Prais

How Emotio

"Fascinating . . . Barrett writes clearly, ~~and~~ ~~helps~~ ~~us~~ ~~sort~~ ~~through~~ ~~our~~ ~~theo~~-
ries, and helps bring us all up to speed on most of the major developments
in the field of emotion studies . . . A thought-provoking journey."
— *Wall Street Journal*

"Most of us make our way through the world without thinking a lot about
what we bring to our encounters with it. Lisa Feldman Barrett does — and
what she has to say about our perceptions and emotions is pretty mind
blowing."
— *Elle*

"We all harbor an intuition about emotions: that the way you experience
joy, fear, or anger happens automatically and is pretty much the same in a
Kalahari hunter-gatherer. In this excellent new book, Lisa Barrett draws on
contemporary research to offer a radically different picture: that the experience of emotion is highly individualized, neurobiologically idiosyncratic,
and inseparable from cognition. This is a provocative, accessible, important book."
— **Robert Sapolsky,** author of
Why Zebras Don't Get Ulcers and *A Primate's Memoir*

"After reading *How Emotions Are Made,* I will never think about emotions
the same way again. Lisa Barrett opens up a whole new terrain for fighting
gender stereotypes and making better policy."
— **Anne-Marie Slaughter,** author of *Unfinished Business*

"*How Emotions Are Made* offers a grand new conception of emotions —
what they are, where they come from, and (most important) what they
aren't. Brain science is the art of the counterintuitive and Lisa Barrett has
a remarkable capacity to make the counterintuitive comprehensible. This
book will have you smacking your forehead wondering why it took so long
to think this way about the brain."
— **Stuart Firestein,** author of
Failure: Why Science Is So Successful and *Ignorance: How It Drives Science*

"Ever wonder where your emotions come from? Lisa Barrett, a world expert in the psychology of emotion, has written the definitive field guide to
feelings and the neuroscience behind them."
— **Angela Duckworth,** best-selling author of *Grit*

"*How Emotions Are Made* is a provocative, insightful, and engaging analysis of the fascinating ways that our brains create our emotional lives, convincingly linking cutting-edge neuroscience studies with everyday emotions. You won't think about emotions in the same way after you read this important book."

— **Daniel L. Schacter,** author of *The Seven Sins of Memory*

"What if everything you thought you knew about lust, anger, grief, and joy was wrong? Lisa Barrett is one of psychology's wisest and most creative scientists and her theory of constructed emotion is radical and fascinating. Through vivid examples and sharp, clear prose, *How Emotions Are Made* defends a bold new vision of the most central aspects of human nature."

— **Paul Bloom,** author of *Against Empathy* and *How Pleasure Works*

"Everything you thought you knew about what you feel and why you feel it turns out to be stunningly wrong. Lisa Barrett illuminates the fascinating new science of our emotions, offering real-world examples of why it matters in realms as diverse as health, parenting, romantic relationships, and national security."

— **Peggy Orenstein,** author of *Girls & Sex*

"This meticulous, well-researched, and deeply thought-out book reveals new insights about our emotions — what they are, where they come from, why we have them. For anyone who has struggled to reconcile brain and heart, this book will be a treasure; it explains the science without shortchanging the humanism of its topic."

— **Andrew Solomon,** best-selling author of *Far from the Tree* and *The Noonday Demon*

"Lisa Barrett masterfully integrates discoveries from affective science, neuroscience, social psychology, and philosophy to make sense of the many instances of emotion that you experience and witness each day. *How Emotions Are Made* will help you remake your life, giving you new lenses to see familiar feelings — from anxiety to love — anew."

— **Barbara Fredrickson,** author of *Positivity* and *Love 2.0*

"Lisa Barrett writes with great clarity about how your emotions are not merely about what you're born with, but also about how your brain pieces your feelings together, and how you can contribute to the process. She tells a compelling story."

— **Joseph LeDoux,** author of *Anxious* and *Synaptic Self*

"A brilliant and original book on the science of emotion, by the deepest thinker about this topic since Darwin."
— **Daniel Gilbert,** best-selling author of *Stumbling on Happiness*

"*How Emotions Are Made* is a tour de force in the quest to understand how we perceive, judge, and decide. It lays the groundwork to address many of the mysteries of human behavior. I look forward to how this more accurate view of emotion will help my clients in athletics and trading."
— **Denise K. Shull,** MA, founder and CEO of the ReThink Group

"With *How Emotions Are Made,* Lisa Feldman Barrett has set the terms of debate for emotion theory in the twenty-first century. In clear, readable prose, she invites us to question both lay and expert understandings of what emotions are — and she musters an impressive body of data to suggest new answers. Barrett's theory of how we construct emotions has major implications for law, including the myth of dispassionate judging. Her 'affective science manifesto for the legal system' deserves to be taken seriously by theorists and practitioners alike."
— **Terry Maroney,** law professor and professor of medicine, health and society, Vanderbilt University

"Every lawyer and judge doing serious criminal trials should read this book. We all grapple with the concepts of free will, emotional impulses, and criminal intent, but here these topics are exposed to a new scrutiny and old assumptions are challenged. The interface of law and brain science is suddenly the area we ought to be debating."
— **Baroness Helena Kennedy,** QC House of Lords, UK

"The extraordinarily powerful writing, logic, and scholarship in this book can't help but arouse admiration, even among those who also experience consternation at its constructivist challenge to simple versions of basic emotions theory." — **Randolph Nesse,** author of *Why We Get Sick*

"Prepare to have your brain twisted around as psychology professor Barrett takes it on a tour of itself . . . Her enthusiasm for her topic brightens every amazing fact and theory about where our emotions come from. Hint: it's not what you think. Indeed, each chapter is chockablock with startling insights . . . Barrett's figurative selfie of the brain is brilliant."
— *Booklist,* starred review

"A well-argued, entertaining disputation of the prevailing view that emotion and reason are at odds . . . As Barrett points out, this has important legal as well as moral implications and leads into the thorny questions surrounding free will. A highly informative, readable, and wide-ranging discussion of 'how psychology, neuroscience, and related disciplines are moving away from the search for emotion fingerprints and instead asking how emotions are constructed.'"   — *Kirkus Reviews,* starred review

"Barrett . . . offers an unintuitive theory that goes against not only the popular understanding but also that of traditional research: emotions don't arise; rather, we construct them on the fly . . . Tracing her own journey from the classical view of emotions, Barrett progressively builds her case, writing in a conversational tone and using down-to-earth metaphors, relegating the heaviest neuroscience to an appendix to keep the book accessible."
      — *Library Journal,* starred review

"Extraordinarily well written, Lisa Barrett's *How Emotions Are Made* chronicles a paradigm shift in the science of emotion. But more than just a chronicle, this book is a brilliant work of translation, translating the new neuroscience of emotion into understandable and readable terms. Since that science has profound implications in areas as disparate as police shootings and TSA profiling, the translation is critical for scientists and citizens, lawmakers and physicians. (For example, what if there is no meaningful scientific difference between premeditated murder, the product of rational thought, which we consider most culpable, and the lesser offense of manslaughter, a 'crime of passion'?) Emotions do not reside in dedicated brain areas, constantly at war with areas charged with cognition or perception, as Pixar caricatured it in *Inside Out,* let alone the brain described by Descartes or Plato or other philosophers. Nor does the brain passively retrieve data from 'outside' to which it reacts. The brain constructs the reality it perceives, and the emotions it (and we) experience, using core brain systems, not specialized circuits. And it does so in concert with other brains, with the culture surrounding it. The implications of this work ('only' challenging two-thousand-year-old assumptions about the brain) and its ambitions are nothing short of stunning. Even more stunning is how extraordinarily well it succeeds."
      — **Nancy Gertner,** senior lecturer on law, Harvard Law School, and former US federal judge for the US District Court of Massachusetts

How Emotions Are Made

How Emotions Are Made

The Secret Life of the Brain

Lisa Feldman Barrett, Ph.D.

HARPER

An Imprint of HarperCollins*Publishers*

Boston New York

First Mariner Books edition 2018
Copyright © 2017 by Lisa Feldman Barrett
Illustrations by Aaron Scott

www.harpercollins.com

Library of Congress Cataloging-in-Publication Data
Names: Barrett, Lisa Feldman, author.
Title: How emotions are made : the secret life of the brain / Lisa Feldman
Barrett.
Description: Boston : HarperCollins Publishers, 2017.
Identifi ers: LCCN 2016038354 (print) | LCCN 2017004323 (ebook) |
ISBN 9780544133310 (hardback) | ISBN 9780544129962 (ebook)
ISBN 9781328915436 (pbk.)
Subjects: LCSH: Emotions. | Emotions — Sociological aspects. | Brain. |
BISAC: PSYCHOLOGY / Emotions. | PSYCHOLOGY / Neuropsychology. |
SCIENCE / Philosophy & Social Aspects. | SCIENCE / Life Sciences / Neuroscience.
Classifica tion: LCC BF561 .B337 2017 (print) | LCC BF561 (ebook) | DDC
152.4 — dc23
LC record available at https://lccn.loc.gov/2016038354

Printed in the United States of America
23 24 25 26 27 LBC 18 17 16 15 14

For Sophia

Contents

Introduction:
The Two-Thousand-Year-Old Assumption

On December 14, 2012, the deadliest school shooting in U.S. history took place at Sandy Hook Elementary School in Newtown, Connecticut. Twenty-six people inside the school, including twenty children, were massacred by a lone gunman. Several weeks after this horror, I watched the governor of Connecticut, Dannel Malloy, give his annual "State of the State" speech on television. He spoke in a strong and animated voice for the first three minutes, thanking individuals for their service. And then he began to address the Newtown tragedy:

> We have all walked a very long and very dark road together. What befell Newtown is not something we thought possible in any of Connecticut's beautiful towns or cities. And yet, in the midst of one of the worst days in our history, we also saw the best of our state. Teachers and a therapist that sacrificed their lives protecting students.[1]

As the governor spoke the last two words, "protecting students," his voice caught in his throat ever so slightly. If you weren't paying close attention, you might have missed it. But that tiny waver *devastated* me. My stomach instantly knotted into a ball. My eyes flooded. The TV camera panned to the crowd where other people had started to sob too. As for Governor Malloy, he stopped speaking and was gazing downward.

Emotions like Governor Malloy's and mine seem primal — hardwired into us, reflexively deployed, shared with all our fellow humans. When triggered, they seem to unleash themselves in each of us in basically the same way. My sadness was like Governor Malloy's sadness was like the crowd's sadness.

Humanity has understood sadness and other emotions in this way for over two thousand years. But at the same time, if humanity has learned anything from centuries of scientific discovery, it's that things aren't always what they appear to be.

The time-honored story of emotion goes something like this: We all have emotions built-in from birth. They are distinct, recognizable phenomena inside us. When something happens in the world, whether it's a gunshot or a flirtatious glance, our emotions come on quickly and automatically, as if someone has flipped a switch. We broadcast emotions on our faces by way of smiles, frowns, scowls, and other characteristic expressions that anyone can easily recognize. Our voices reveal our emotions through laughter, shouts, and cries. Our body posture betrays our feelings with every gesture and slouch.

Modern science has an account that fits this story, which I call the *classical view of emotion*. According to this view, the waver in Governor Malloy's voice launched a chain reaction that began in my brain. A particular set of neurons — call it the "sadness circuit" — leaped into action and caused my face and body to respond in a certain, specific way. My brow furrowed, I frowned, my shoulders stooped, and I cried. This proposed circuit also triggered physical changes inside my body, causing my heart rate and breathing to speed up, my sweat glands to activate, and my blood vessels to constrict.* This collection of movements on the inside and outside of my body are said to be like a "fingerprint" that uniquely identifies sadness, much like your own fingerprints uniquely identify you.

The classical view of emotion holds that we have many such emotion circuits in our brains, and each is said to cause a distinct set of changes, that is, a fingerprint. Perhaps an annoying coworker triggers your "anger neurons," so your blood pressure rises; you scowl, yell, and feel the heat of fury. Or an alarming news story triggers your "fear neurons," so your heart races; you freeze and feel a flash of dread. Because we experience anger, happiness, surprise, and other emotions as clear and identifiable states of being,

---

* When I use the word "body" in this book, I am excluding the brain, as in the sentence, "Your brain tells your body to move." To refer to the entire body including the brain, I write "the anatomical body."

it seems reasonable to assume that each emotion has a defining underlying pattern in the brain and body.

Our emotions, according to the classical view, are artifacts of evolution, having long ago been advantageous for survival, and are now a fixed component of our biological nature. As such, they are universal: people of every age, in every culture, in every part of the world should experience sadness more or less as you do — and more or less as did our hominin ancestors who roamed the African savanna a million years ago. I say "more or less" because no one believes that faces, bodies, and brain activity look *exactly* the same each time someone is sad. Your heart rate and breathing and blood flow won't always change by the same amount. Your brow might furrow slightly less by chance or by custom.[2]

Emotions are thus thought to be a kind of brute reflex, very often at odds with our rationality. The primitive part of your brain wants you to tell your boss he's an idiot, but your deliberative side knows that doing so would get you fired, so you restrain yourself. This kind of internal battle between emotion and reason is one of the great narratives of Western civilization. It helps define you as human. Without rationality, you are merely an emotional beast.

This view of emotions has been around for millennia in various forms. Plato believed a version of it. So did Hippocrates, Aristotle, the Buddha, René Descartes, Sigmund Freud, and Charles Darwin. Today, prominent thinkers such as Steven Pinker, Paul Ekman, and the Dalai Lama also offer up descriptions of emotions rooted in the classical view. The classical view is found in virtually every introductory college textbook on psychology, and in most magazine and newspaper articles that discuss emotion. Preschools throughout America hang posters displaying the smiles, frowns, and pouts that are supposed to be the universal language of the face for recognizing emotions. Facebook even commissioned a set of emoticons inspired by Darwin's writings.[3]

The classical view is also entrenched in our culture. Television shows like *Lie to Me* and *Daredevil* are predicated on the assumption that your innermost feelings are exposed by your heart rate or facial movements. *Sesame Street* teaches children that emotions are distinct things inside us seeking expression in the face and body, as does the Pixar movie *Inside Out*. Companies like Affectiva and Realeyes offer to help businesses detect their customers' feelings through "emotion analytics." In the NBA draft, the Milwaukee Bucks evaluate a player's "psychological, character and personality issues" and assess "team chemistry" from facial expressions. And for several decades, the U.S. Federal Bureau of Investigation (FBI) based some of its advanced agent training on the classical view.[4]

More significantly, the classical view of emotion is embedded in our social institutions. The American legal system assumes that emotions are part of an inherent animal nature and cause us to perform foolish and even violent acts unless we control them with our rational thoughts. In medicine, researchers study the health effects of anger, supposing that there is a single pattern of changes in the body that goes by that name. People suffering from a variety of mental illnesses, including children and adults diagnosed with autism spectrum disorder, are taught how to recognize facial configurations for specific emotions, ostensibly to help them communicate and relate to others.

And yet . . . despite the distinguished intellectual pedigree of the classical view of emotion, and despite its immense influence in our culture and society, there is abundant scientific evidence that this view cannot possibly be true. Even after a century of effort, scientific research has not revealed a consistent, physical fingerprint for even a single emotion. When scientists attach electrodes to a person's face and measure how facial muscles actually move during the experience of an emotion, they find tremendous variety, not uniformity. They find the same variety — the same absence of fingerprints — when they study the body and the brain. You can experience anger with or without a spike in blood pressure. You can experience fear with or without an amygdala, the brain region historically tagged as the home of fear.

To be sure, hundreds of experiments offer some evidence for the classical view. But *hundreds more* cast that evidence into doubt. The only reasonable scientific conclusion, in my opinion, is that emotions are not what we typically think they are.

So what are they, really? When scientists set aside the classical view and just look at the data, a radically different explanation for emotion comes to light. In short, we find that your emotions are not built-in but made from more basic parts. They are not universal but vary from culture to culture. They are not triggered; you create them. They emerge as a combination of the physical properties of your body, a flexible brain that wires itself to whatever environment it develops in, and your culture and upbringing, which provide that environment. Emotions are real, but not in the objective sense that molecules or neurons are real. They are real in the same sense that money is real — that is, hardly an illusion, but a product of human agreement.[5]

This view, which I call the *theory of constructed emotion*, offers a very different interpretation of the events during Governor Malloy's speech. When Malloy's voice caught in his throat, it did not trigger a brain circuit for sadness inside me, causing a distinctive set of bodily changes. Rather, I felt sadness in that moment because, having been raised in a certain culture, I learned long ago that "sadness" is something that may occur when certain bodily feelings coincide with terrible loss. Using bits and pieces of past experience, such as my knowledge of shootings and my previous sadness about them, my brain rapidly predicted what my body should do to cope with such tragedy. Its predictions caused my thumping heart, my flushed face, and the knots in my stomach. They directed me to cry, an action that would calm my nervous system. And they made the resulting sensations meaningful as an instance of sadness.

In this manner, my brain *constructed* my experience of emotion. My particular movements and sensations were not a fingerprint for sadness. With different predictions, my skin would cool rather than flush and my stomach would remain unknotted, yet my brain could still transform the resulting sensations into sadness. Not only that, but my original thumping heart, flushed face, knotted stomach, and tears could become meaningful as a different emotion, such as anger or fear, instead of sadness. Or in a very different situation, like a wedding celebration, those same sensations could become joy or gratitude.

If this explanation doesn't make complete sense or even sounds counterintuitive so far, believe me, I am right there with you. After Governor Malloy's speech, as I came back to myself, wiping my tears, I was reminded that no matter what I *know* about emotions as a scientist, I *experience* them much as the classical view conceives them. My sadness felt like an instantly recognizable wave of bodily changes and feelings that overwhelmed me as a reaction to tragedy and loss. If I were not a scientist using experiments to reveal that emotions are in fact made and not triggered, I too would trust my immediate experience.

The classical view of emotion remains compelling, despite the evidence against it, precisely because it's intuitive. The classical view also provides reassuring answers to deep, fundamental questions like: Where do you come from, evolutionarily speaking? Are you responsible for your actions when you get emotional? Do your experiences accurately reveal the world outside you?

The theory of constructed emotion answers such questions differently. It's a different theory of human nature that helps you see yourself and others in a new and more scientifically justified light. The theory of constructed emotion might not fit the way you typically experience emotion and, in fact, may well violate your deepest beliefs about how the mind works, where humans come from, and why we act and feel as we do. But the theory consistently predicts and explains the scientific evidence on emotion, including plenty of evidence that the classical view struggles to make sense of.

Why should you care which theory of emotion is correct? Because belief in the classical view affects your life in ways you might not realize. Think about the last time you went through airport security, where taciturn agents of the Transportation Security Administration (TSA) X-rayed your shoes and evaluated your likelihood as a terrorist threat. Not long ago, a training program called SPOT (Screening Passengers by Observation Techniques) taught those TSA agents to detect deception and assess risk based on facial and bodily movements, on the theory that such movements reveal your innermost feelings. It didn't work, and the program cost taxpayers $900 million. We need to understand emotion scientifically so government agents won't detain us — or overlook those who actually do pose a threat — based on an incorrect view of emotion.[6]

Now imagine that you're in a doctor's office, complaining of chest pressure and shortness of breath, which may be heart attack symptoms. If you're a woman, you're more likely to be diagnosed with anxiety and sent home, whereas if you're a man, you're more likely to be diagnosed with heart disease and receive lifesaving preventive treatment. As a result, women over age sixty-five die more frequently of heart attacks than men do. The perceptions of doctors, nurses, and the female patients themselves are shaped by classical view beliefs that they can detect emotions like anxiety, and that women are inherently more emotional than men . . . with fatal consequences.[7]

Belief in the classical view can even start wars. The Gulf War in Iraq was launched, in part, because Saddam Hussein's half-brother thought he could read the emotions of the American negotiators and informed Saddam that the United States wasn't serious about attacking. The subsequent war claimed the lives of 175,000 Iraqis and hundreds of coalition forces.[8]

We are, I believe, in the midst of a revolution in our understanding of emotion, the mind, and the brain — a revolution that may compel us to radically rethink such central tenets of our society as our treatments for mental and physical illness, our understanding of personal relationships, our approaches to raising children, and ultimately our view of ourselves. Other scientific disciplines have seen revolutions of this kind, each one a momentous shift away from centuries of common sense. Physics moved from Isaac Newton's intuitive ideas about time and space to Albert Einstein's more relative ideas, and eventually to quantum mechanics. In biology, scientists carved up the natural world into fixed species, each having an ideal form, until Charles Darwin introduced the concept of natural selection.

Scientific revolutions tend to emerge not from a sudden discovery but by asking better questions. How are emotions made, if they aren't simply triggered reactions? Why do they vary so much, and why have we believed for so long that they have distinctive fingerprints? These questions in and of themselves can be delightfully interesting to ponder. But taking pleasure in the unknown is more than just a scientific indulgence. It's part of the spirit of adventure that makes us human.

In the pages that follow, I invite you to share that adventure with me. Chapters 1–3 introduce the new science of emotion: how psychology, neuroscience, and related disciplines are moving away from the search for emotion fingerprints and instead asking how emotions are constructed. Chapters 4–7 explain how, exactly, emotions are made. And chapters 8–12 explore the practical, real-world implications of this new theory of emotions on our approaches to health, emotional intelligence, child-rearing, personal relationships, systems of law, and even human nature itself. To close the book, chapter 13 reveals how the science of emotion illuminates the age-old mystery of how a human brain creates a human mind.

1

The Search for
Emotion's "Fingerprints"

Once upon a time, in the 1980s, I thought I would be a clinical psychologist. I headed into a Ph.D. program at the University of Waterloo, expecting to learn the tools of the trade as a psychotherapist and one day treat patients in a stylish yet tasteful office. I was going to be a consumer of science, not a producer. I certainly had no intention of joining a revolution to unseat basic beliefs about the mind that have existed since the days of Plato. But life sometimes tosses little surprises in your direction.

It was in graduate school that I felt my first tug of doubt about the classical view of emotion. At the time, I was researching the roots of low self-esteem and how it leads to anxiety or depression. Numerous experiments showed that people feel depressed when they fail to live up to their own ideals, but when they fall short of a standard set by others, they feel anxious. My first experiment in grad school was simply to replicate this well-known phenomenon before building on it to test my own hypotheses. In the course of this experiment, I asked a large number of volunteers if they felt anxious or depressed using well-established checklists of symptoms.[1]

I'd done more complicated experiments as an undergraduate student, so this one should have been a piece of cake. Instead, it crashed and burned. My volunteers did not report anxious or depressed feelings in the expected pattern. So I tried to replicate a second published experiment, and it failed too. I tried again, over and over, each experiment taking months. After three years, all I'd achieved was the same failure *eight times in a row*. In science, experiments often don't replicate, but eight consecutive failures is an impressive record. My internal critic taunted me: *not everyone is cut out to be a scientist.*

When I looked closely at all the evidence I had collected, however, I noticed something consistently odd across all eight experiments. Many of my subjects appeared to be unwilling, or unable, to distinguish between feeling anxious and feeling depressed. Instead, they had indicated feeling both or neither; rarely did a subject report feeling just one. This made no sense. Everybody knows that anxiety and depression, when measured as emotions, are decidedly different. When you're anxious, you feel worked up, jittery, like you're worried something bad will happen. In depression you feel miserable and sluggish; everything seems horrible and life is a struggle. These emotions should leave your body in completely opposite physical states, and so they should feel different and be trivial for any healthy person to tell apart. Nevertheless, the data declared that my test subjects weren't doing so. The question was . . . why?

As it turned out, my experiments weren't failing after all. My first "botched" experiment actually revealed a genuine discovery — that people often did not distinguish between feeling anxious and feeling depressed. My next seven experiments hadn't failed either; they'd replicated the first one. I also began noticing the same effect lurking in other scientists' data. After completing my Ph.D. and becoming a university professor, I continued pursuing this mystery. I directed a lab that asked hundreds of test subjects to keep track of their emotional experiences for weeks or months as they went about their lives. My students and I inquired about a wide variety of emotional experiences, not just anxious and depressed feelings, to see if the discovery generalized.

These new experiments revealed something that had never been documented before: everyone we tested used the same emotion words like "angry," "sad," and "afraid" to communicate their feelings but not necessarily to mean the same thing. Some test subjects made fine distinctions with their word use: for example, they experienced sadness and fear as qualitatively different. Other subjects, however, lumped together words like "sad" and "afraid" and "anxious" and "depressed" to mean "I feel crappy" (or, more scientifically, "I feel unpleasant"). The effect was the same for pleasant emotions like happiness, calmness, and pride. After testing over seven hundred American subjects, we discovered that people vary tremendously in how they differentiate their emotional experiences.

A skilled interior designer can look at five shades of blue and distinguish azure, cobalt, ultramarine, royal blue, and cyan. My husband, on the other hand, would call them all blue. My students and I had discovered a similar phenomenon for emotions, which I described as *emotional granularity*.[2]

Here's where the classical view of emotion entered the picture. Emotional granularity, in terms of this view, must be about accurately reading your internal emotional states. Someone who distinguished among different feelings using words like "joy," "sadness," "fear," "disgust," "excitement," and "awe" must be detecting physical cues or reactions for each emotion and interpreting them correctly. A person exhibiting lower emotional granularity, who uses words like "anxious" and "depressed" interchangeably, must be failing to detect these cues.

I began wondering if I could teach people to improve their emotional granularity by coaching them to recognize their emotional states accurately. The key word here is "accurately." How can a scientist tell if someone who says "I'm happy" or "I'm anxious" is accurate? Clearly, I needed some way to *measure an emotion objectively* and then compare it to what the person reports. If a person reports feeling anxious, and the objective criteria indicate that he is in a state of anxiety, then he is accurately detecting his own emotion. On the other hand, if the objective criteria indicate that he is depressed or angry or enthusiastic, then he's inaccurate. With an objective test in hand, the rest would be simple. I could ask a person how he feels and compare his answer to his "real" emotional state. I could correct any of his apparent mistakes by teaching him to better recognize the cues that distinguish one emotion from another and improve his emotional granularity.

Like most students of psychology, I had read that each emotion is supposed to have a distinct pattern of physical changes, roughly like a fingerprint. Each time you grasp a doorknob, the fingerprints that you leave behind may vary depending on the firmness of your grip, how slippery the surface is, or how warm and pliable your skin is at that moment. Nevertheless, your fingerprints look similar enough each time to identify you uniquely. The "fingerprint" of an emotion is likewise assumed to be similar enough from one instance to the next, and in one person to the next, regardless of age, sex, personality, or culture. In a laboratory, scientists should be able to tell whether someone is sad or happy or anxious just by looking at physical measurements of a person's face, body, and brain.

I felt confident that these emotion fingerprints could provide the objective criteria I needed to measure emotion. If the scientific literature was correct, then assessing people's emotional accuracy would be a breeze. But things did not turn out quite as I expected.

•   •   •

According to the classical view of emotion, our faces hold the key to assessing emotions objectively and accurately. A primary inspiration for this idea is Charles Darwin's book *The Expression of the Emotions in Man and Animals,* where he claimed that emotions and their expressions were an ancient part of universal human nature. All people, everywhere in the world, are said to exhibit and recognize facial expressions of emotion without any training whatsoever.[3]

So, I thought that my lab should be able to measure facial movements, assess our test subjects' true emotional state, compare it to their verbal reports of emotion, and calculate their accuracy. If subjects made a pouting expression in the lab, for instance, but did not report feeling sad, we could train them to recognize the sadness they must be feeling. Case closed.

The human face is laced with forty-two small muscles on each side. The facial movements that we see each other make every day — winks and blinks, smirks and grimaces, raised and wrinkled brows — occur when combinations of facial muscles contract and relax, causing connective tissue and skin to move. Even when your face seems completely still to the naked eye, your muscles are still contracting and relaxing.[4]

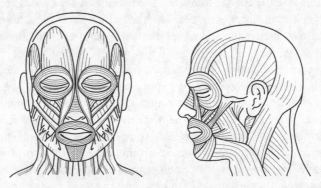

Figure 1-1: Muscles of the human face

According to the classical view, each emotion is displayed on the face as a particular pattern of movements — a "facial expression." When you're happy, you're supposed to smile. When you're angry, you're supposed to furrow your brow. These movements are said to be part of the fingerprint of their respective emotions.

Back in the 1960s, the psychologist Silvan S. Tomkins and his protégés Carroll E. Izard and Paul Ekman decided to test this in the lab. They created sets of meticulously posed photographs, such as those in figure 1-2, to represent six so-called basic emotions they believed had biological fingerprints: anger, fear, disgust, surprise, sadness, and happiness. These photos, which featured actors who were carefully coached, were supposed to be the clearest examples of facial expressions for these emotions. (They might look exaggerated or artificial to you, but they were designed this way on purpose, because Tomkins believed they gave the strongest, clearest signals for emotion.)[5]

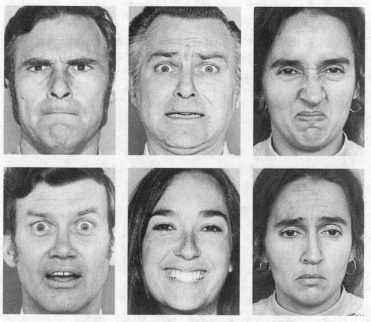

Figure 1-2: Some facial photographs from basic emotion method studies

Using posed photos like these, Tomkins and his crew applied an experimental technique to study how well people "recognize" emotional expressions, or, more precisely, how well they perceive facial movements as expressions of emotion. Hundreds of published experiments have used this method, and it's still considered the gold standard today. A test subject is given a photograph and a set of emotion words, as in figure 1-3.

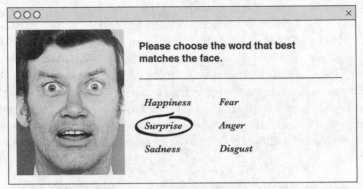

Figure 1-3: Basic emotion method: picking a word to match the face

The subject then chooses the word that best matches the face. In this case, the intended word is "Surprise." Or, using a slightly different setup, a test subject is given two posed photos and a brief story, as in figure 1-4, and then picks which face best matches the story. In this case, the intended face is on the right.[6]

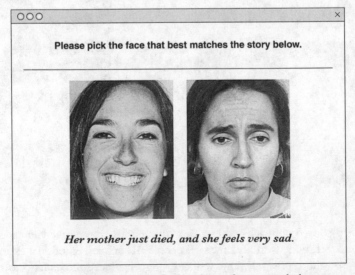

Figure 1-4: Basic emotion method: picking a face to match the story

This research technique — let's call it the basic emotion method — revolutionized the scientific study of what Tomkins's group called "emotion recognition." Using this method, scientists showed that people from around the world could consistently match the same emotion words (translated into the local language) to posed faces. In one famous study, Ekman and his colleagues traveled to Papua New Guinea and ran experiments with a local population, the Fore people, who had little contact with the Western world. Even this remote tribe could consistently match the faces to the expected emotion words and stories. In later years, scientists ran similar studies in many other countries such as Japan and Korea. In each case, subjects easily matched the posed scowls, pouts, smiles, and so on to the provided emotion words or stories.[7]

From this evidence, scientists concluded that emotion recognition is universal: no matter where you are born or grow up, you should be able to recognize American-style facial expressions like those in the photos. The only way expressions could be universally recognized, the reasoning went, is if they are universally produced: thus, facial expressions must be reliable, diagnostic fingerprints of emotion.[8]

Other scientists, however, worried that the basic emotion method was too indirect and subjective to reveal emotion fingerprints because it involves human judgment. A more objective technique, called facial electromyography (EMG), removes human perceivers altogether. Facial EMG places electrodes on the surface of the skin to detect the electrical signals that make facial muscles move. It precisely identifies the parts of the face as they move, how much, and how often. In a typical study, test subjects wear electrodes over their eyebrows, forehead, cheeks, and jaw as they view films or photos, or as they remember or imagine situations, to evoke a variety of emotions. Scientists record the electrical changes in muscle activity and calculate the degree of movement in each muscle during each emotion. If people move the same facial muscles in the same pattern each time they experience a given emotion — scowling in anger, smiling in happiness, pouting in sadness, and so on — and *only* when they experience that emotion, then the movements might be a fingerprint.[9]

As it turns out, facial EMG presents a serious challenge to the classical view of emotion. In study after study, the muscle movements do not reliably indicate when someone is angry, sad, or fearful; they don't form predictable fingerprints for each emotion. At best, facial EMG reveals that these movements distinguish pleasant versus unpleasant feeling. Even more damning,

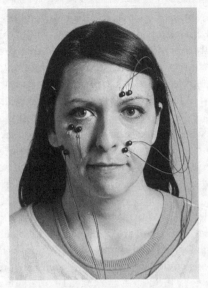

Figure 1-5: Facial electromyography

the facial movements recorded in these studies do not reliably match the posed photos created for the basic emotion method.[10]

Let's take a moment and consider the implications of these findings. Hundreds of experiments have shown that people worldwide can match emotion words to so-called expressions of emotion, posed by actors who aren't actually feeling those emotions. However, those expressions can't be consistently and specifically detected by objective measures of facial muscle movements when people are *actually feeling* emotion. We all move our facial muscles all the time, of course, and when we look at each other, we effortlessly see emotion in some of these movements. Nevertheless, from a purely objective standpoint, when scientists measure *just the muscle movements themselves,* those movements do not conform to the photographs.

It's conceivable that facial EMG is too limited to capture all the meaningful actions in a face during an emotional experience. A scientist can place about six electrodes on each side of the face before a test subject starts to feel uncomfortable, too few to capture all forty-two facial muscles meaningfully. So scientists also employ an alternative technique called facial action coding (FACS), in which trained observers laboriously classify a subject's individual facial movements as they occur. It's less objective than facial EMG, since it relies on human perceivers, but presumably more objective than matching words to posed faces in the basic emotion method. Nevertheless, the movements observed during facial action coding also don't consistently match the posed photos.[11]

These same inconsistencies show up in infants. If facial expressions are universal, then babies should be even more likely than adults to express anger with a scowl and sadness with a pout, because they're too young to learn rules of social appropriateness. And yet when scientists observe infants in situations that should evoke emotion, the infants do not make the expected expressions. For example, the developmental psychologists Linda A. Camras and Harriet Oster and their colleagues videotaped babies from various cultures, employing a growling gorilla toy to startle them (to induce fear) or restraining their arm (to induce anger). Camras and Oster found, using FACS, that the range of babies' facial movements in the two situations was indistinguishable. Nevertheless, when adults watched these videos, they somehow identified the infants in the gorilla film as afraid and infants in the arm restraint film as angry, even when Camras and Oster blanked out the babies' faces electronically! The adults were distinguishing fear from anger based on the context, without seeing facial movements at all.[12]

Don't get me wrong: newborns and young infants move their faces in meaningful ways. They make many distinctive facial movements when the situation implies that they might be interested or puzzled, or when they feel distress in response to pain or distaste in response to offending smells and tastes. But newborns don't show differentiated, adult-like expressions like the photographs from the basic emotion method.[13]

Other scientists also have demonstrated, as Camras and Oster did, that you take tremendous information from the surrounding context. They graft photographs of faces and bodies that don't belong together, like an angry scowling face attached to a body that's holding a dirty diaper, and their test subjects nearly always identify the emotion appropriate to the body, not the face — in this case, disgust rather than anger. Faces are constantly moving, and your brain relies on many different factors at once — body posture, voice, the overall situation, your lifetime of experience — to figure out which movements are meaningful and what they mean.[14]

When it comes to emotion, a face doesn't speak for itself. In fact, the poses of the basic emotion method were not discovered by observing faces in the real world. Scientists *stipulated* those facial poses, inspired by Darwin's book, and asked actors to portray them. And now these faces are simply assumed to be the universal expressions of emotion.[15]

But they aren't universal. To further demonstrate this, my lab conducted a study using photos from a group of emotion experts — accomplished actors. The photos came from the book *In Character: Actors Acting*, in which actors portray emotions by posing their faces to match written scenarios. We divided our U.S. test subjects into three groups. The first group read only the scenarios, for example, "He just witnessed a shooting on his quiet, tree-shaded block in Brooklyn." A second group saw only the facial configurations, such as Martin Landau's pose for the shooting scenario (figure 1-6, center). A third group saw the scenarios and the faces. In each case, we handed subjects a short list of emotion words to categorize whatever emotion they saw.[16]

For the shooting scenario I just mentioned, 66 percent of subjects who read the scenario alone or with Landau's face rated the scenario as a fearful situation. But for subjects who saw Landau's face alone, devoid of context, only 38 percent of them rated it as fear and 56 percent rated it as surprise. (Figure 1-6 compares Landau's facial configuration to basic emotion method photos for "fear" and "surprise." Does Landau look afraid or surprised? Or both?)

Figure 1-6: Actor Martin Landau (center) flanked by
basic emotion method faces for fear (left) and surprise (right)

Other actors' poses for fear were strikingly different from Landau's. In one case, the actress Melissa Leo portrayed fear for the scenario: "She is trying to decide if she should tell her husband about a rumor going around that she is gay before he hears it from someone else." Her mouth is closed and downturned, and her brow is slightly knitted. Nearly three quarters of our test subjects who saw her face alone rated it as sad, but when presented with the scenario, 70 percent of subjects rated her face as displaying fear.[17]

This sort of variation held true for every emotion that we studied. An emotion like "Fear" does not have a single expression but a *diverse population of facial movements* that vary from one situation to the next.* (Think about it: When is the last time an actor won an Academy Award for pouting when sad?)

This may seem obvious once you pause to consider your own emotional experiences. When you experience an emotion such as fear, you might move your face in a variety of ways. While cowering in your seat at a horror movie, you might close your eyes or cover them with your hands. If you're uncertain whether a person directly in front of you could harm you, you might narrow your eyes to see the person's face better. If danger is potentially lurking around the next corner, your eyes might widen to improve your peripheral vision. "Fear" takes no single physical form. Variation is the norm. Likewise, happiness, sadness, anger, and every other emotion you know is a diverse *category,* with widely varying facial movements.[18]

If facial movements have so much variation within an emotion category like "Fear," you might wonder why we find it so natural to believe that a wide-eyed face is the universal fear expression. The answer is that it's a stereotype, a symbol that fits a well-known theme for "Fear" within our culture. Preschools teach these stereotypes to children: "People who scowl are angry. People who pout are sad." They are cultural shorthands or conventions. You see them in cartoons, in advertisements, in the faces of dolls, in emojis — in an endless array of imagery and iconography. Textbooks teach these stereotypes to psychology students. Therapists teach them to their patients. The media spreads them widely throughout the Western world. "Now, wait just a minute," you might be thinking. "Is she saying that our culture has *created* these expressions, and we all have *learned* them?" Well . . . yes. And the classical view perpetuates these stereotypes as if they are authentic fingerprints of emotion.

To be sure, faces are instruments of social communication. Some facial movements have meaning, but others do not, and right now, we know pre-

---

* In this book, I use initial capitals and double quotation marks to denote an emotion in general, such as "Fear," as opposed to a single instance of fear.

cious little about how people figure out which is which, other than that context is somehow crucial (body language, social situation, cultural expectation, etc.). When facial movements do convey a psychological message — say, raising an eyebrow — we don't know if the message is always emotional, or even if its meaning is the same each time. If we put all the scientific evidence together, we cannot claim, with any reasonable certainty, that each emotion has a diagnostic facial expression.[19]

• • •

In my search for unique fingerprints of emotion, I clearly needed a more reliable source than the human face, so next I looked to the human body. Perhaps some telling changes in heart rate, blood pressure, and other body functions would provide the necessary fingerprints to teach people to recognize their emotions more accurately.

Some of the strongest experimental support for bodily fingerprints comes from a famous study by Paul Ekman, the psychologist Robert W. Levenson, and their colleague Wallace V. Friesen, published in the journal *Science* in 1983. They hooked up test subjects to machines to measure changes in the autonomic nervous system: variations in heart rate, temperature, and skin conductance (a measure of sweat). They also measured variations in arm tension, rooted in the skeletomotor nervous system. They then used an experimental technique to evoke anger, sadness, fear, disgust, surprise, and happiness, and observed the physical changes during each emotion. After analyzing the data, Ekman and his colleagues concluded that they had measured clear and consistent changes in these bodily responses, relating them to particular emotions. This study seemingly established objective, biological fingerprints in the body for each of the studied emotions, and today it remains a classic in the scientific literature.[20]

The famous 1983 study evoked emotion in a curious way — by having test subjects make and hold a facial pose from the basic emotion method. To evoke sadness, for example, a subject would frown for ten seconds. To evoke anger, a subject would scowl. While face-posing, subjects could use a mirror and were coached by Ekman himself to move particular facial muscles.[21]

The idea that a posed, so-called facial expression can trigger an emotional state is known as the facial feedback hypothesis. Allegedly, contorting your face into a particular configuration causes the specific physiological changes associated with that emotion in your body. Try it yourself. Knit your brows and pout for ten seconds — do you feel sad? Smile broadly. Do you feel happier? The facial feedback hypothesis is highly controversial —

there is wide disagreement on whether a full-blown emotional experience can be evoked this way.[22]

The 1983 study did, in fact, observe bodily changes as people posed the required facial configurations. This is a remarkable finding: just posing a particular facial configuration changed the test subjects' peripheral nervous system activity, even while they were comfortably motionless in a chair. Their fingertips were warmer when posing a scowl (anger pose). Their heartbeats were faster when posing scowls, wide-eyed startle (fear pose), and pouts (sad pose) when compared to the poses for happiness, surprise, and disgust. The remaining two measures, skin conductance and arm tension, did not distinguish one facial configuration from another.[23]

Even so, you must take some additional steps before you can claim that you've found a bodily fingerprint for an emotion. For one thing, you must show that the response during one emotion, say, anger, is different from that of other emotions — that is, it's specific to instances of anger. Here, the 1983 study starts having some difficulty. It showed some specificity for anger but not for the other emotions tested. That means the bodily responses for different emotions were too similar to be distinct fingerprints.

In addition, you must show that no other explanations can account for your results. Then, and only then, can you claim to have found physical fingerprints for anger, sadness, and the rest. The 1983 study is, for this reason, subject to an alternative explanation, because the test subjects were given instructions for how to pose their faces. Western subjects could conceivably identify most of the target emotions from these instructions. This understanding can actually produce the heart rate and other physical changes Ekman and colleagues observed, a fact that was unknown when these studies were conducted. This alternative explanation is borne out by their later experiment with an Indonesian tribe, the Minangkabau of West Sumatra. These volunteers had less understanding of Western emotions and did not show the same physical changes as Western test subjects; they also reported feeling the expected emotion much less frequently than the Western subjects did.[24]

Other subsequent research has evoked emotions using a variety of different methods but has not replicated the original physiological differences observed in the 1983 paper. Quite a few studies employ horror movies, tearful chick flicks, and other evocative material to bring on particular emotions, while scientists measure subjects' heart rate, respiration, and other bodily functions. Many such studies found great variability in physical measurements, meaning no clear pattern of bodily changes that distinguished emotions. In other studies, scientists did find distinguishing patterns, but different studies often found *different* patterns, even when using exactly the same film clips. In other words, when studies distinguished anger from sadness from fear, they did not always replicate one another, implying that the instances of anger, sadness, and fear cultivated in one study were different from those cultivated in another.[25]

When faced with a large collection of diverse experiments like this, it's hard to extract a consistent story. Fortunately, scientists have a technique to analyze all the data together and reach a unified conclusion. It's called a "meta-analysis." Scientists comb through large numbers of experiments conducted by different researchers, combining their results statistically. As a simple example, suppose you wanted to check if increased heart rate is part of the bodily fingerprint of happiness. Rather than run your own experiment, you could do a meta-analysis of other experiments that measured heart rate during happiness, even incidentally (e.g., the study could be about the relationship between sex and heart attacks and have nothing centrally to do with emotion). You would search for all the relevant scientific papers, collect the relevant statistics from them, and analyze them *en masse* to test the hypothesis.

Where emotions and the autonomic nervous system are concerned, four significant meta-analyses have been conducted in the last two decades, the largest of which covered more than 220 physiology studies and nearly 22,000 test subjects. None of these four meta-analyses found consistent and specific emotion fingerprints in the body. Instead, the body's orchestra of internal organs can play many different symphonies during happiness, fear, and the rest.[26]

You can see this variation easily in an experimental procedure used by laboratories around the world, where test subjects perform a difficult task such as counting backward by thirteen as fast as possible, or speaking about a polarizing topic like abortion or religion, while being ridiculed. As they struggle, the experimenter berates them for poor performance, making critical and even insulting remarks. Do all the test subjects get angry? No, they don't. More importantly, those who do feel angry show different patterns of bodily changes. Some people fume in anger, but some cry. Others become quiet and cunning. Still others just withdraw. Each behavior (fuming, crying, planning, withdrawing) is supported by a different physiological pattern in the body, a detail long known by physiologists who study the body for its own sake. Even small changes in body posture, like lying back versus leaning forward with arms crossed, can completely alter an angry person's physiological response.[27]

When I address audiences at conferences and present these meta-analyses, some people become incredulous: "Are you saying that in a frustrating, humiliating situation, not everyone will get angry so that their blood boils and their palms sweat and their cheeks flush?" And my answer is yes, that is exactly what I am saying. As a matter of fact, earlier in my career, when I was giving my first talks about these ideas, you could see variations in anger firsthand in audience members who *really* didn't like the evidence. Sometimes they would shift around in their seats. Other times they shook their head in a silent "no." Once a colleague yelled at me while his face turned red and he stabbed his finger in the air. Another colleague asked me, in a sympathetic tone, if I had ever felt real fear, because if I'd ever been seriously harmed, I would never be suggesting such a preposterous idea. Yet another colleague said he would tell my brother-in-law (a sociologist of his acquaintance) that I was damaging the science of emotion. My favorite example involved a much more senior colleague, built like a linebacker and towering a foot above me, who cocked his fist and offered to punch me in the face to demonstrate what real anger looks like. (I smiled and thanked him for the thoughtful offer.) In these examples, my colleagues demonstrated the variability of anger far more handily than my presentation did.

What does it mean that four meta-analyses, summarizing hundreds of experiments, revealed no consistent, specific fingerprints in the autonomic nervous system for different emotions? It doesn't mean that emotions are an illusion, or that bodily responses are random. It means that on different occasions, in different contexts, in different studies, within the same individual and across different individuals, *the same emotion category involves different bodily responses*. Variation, not uniformity, is the norm. These results are consistent with what physiologists have known for over fifty years: different behaviors have different patterns of heart rate, breathing, and so on to support their unique movements.[28]

Despite tremendous time and investment, research has not revealed a consistent bodily fingerprint for even a single emotion.

· · ·

My first two attempts to find objective fingerprints of emotion — in the face and body — had led me smack into a closed door. But as they say, when a door closes, sometimes a window opens. My window was the unexpected realization that an emotion is not a *thing* but a category of instances, and any emotion category has tremendous variety. Anger, for example, varies far more than the classical view of emotion predicts or can explain. When you're angry at someone, do you shout and swear or do you seethe quietly? Do you tease back in reproach? How about widening your eyes and raising your eyebrows? During these times, your blood pressure might go up or down or stay the same. You might feel your heart beating in your chest, or not. Your hands might become clammy, or they might remain dry . . . whatever best prepares your body for action in that situation.

How does your brain create and keep track of all these diverse angers? How does it know which one fits the situation best? If I asked how you felt in each of these situations, would you give a detailed answer like "aggravated," "irritated," "outraged," or "vengeful" automatically with little effort? Or would you answer "angry" in each case, or simply, "I feel bad"? How do you even know the answer? These are mysteries that the classical view of emotion doesn't acknowledge.

I didn't know it at the time, but as I considered emotion categories in all their diversity, I was unwittingly applying a standard way of thinking in biology called *population thinking,* which was proposed by Darwin. A category, such as a species of animal, is a population of unique members who vary from one another, with no fingerprint at their core. The category can be described at the group level only in abstract, statistical terms. Just as no American family consists of 3.13 people, no instance of anger must include an average anger pattern (should we be able to identify one). Nor will any instance necessarily resemble the elusive fingerprint of anger. What we have been calling a fingerprint might just be a stereotype.[29]

Once I adopted a mindset of population thinking, my whole landscape shifted, scientifically speaking. I began to see variation not as error but as normal and even desirable. I continued my quest for an objective way to distinguish one emotion from another, but it wasn't quite the same quest anymore. With growing skepticism, I had only one place left to look for fingerprints. It was time to turn to the brain.*

Scientists have long studied people with brain damage (brain lesions) to try to locate an emotion in a specific area of the brain. If someone with a lesion in a particular area of the brain has difficulty experiencing or perceiv-

---

* For a quick overview of brain terminology — neurons, lobes, and so on — see appendix A.

ing a particular emotion, and only that emotion, then this would be considered evidence that the emotion specifically depends on the neurons in that region. It's a bit like finding out which circuit breakers in your house control which parts of your electrical system. Initially, all breakers are on and your house runs normally. When you shut off one breaker (giving your electrical system a lesion of sorts) and observe that your kitchen lights no longer function, you've discovered a purpose of the breaker.

The search for fear in the brain is an instructive example because for many years, scientists have considered it a textbook case of localizing emotion to a single brain area — namely, the amygdala, a group of nuclei found deep in the brain's temporal lobe.* The amygdala was first linked to fear in the 1930s when two scientists, Heinrich Klüver and Paul C. Bucy, removed the temporal lobes of rhesus monkeys. Lacking an amygdala, these monkeys approached objects and animals that would normally frighten them, like snakes, unfamiliar monkeys, or others that they'd avoided before the surgery, without hesitation. Klüver and Bucy attributed these deficits to an "absence of fear."[30]

Not long afterward, other scientists began studying humans with amygdala damage to see if those patients continued to experience and perceive fear. The most intensively studied case is a woman known as "SM," afflicted with a genetic disease that gradually obliterates the amygdala during childhood and adolescence, called Urbach-Wiethe disease. Overall, SM was (and still is) mentally healthy and of normal intelligence, but her relationship to fear seemed quite unusual in laboratory tests. Scientists showed her horror movies like *The Shining* and *The Silence of the Lambs*, exposed her to live snakes and spiders, and even took her through a haunted house, but she reported no strong feelings of fear. When SM was shown wide-eyed facial configurations from the basic emotion method's set of photos, she had difficulty identifying them as fearful. SM experienced and perceived other emotions normally.[31]

Scientists tried unsuccessfully to teach SM to feel fear, using a procedure commonly called fear learning. They showed her a picture and then immediately blasted a boat horn at one hundred decibels to startle her. This sound was meant to trigger SM's fear response if she had one. At the same time, they measured SM's skin conductance, which many scientists believe to be a measure of fear and is related to amygdala activity. After many repe-

_____

* Actually, we have two amygdalae, one each in the left and right temporal lobes.

titions of the picture followed by the horn blast, they showed SM the picture alone and measured her response. People with intact amygdalae would have learned to associate the picture with the startling sound, so if just shown the picture, their brain would predict the horn blast and their skin conductance would jump. But no matter how many times scientists paired the picture and the loud sound, SM's skin conductance didn't increase when viewing the picture alone. The experimenters concluded that SM could not learn to fear new objects.[32]

Overall, SM seemed fearless, and her damaged amygdalae seemed to be the reason. From this and other similar evidence, scientists concluded that a properly functioning amygdala was the brain center for fear.

But then, a funny thing happened. Scientists found that SM could see fear in body postures and hear fear in voices. They even found a way to make SM feel terror, by asking her to breathe air that was loaded with extra carbon dioxide. Lacking the normal degree of oxygen, SM panicked. (Don't worry, she was not in danger.) So SM could clearly feel and perceive fear under some circumstances, even without her amygdalae.[33]

As brain lesion research progressed, other people with amygdala damage were discovered and tested, and the clear and specific link between fear and the amygdala dissolved. Perhaps the most important counterevidence came from a pair of identical twins who lost the supposed fear-related parts of their amygdalae to Urbach-Wiethe disease. Both were diagnosed at the age of twelve, have normal intelligence, and have a high school education. Despite their identical DNA, equivalent brain damage, and a common environment both as children and adults, the twins have very different profiles regarding fear. One twin, BG, is much like SM: she has similar fear-related deficits yet experiences fear when breathing carbon dioxide–loaded air. The other twin, AM, has basically normal responses during fear: other brain networks are compensating for her missing amygdalae. So we have identical twins, with identical DNA, suffering from identical brain damage, living in highly similar environments, but one has some fear-related deficits while the other has none.[34]

These findings undermine the idea that the amygdala contains the circuit for fear. They point instead to the idea that the brain must have multiple ways of creating fear, and therefore the emotion category "Fear" cannot be necessarily localized to a specific region. Scientists have studied other emotion categories in lesion patients besides fear, and the results have been similarly variable. Brain regions like the amygdala are routinely important to emotion, but they are neither necessary nor sufficient for emotion.[35]

This is one of the most surprising things I learned as I began to study neuroscience: a mental event, such as fear, is not created by only one set of neurons. Instead, combinations of different neurons can create instances of fear. Neuroscientists call this principle *degeneracy.* Degeneracy means "many to one": many combinations of neurons can produce the same outcome. In the quest to map emotion fingerprints in the brain, degeneracy is a humbling reality check.[36]

My lab has observed degeneracy while performing brain scans on volunteers. We showed them evocative photos, with subject matter like skydiving and bloody corpses, and asked them how much bodily arousal they felt. Men and women reported equivalent feelings of arousal, and both had increased activity in two brain areas, the anterior insula and early visual cortex. However, women's feelings of arousal were more strongly linked to the anterior insula, while men's were more strongly linked to visual cortex. This is evidence that the same experience — feelings of arousal — was associated with different patterns of neural activity, an example of degeneracy.[37]

Another surprising thing I learned while training to be a neuroscientist, along with degeneracy, is that many parts of the brain serve more than one purpose. The brain contains *core systems* that participate in creating a wide variety of mental states. A single core system can play a role in thinking, remembering, decision-making, seeing, hearing, and experiencing and perceiving diverse emotions. A core system is "one to many": a single brain area or network contributes to many different mental states. The classical view of emotion, in contrast, considers particular brain areas to have dedicated psychological functions, that is, they are "one to one." Core systems are therefore the antithesis of neural fingerprints.[38]

To be clear, I'm not saying that every neuron in the brain does exactly the same thing, nor that every neuron can stand in for every other. (That view is called equipotentiality, and it's been long disproved.) I am saying that most neurons are multipurpose, playing more than one part, much as flour and eggs in your kitchen can participate in many recipes.

The reality of core systems has been established through virtually every experimental method in neuroscience, but it's most easily seen with brain-imaging techniques that observe the brain in action. The most common method is called functional magnetic resonance imaging (fMRI), which can peer harmlessly into the heads of living people who are experiencing emotion or perceiving emotion in others, recording the changes in magnetic signals related to firing neurons.[39]

Even so, scientists employ fMRI to search for emotion fingerprints throughout the brain. If a particular blob of brain circuitry shows increased activation during a particular emotion, researchers reason, that would be evidence that the blob computes the emotion. Scientists initially focused their scanners on the amygdala and whether it contains the neural fingerprint for fear. One key piece of evidence came from test subjects who looked at photos of so-called fear poses from the basic emotion method while in the scanner. Their amygdalae increased in activity compared to when they viewed faces with neutral expressions.[40]

As research continued, however, anomalies emerged. Yes, the amygdala was showing an increase in activity, but only in certain situations, like when the eyes of a face were staring directly at the viewer. If the eyes were gazing off to the side, the neurons in the amygdala barely changed their firing rates. Also, if test subjects viewed the same stereotyped fear pose over and over again, their amygdala activation rapidly tapered off. If the amygdala truly housed the circuit for fear, then this habituation should not occur — the circuit should fire in an obligatory way whenever it is presented with a triggering "fear" stimulus. From these contrary results, it became clear to me — and ultimately to many other scientists — that the amygdala is not the home of fear in the brain.[41]

In 2008, my lab along with neurologist Chris Wright demonstrated why the amygdala increases in activity in response to the basic emotion fear faces. The activity increases in response to *any* face — whether fearful or neutral — *as long as it is novel* (i.e., the test subjects have not seen it before). Since the wide-eyed, fearful facial configurations of the basic emotion method occur rarely in everyday life, they are novel when test subjects view them in brain-imaging experiments. These findings, and others like them, provide an alternative explanation for the original experiments that don't require the amygdala to be the brain locus of fear.[42]

Over the past two decades, this back-and-forth trajectory, with evidence followed by counterevidence, has occurred in research on every brain region that has ever been identified as the neural fingerprint of an emotion. So my lab set out to settle the question of whether brain blobs are really emotion fingerprints once and for all. We examined *every* published neuroimaging study on anger, disgust, happiness, fear, and sadness, and combined those that were usable statistically in a meta-analysis. Altogether, this comprised nearly 100 published studies involving nearly 1,300 test subjects across almost 20 years.[43]

To make sense of this large amount of data, we divided the human brain virtually into tiny cubes called voxels, the 3-D version of pixels. Then, for every voxel in the brain during every emotion studied in every experiment, we recorded whether or not an increase in activation was reported. Now we could compute the probability that each voxel would show an increase in activation during the experience or perception of each emotion. When the probability was greater than chance, we called it statistically significant.

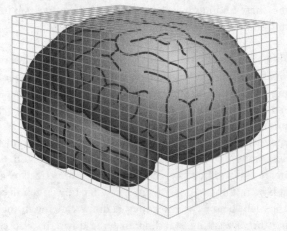

Figure 1-7: The human brain divided into voxels

Our comprehensive meta-analysis found little to support the classical view of emotion. The amygdala, for example, did show a consistent increase in activity for studies of fear, more than what you'd expect by chance, but only in a quarter of fear experience studies and about 40 percent of fear perception studies. These numbers fall short of what you'd expect for a neural fingerprint. Not only that, but the amygdala also showed a consistent increase during studies of anger, disgust, sadness, and happiness, indicating that whatever functions the amygdala was performing in some instances of fear, it was also performing those functions during some instances of those other emotions.

Interestingly, amygdala activity likewise increases during events usually considered non-emotional, such as when you feel pain, learn something new, meet new people, or make decisions. It's probably increasing now as you read these words. In fact, every supposed emotional brain region has also been implicated in creating non-emotional events, such as thoughts and perceptions.

Overall, we found that *no brain region contained the fingerprint for any single emotion.* Fingerprints are also absent if you consider multiple connected regions at once (a brain network), or stimulate individual neurons with electricity. The same results hold in experiments with other animals that allegedly have emotion circuits, such as monkeys and rats. Emotions arise from firing neurons, but no neurons are exclusively dedicated to emotion. For me, these findings have been the final, definitive nail in the coffin for localizing emotions to individual parts of the brain.[44]

· · ·

By now, I hope you see that for a very long time, people have held a mistaken view of emotions. Many research studies claim to have identified physical fingerprints that distinguish one emotion from another. Nevertheless, these supportive studies are found within a *much larger* scientific context that doesn't support the classical view.*

Some scientists might say that the contrary studies are simply wrong; after all, experiments on emotion can be pretty tricky to pull off. Some areas of the brain are really difficult to see. Heart rate is influenced by all kinds of factors that have nothing to do with emotion, like how much sleep test subjects had the night before, whether they had any caffeine in the last hour, and whether they are sitting, standing, or lying down. It's also challenging to make test subjects experience emotion on cue. Trying to evoke blood-curdling fear or brain-boiling anger is against the rules: all universities have Institutional Review Boards that prevent people like me from inflicting too much emotional agony on innocent volunteers.[45]

But even considering all these caveats, far more experiments call the classical view into doubt than we would expect by chance, or even due to inad-

---

* I sometimes hear comments from emotion researchers who subscribe to the classical view: "What about these other fifty studies, with these thousands of subjects, that show incontrovertible evidence for emotion fingerprints?" Yes, there are many such confirmatory studies, but a theory of emotion must explain all the evidence, not just the portion that supports the theory. One must not point to fifty thousand black dogs as proof that all dogs are black.

equate experimental methods. Facial EMG studies demonstrate that people move their facial muscles in many different ways, not one consistent way, when feeling an instance of the same emotion category. Large meta-analyses conclude that a single emotion category involves different bodily responses, not a single, consistent response. Brain circuitry operates by the many-to-one principle of degeneracy: instances of a single emotion category, such as fear, are handled by different brain patterns at different times and in different people. Conversely, the same neurons can participate in creating different mental states (one-to-many).

I hope you've caught the pattern emerging here: *variation is the norm.* Emotion fingerprints are a myth.

If we want to truly understand emotions, we must start taking that variation seriously. We must consider that an emotion word, like "anger," does not refer to a specific response with a unique physical fingerprint but to a group of highly variable instances that are tied to specific situations. What we colloquially call emotions, such as anger, fear, and happiness, are better thought of as emotion categories, because each is a collection of diverse instances. Just as instances of the category "Cocker Spaniel" vary in their physical attributes (tail length, nose length, coat thickness, running speed, and so on) more than genes alone can account for, so might instances of "Anger" vary in their physical manifestations (facial movements, heart rate, hormones, vocal acoustics, neural activity, and so on), and this variation might be related to the environment or context.[46]

When you adopt a mindset of variation and population thinking, so-called emotion fingerprints give way to better explanations. Here's an example of what I mean. Some scientists, using techniques from artificial intelligence, can train a software program to recognize many, many brain scans of people experiencing different emotions (say, anger and fear). The program computes a statistical pattern that summarizes each emotion category and then — here's the cool part — can actually analyze new scans and determine if they are closer to the summary pattern for anger or fear. This technique, called pattern classification, works so well that it's sometimes called "neural mind-reading."

Some of these scientists claim that the statistical summaries depict neural fingerprints for anger and fear. But that's a gigantic logical error. The statistical pattern for fear is not an actual brain state, just an abstract summary of many instances of fear. These scientists are mistaking a mathematical average for the norm.[47]

My collaborators and I applied pattern classification to our meta-analysis of brain-imaging studies of emotion. Our computer program learned to classify scans from about 150 different studies. We found patterns across the brain that predict better than chance whether the test subjects in a specific study were experiencing anger, disgust, fear, happiness, or sadness. These patterns are not emotion fingerprints, however. The pattern for anger, for example, consists of a set of voxels across the brain, but that pattern need not appear in any individual brain scan for anger. The pattern is an abstract summary. In fact, no individual voxel appeared in all the scans of anger.[48]

When properly applied, pattern classification is an example of population thinking. A species, you may recall, is a collection of diverse individuals, so it can be summarized only in statistical terms. The summary is an abstraction that does not exist in nature — it does not describe any individual member of the species. Where emotion is concerned, on different occasions and in different people, different combinations of neurons can create instances of an emotion category like anger. Even when two experiences of anger feel the same to you, they can have different brain patterns via degeneracy. But we can still summarize many varying instances of anger to describe how, in abstract terms, they might be distinguishable from all the varying instances of fear. (Analogy: no two Labrador Retrievers are identical, but they're all distinguishable from Golden Retrievers.)

My long search for fingerprints in the face, body, and brain brought me to a realization that I had not expected — that we need a new theory of what emotions are and where they come from. In the chapters that follow, I introduce you to this new theory, which accounts for all the findings of the classical view as well as all the inconsistencies you've just seen. By moving beyond fingerprints and following the evidence, we will seek a better and more scientifically justified understanding, not only of emotion but also of ourselves.

2

Emotions Are Constructed

P lease take a look at the black splotches in figure 2-1.

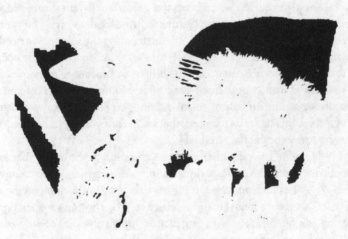

Figure 2-1: Mystery blobs

If this is your first time viewing these blobs, your brain is working hard to make sense of them. Neurons in your visual cortex are processing the lines and edges. Your amygdala is firing rapidly because the input is novel. Other brain regions are sifting through your past experiences to determine if you've encountered anything like this input before and are conversing with your body to prepare it for an as-yet-undetermined action. Most likely, you are in a state called *experiential blindness,* seeing only black blobs of unknown origin.

To cure your experiential blindness, look at the image on page 308 (appendix B). Then come back to this page. You should no longer see formless blobs but a familiar object.

What just happened in your brain to change your perception of these blobs? Your brain added stuff from the full photograph into its vast array of prior experiences and *constructed* the familiar object you now see in the blobs. Neurons in your visual cortex changed their firing to create lines that aren't present, linking the blobs into a shape that isn't physically there. You are, in a manner of speaking, hallucinating. Not the scary "I'd better get to the hospital" kind of hallucination, but the everyday "my brain is built to work like this" hallucination.

Your experience with figure 2-1 reveals a couple of insights. Your past experiences — from direct encounters, from photos, from movies and books — give meaning to your present sensations. Additionally, the entire process of construction is invisible to you. No matter how hard you try, you cannot observe yourself or experience yourself constructing the image. We needed a specially designed example to unmask the fact that construction is occurring. You consciously experienced the shift from unknown to known because you saw figure 2-1 both before and after you had the relevant knowledge to draw on. The process of construction is so habitual that you might never again see this figure as formless shapes, even if you try hard to un-see it and recapture experiential blindness.

This little magic trick of the brain is so common and normal that psychologists discovered it time and time again before they understood how it worked. We will call it *simulation.* It means that your brain changed the firing of its own sensory neurons in the absence of incoming sensory input. Simulation can be visual, as with our picture, or involve any of your other senses. Ever have a song playing in your head that you can't get rid of? That audio hallucination is also a simulation.[1]

Think of the last time someone handed you a red, juicy apple. You reached out for it, took a bite, and experienced the tart flavor. During those moments, neurons were firing in the sensory and motor regions of your brain. Motor neurons fired to produce your movements, and sensory neurons fired to process your sensations of the apple, like its red color with a blush of green; its smoothness against your hand; its crisp, floral scent; the audible crunch when you bit into it; and its tangy taste with a hint of sweetness. Other neurons made your mouth water to release enzymes and begin digestion, released cortisol to prepare your body to metabolize the sugars in the apple, and perhaps made your stomach churn a bit. But here's the cool thing: just now, when you read the word "apple," your brain responded to a certain extent as if an apple were actually present. Your brain combined bits and pieces of knowledge of previous apples you've seen and tasted, and changed the firing of neurons in your sensory and motor regions to construct a mental instance of the concept "Apple." Your brain simulated a nonexistent apple using sensory and motor neurons. Simulation happens as quickly and automatically as a heartbeat.[2]

For my daughter's twelfth birthday, we exploited the power of simulation (and had some fun) by throwing a "gross foods" party. When her guests arrived, we served them pizza doctored with green food coloring so the cheese looked like fuzzy mold, and peach gelatin laced with bits of vegetables to look like vomit. For drinks, we served white grape juice in medical urine sample cups. Everybody was exuberantly disgusted (it was perfect twelve-year-old humor), and several guests could not bring themselves to touch the food as they involuntarily simulated vile tastes and smells. The *pièce de résistance,* however, was the party game we played after lunch: a simple contest to identify foods by their smell. We used mashed baby food — peaches, spinach, beef, and so on — and artfully smeared it on diapers, so it looked exactly like baby poo. Even though the guests knew that the smears were food, several actually gagged from the simulated smell.[3]

Simulations are your brain's guesses of what's happening in the world. In every waking moment, you're faced with ambiguous, noisy information from your eyes, ears, nose, and other sensory organs. Your brain uses your past experiences to construct a hypothesis — the simulation — and compares it to the cacophony arriving from your senses. In this manner, simulation lets your brain impose meaning on the noise, selecting what's relevant and ignoring the rest.

The discovery of simulation in the late 1990s ushered in a new era in psychology and neuroscience. Scientific evidence shows that what we see, hear, touch, taste, and smell are largely simulations of the world, not reactions to it. Forward-looking thinkers speculate that simulation is a common mechanism not only for perception but also for understanding language, feeling empathy, remembering, imagining, dreaming, and many other psychological phenomena. Our common sense might declare that thinking, perceiving, and dreaming are different mental events (at least to those of us in Western cultures), yet one general process describes them all. Simulation is the default mode for all mental activity. It also holds a key to unlocking the mystery of how the brain creates emotions.[4]

Outside your brain, simulation can cause tangible changes in your body. Let's try a little creative simulation with our bee. In your mind's eye, see the bee bouncing lightly on the petal of a fragrant white flower, buzzing around as it searches for pollen. If you're fond of bees, then the flutter of imaginary wings is right now causing other neurons to prepare your body to move in for a closer look — preparing your heart to beat faster, your sweat glands to fill, and your blood pressure to decrease. Or if you have been badly stung in the past, your brain may ready your body to run away or make a swatting motion, formulating some other pattern of physical changes. Each time your brain simulates sensory input, it prepares automatic changes in your body that have the potential to change your feeling.

Your bee-related simulations are rooted in your mental *concept* of what a "Bee" is. This concept not only includes information about the bee itself (what it looks and sounds like, how you act on it, what changes in your autonomic nervous system allow your action, etc.), but also information contained in other concepts related to bees ("Meadow," "Flower," "Honey," "Sting," "Pain," etc.). All this information is integrated with your concept "Bee," guiding how you simulate the bee in this particular context. So, a concept like "Bee" is actually a collection of neural patterns in your brain, representing your past experiences. Your brain combines these patterns in different ways to perceive and flexibly guide your action in new situations.[5]

Using your concepts, your brain groups some things together and separates others. You can look at three mounds of dirt and perceive two of them as "Hills" and one as a "Mountain," based on your concepts. Construction treats the world like a sheet of pastry, and your concepts are cookie cutters that carve boundaries, not because the boundaries are natural, but because they're useful or desirable. These boundaries have physical limitations of course; you'd never perceive a mountain as a lake. Not everything is relative.[6]

Your concepts are a primary tool for your brain to guess the meaning of incoming sensory inputs. For example, concepts give meaning to changes in sound pressure so you hear them as words or music instead of random noise. In Western culture, most music is based on an octave divided into twelve equally spaced pitches, an arrangement known as the twelve tone equal-tempered scale. All people of Western culture with normal hearing have a concept for this ubiquitous scale, even if they can't explicitly describe it. Not all music uses this scale, however. When Westerners hear Indonesian gamelan music for the first time, which is based on seven pitches per octave with varied tunings, it's more likely to sound like noise. A brain that's been wired by listening to twelve-tone scales doesn't have a concept for that music. Personally, I am experientially blind to dubstep, although my teenage daughter clearly has that concept.

Concepts also give meaning to the chemicals that create tastes and smells. If I served you pink ice cream, you might expect (simulate) the taste of strawberry, but if it tasted like fish, you would find it jarring, perhaps even disgusting. If I instead introduced it as "chilled salmon mousse" to give your brain fair warning, you might find the same taste delicious (assuming you enjoy salmon). You might think of food as existing in the physical world, but in fact the concept "Food" is heavily cultural. Obviously, there are some biological constraints; you can't eat razor blades. But there are some perfectly edible substances that we don't all perceive as food, such as hachinoko, a Japanese delicacy made of baby bees, which most Americans would vigorously avoid. This cultural difference is due to concepts.[7]

Every moment that you are alive, your brain uses concepts to simulate the outside world. Without concepts, you are experientially blind, as you were with the blobby bee. With concepts, your brain simulates so invisibly and automatically that vision, hearing, and your other senses seem like reflexes rather than constructions.

Now consider this: what if your brain uses this same process to make meaning of the sensations from *inside your body* — the commotion arising from your heartbeat, breathing, and other internal movements?

From your brain's perspective, your body is just another source of sensory input. Sensations from your heart and lungs, your metabolism, your changing temperature, and so on, are like the ambiguous blobs of figure 2-1. These purely physical sensations inside your body have no objective psychological meaning. Once your concepts enter the picture, however, those sensations may take on additional meaning. If you feel an ache in your stomach while sitting at the dinner table, you might experience it as hunger. If flu season is just around the corner, you might experience that same ache as nausea. If you are a judge in a courtroom, you might experience the ache as a gut feeling that the defendant cannot be trusted. In a given moment, in a given context, your brain uses concepts to give meaning to internal sensations as well as to external sensations from the world, all simultaneously. From an aching stomach, your brain constructs an instance of hunger, nausea, or mistrust.[8]

Now consider that same stomachache if you're sniffing a diaper heavy with pureed lamb, as my daughter's friends did at her gross foods birthday party. You might experience the ache as disgust. Or if your lover has just walked into the room, you might experience the ache as a pang of longing. If you're in a doctor's office waiting for the results of a medical test, you might experience that same ache as an anxious feeling. In these cases of disgust, longing, and anxiety, the concept active in your brain is an *emotion concept*. As before, your brain makes meaning from your aching stomach, together with the sensations from the world around you, by constructing an instance of that concept.

An instance of *emotion*.

And that just might be how emotions are made.

• • •

Back when I was in graduate school, a guy in my psychology program asked me out on a date. I didn't know him very well and was reluctant to go because, honestly, I wasn't particularly attracted to him, but I had been cooped up too long in the lab that day, so I agreed. As we sat together in a coffee shop, to my surprise, I felt my face flush several times as we spoke. My stomach fluttered and I started having trouble concentrating. Okay, I realized, I was wrong. I am clearly attracted to him. We parted an hour later — after I agreed to go out with him again — and I headed home, intrigued. I walked into my apartment, dropped my keys on the floor, threw up, and spent the next seven days in bed with the flu.

The same neural process of construction that simulates a bee from blobs also constructs feelings of attraction from a fluttering stomach and a flushing face. An emotion is your brain's *creation* of what your bodily sensations mean, in relation to what is going on around you in the world. Philosophers have long proposed that your mind makes sense of your body in the world, from René Descartes in the seventeenth century to William James (considered the father of American psychology) in the nineteenth; as you will learn, however, neuroscience now shows us how this process — and much more — occurs in the brain to make an emotion on the spot. I call this explanation the *theory of constructed emotion*:[9]

In every waking moment, your brain uses past experience, organized as concepts, to guide your actions and give your sensations meaning. When the concepts involved are emotion concepts, your brain constructs instances of emotion.

If a swarm of buzzing bees is squeezing underneath your front door while your heart is pounding in your chest, your brain's prior knowledge of stinging insects gives meaning to the sensations from your body and to the sights, sounds, smells, and other sensations from the world, simulating the swarm, the door, and an instance of fear. The exact same bodily sensations in another context, like watching a fascinating film about the hidden lives of bees, might construct an instance of excitement. Or if you see a picture of a smiling cartoon bee in a children's book, reminding you of a beloved niece whom you took to a Disney movie, you could mentally construct the bee, the niece, and an instance of pleasant nostalgia.

My experience in the coffee shop, where I felt attraction when I had the flu, would be called an error or misattribution in the classical view, but it's no more a mistake than seeing a bee in a bunch of blobs. An influenza virus in my blood contributed to fever and flushing, and my brain made meaning from the sensations in the context of a lunch date, constructing a genuine feeling of attraction, in the normal way that the brain constructs any other mental state. If I'd had exactly the same bodily sensations while at home in bed with a thermometer, my brain might have constructed an instance of "Feeling Sick" using the same manufacturing process. (The classical view, in contrast, would require feelings of attraction and malaise to have different bodily fingerprints triggered by different brain circuitry.)[10]

Emotions are not reactions to the world. You are not a passive receiver of sensory input but an active constructor of your emotions. From sensory input and past experience, your brain constructs meaning and prescribes action. If you didn't have concepts that represent your past experience, all your sensory inputs would just be noise. You wouldn't know what the sensations are, what caused them, nor how to behave to deal with them. With concepts, your brain makes meaning of sensation, and sometimes that meaning is an emotion.

The theory of constructed emotion and the classical view of emotion tell vastly different stories of how we experience the world. The classical view is intuitive — events in the world trigger emotional reactions inside of us. Its story features familiar characters like thoughts and feelings that live in distinct brain areas. The theory of constructed emotion, in contrast, tells a story that doesn't match your daily life — your brain invisibly constructs everything you experience, including emotions. Its story features unfamiliar characters like simulation and concepts and degeneracy, and it takes place throughout the whole brain at once.

This unfamiliar story creates a challenge because people expect stories with familiar structures. Every superhero story is assumed to have a villain. Every romantic comedy requires an attractive couple faced with a humorous misunderstanding that turns out all right in the end. Our challenge here is that the dynamics of the brain, and how emotions are made, do not follow a linear, cause-and-effect sort of story. (This challenge is common in science; for example, in quantum mechanics, the distinction between a cause and an effect is not meaningful.) Nevertheless, every book must tell a story, even for a nonlinear subject like brain function. Mine will occasionally have to defy the usual linear framework of human storytelling.

For now, my aim is simply to give you some intuition about the construction of emotion and why this scientific explanation makes sense. We'll see later that this theory incorporates the most up-to-date, neuroscientific understanding of how the brain works, and it explains the great variation in emotional experiences and perceptions in everyday life. It can help us figure out how instances of happiness, sadness, anger, fear, and other emotion categories are constructed by the same brain mechanism that constructed the blobby bee, the juicy apple, and the smell of poo from mashed baby food, with no need for emotion circuits or other biological fingerprints.

. . .

I'm not the first person to propose that emotions are made. The theory of constructed emotion belongs to a broader scientific tradition called *construction*, which holds that your experiences and behaviors are created in the moment by biological processes within your brain and body. Construction is based on a very old set of ideas that date back to Ancient Greece, when the philosopher Heraclitus famously wrote, "No man ever steps in the same river twice," because only a mind perceives an ever-changing river as a distinct body of water. Today, constructionism spans many topics including memory, perception, mental illness, and, of course, emotion.[11]

A constructionist approach to emotion has a couple of core ideas. One idea is that an emotion category such as anger or disgust does not have a fingerprint. One instance of anger need not look or feel like another, nor will it be caused by the same neurons. Variation is the norm. Your range of angers is not necessarily the same as mine, although if we were raised in similar circumstances, we will likely have some overlap.

Another core idea is that the emotions you experience and perceive are not an inevitable consequence of your genes. What's inevitable is that you'll have *some kinds* of concepts for making sense of sensory input from your body in the world because, as we learn in chapter 5, your brain has wiring for this purpose. Even single-celled animals can make sense of changes in their environment. But *particular* concepts like "Anger" and "Disgust" are not genetically predetermined. Your familiar emotion concepts are built-in only because you grew up in a particular social context where those emotion concepts are meaningful and useful, and your brain applies them outside your awareness to construct your experiences. Heart rate changes are inevitable; their emotional meaning is not. Other cultures can and do make other kinds of meaning from the same sensory input.[12]

The theory of constructed emotion incorporates ideas from several flavors of construction. One flavor, called social construction, studies the role of social values and interests in determining how we perceive and act in the world. An example would be whether or not Pluto is a planet, which is a decision not based in astrophysics but in culture. Spherical rocks in space are objectively real and come in various sizes, but the idea of a "Planet," representing a particular combination of features of interest, is made up by people. Each of us understands the world in a way that is useful but not necessarily true in some absolute, objective sense. Where emotion is concerned, social construction theories ask how feelings and perceptions are influenced by our social roles or beliefs. For example, my perceptions are influenced by the fact that I am a woman, a mother, an atheist who is culturally Jewish, and a rather pale person living in a country that once enslaved people for having more melanin in their skin than I do. Social construction tends to ignore biology, however, as irrelevant to emotion. Instead, the theories suggest that emotions are triggered differently depending on your social role. Social constructionist theories, then, are primarily concerned with social circumstances in the world outside you, without considering how those circumstances affect the brain's wiring.[13]

Another flavor of construction, known as psychological construction, turns this focus inward. It proposes that your perceptions, thoughts, and feelings are themselves constructed from more basic parts. Some nineteenth-century philosophers viewed the mind like a big chemistry set, combining simpler sensations into thoughts and emotions the way that atoms combine to make molecules. Others saw the mind as a set of all-purpose parts, like Lego blocks, that contribute to various mental states like cognitions and emotions. William James proposed that our incredibly varied emotional experiences are constructed from common ingredients. "Emotional brain processes," he wrote, "not only resemble the ordinary sensorial brain-processes, but in very truth are nothing but such processes variously combined." In the 1960s, the psychologists Stanley Schachter and Jerome Singer famously injected test subjects with adrenaline — without the subjects' knowledge — and saw them experience this mysterious arousal as anger or euphoria, depending on the context surrounding them. In all these views, an instance of anger or elation does not reveal its causal mechanisms — a marked contrast to the classical view, where each emotion has a dedicated mechanism in the brain, and the same word (e.g., "sadness") names the mechanism and its product. In recent years, a new generation of scientists has been crafting psychological construction-based theories for understanding emotions and how they work. Not every theory agrees on every assumption, but together they assert that emotions are made, not triggered; emotions are highly variable, without fingerprints; and emotions are not, in principle, distinct from cognitions and perceptions.[14]

You might be surprised to learn that these same principles of construction appear to hold for the brain's physical architecture, an idea called neuroconstruction. Consider two neurons that are connected by a synapse. Clearly these brain cells exist in an objective sense. But there is no objective way to tell whether the two neurons are part of a unit called a "circuit" or "system," or whether each neuron belongs to a separate circuit where one "regulates" the other. The answer depends entirely on human perspective. Similarly, your brain's interconnections are not inevitable consequences of your genes alone. We know today that experience is a contributing factor. Your genes turn on and off in different contexts, including the genes that shape your brain's wiring. (Scientists call this phenomenon plasticity.) That means some of your synapses literally come into existence because other people talked to you or treated you in a certain way. In other words, construction extends all the way down to the cellular level. The macro structure of your brain is largely predetermined, but the microwiring is not. As a consequence, past experience helps determine your future experiences and perceptions. Neuroconstruction explains how human infants are born without the ability to recognize a face but can develop that capacity within the first few days after birth. It also explains how early cultural experiences — for instance, how often your caregivers were in physical contact with you, and whether you slept alone in a crib or in a family bed — differentially shape the wiring of the brain.[15]

The theory of constructed emotion incorporates elements of all three flavors of construction. From social construction, it acknowledges the importance of culture and concepts. From psychological construction, it considers emotions to be constructed by core systems in the brain and body. And from neuroconstruction, it adopts the idea that experience wires the brain.

• • •

The theory of constructed emotion tosses away the most basic assumptions of the classical view. For instance, the classical view assumes that happiness, anger, and other emotion categories each have a distinctive bodily fingerprint. In the theory of constructed emotion, *variation* is the norm. When you are angry, you might scowl, frown mildly or severely, shout, laugh, or even stand in eerie calmness, depending on what works best in the situation. Your heart rate likewise might increase, decrease, or stay the same, whatever is necessary to support the action you are performing. When you perceive someone else as angry, your perceptions are similarly varied. An emotion word such as "anger," therefore, names a population of diverse instances, each one constructed to best guide action in the immediate circumstance. There is no single difference between anger and fear, because there's no single "Anger" and no single "Fear." These ideas are inspired by William James, who wrote at length on the variability of emotional life, and by Charles Darwin's revolutionary idea that a biological category, such as a species, is a population of unique individuals.[16]

You can think about emotion categories like cookies. There are crisp ones, chewy ones, sweet ones, savory ones, large, small, flat, rounded, rolled, sandwiched, floured, flourless, and more. The members of the category "Cookie" vary tremendously but are deemed equivalent for some purpose: to be a tasty snack or dessert. Cookies need not look the same or be created with the same recipe; they are a population of diverse instances. Even within a more fine-grained category like "Chocolate Chip Cookie," there is still diversity created by the type of chocolate, the amount of flour, the ratio of brown sugar to white sugar, the fat content of the butter, and the time spent chilling the dough. Likewise, any category of emotion such as "Happiness" or "Guilt" is filled with variety.[17]

The theory of constructed emotion dispenses with fingerprints not only in the body but also in the brain. It avoids questions that imply a neural fingerprint exists, like "Where are the neurons that trigger fear?" The word "where" has a built-in assumption that a particular set of neurons activates every time you and everyone else on the planet feel afraid. In the theory of constructed emotion, a category of emotion such as sadness, fear, or anger has no distinct brain location, and each instance of emotion is a whole-brain state to be studied and understood. Therefore we ask how, not where, emotions are made. The more neutral question, "How does the brain create an instance of fear?" does not presume a neural fingerprint behind the scenes, only that experiences and perceptions of fear are real and worthy of study.

If instances of emotion are like cookies, then the brain is like a kitchen, stocked with common ingredients such as flour, water, sugar, and salt. Beginning with these ingredients, we can create diverse foods such as cookies, bread, cake, muffins, biscuits, and scones. Likewise, your brain has core "ingredients," which we called core systems in chapter 1. They combine in complex ways, roughly analogous to recipes, to produce diverse instances of happiness, sadness, anger, fear, and so on. The ingredients themselves are multipurpose, not dedicated to emotions but participating in their construction. Instances of two different emotion categories, such as fear and anger, can be made from similar ingredients, just as cookies and bread both contain flour. Conversely, two instances of the same emotion category, like fear, will have some variation in their ingredients, just as some cookies have nuts and others do not. This phenomenon is our old friend degeneracy at work: different instances of fear are constructed by different combinations of the core systems throughout the brain. We can describe the instances of fear together by a pattern of brain activity, but this pattern is a statistical summary and need not describe any actual instance of fear.[18]

My kitchen analogy, like all analogies in science, has its limits. A brain network, as a core system, is not a "thing" like flour or salt. It's a collection of neurons that we view as a unit, statistically speaking, but only a subset of those neurons participate at any given time. If you have ten feelings of fear that involve a particular brain network, each feeling can involve different neurons from the network.* This is degeneracy at the network level. Additionally, cookies and bread are discrete, physical objects, whereas instances

---

* If you prefer sports analogies, a network is like a baseball team. In a given moment, only nine out of the team's twenty-five players participate, and the nine may change at any time, yet we say that "the team" won or lost the game.

of emotion are momentary snapshots of continuous brain activity, and we merely perceive these snapshots as discrete events. Nevertheless, you may find the kitchen analogy useful to imagine how interacting networks produce diverse mental states.[19]

The core systems that construct the mind interact in complex ways, without any central manager or chef to run the show. However, these systems cannot be understood independently like the disassembled parts of a machine, or like so-called emotion modules or organs. That's because their interactions produce new properties that are not present in the parts alone. By analogy, when you bake bread with flour, water, yeast, and salt, a new product emerges from the complex, chemical interplay of the ingredients. Bread has its own emergent properties, like "crustiness" and "chewiness," that are not present in its ingredients alone. In fact, if you try to identify all the ingredients by tasting the finished bread, you are in for a difficult time. Consider the salt: bread doesn't taste salty even though salt is absolutely essential. Similarly, an instance of fear cannot be reduced to mere ingredients. Fear is not a bodily pattern — just as bread is not flour — but emerges from the interactions of core systems. An instance of fear has irreducible, emergent properties not found in the ingredients alone, such as unpleasantness (as your car skids out of control on a slippery highway) or pleasantness (on an undulating rollercoaster). You cannot reverse-engineer a recipe for an instance of fear from a feeling of fear.[20]

Even if we did know the ingredients of emotion but studied them only in isolation, we'd get an inaccurate understanding of how they work together to construct emotion. If we study salt in isolation by tasting and weighing it, we will not understand how it contributes to the creation of bread. That's because salt interacts chemically with the other ingredients during baking: controlling yeast growth, shoring up the gluten in the dough, and, most importantly, enhancing flavor. To understand how salt transforms a recipe of bread, you must watch it work in context. Likewise, each ingredient of emotion must be studied in the context of the rest of the brain that influences it. This philosophy, known as holism, explains why I get different results each time I bake bread in my own kitchen, even using exactly the same recipe. I weigh every ingredient. I knead the dough for the same amount of time. I set the oven to the same temperature. I count the number of sprays of water I spritz into the oven to make the bread crusty. It's all very systematic, and yet, the result is sometimes lighter, sometimes heavier, sometimes sweeter. That's because baking has additional context that the recipe doesn't mention, like the amount of force I use in kneading, the humidity in the kitchen, and the precise temperature at which the dough rises. Holism explains why bread baked in my home in Boston is never as tasty as bread baked at my friend Ann's house in Berkeley, California. The Berkeley loaf has a superior flavor because of the different yeasts floating naturally in the air and the elevation above sea level. These additional variables can dramatically impact the end product, and expert bakers know this. Holism, emergent properties, and degeneracy are the very antithesis of fingerprints.[21]

After bodily and neural fingerprints, the next core assumption of the classical view we discard is how emotions evolved. The classical view proposes that we have a gift-wrapped animal brain — ancient emotion circuits passed down from ancestral animals, wrapped in uniquely human circuitry for rational thought — like icing on an already-baked cake. This view is often touted as "the" evolutionary theory of emotion, when in reality it is just one evolutionary theory.

Construction incorporates the latest scientific findings about Darwinian natural selection and population thinking. For example, the many-to-one principle of degeneracy — many different sets of neurons can produce the same outcome — brings about greater robustness for survival. The one-to-many principle — any single neuron can contribute to more than one outcome — is metabolically efficient and increases the computational power of the brain. This kind of brain creates a flexible mind without fingerprints.[22]

The final major assumption of the classical view is that certain emotions are inborn and universal: all healthy people around the world are supposed to display and recognize them. The theory of constructed emotion, in contrast, proposes that emotions are not inborn, and if they are universal, it's due to shared concepts. What's universal is the ability to form concepts that make our physical sensations meaningful, from the Western concept "Sadness" to the Dutch concept *Gezellig* (a specific experience of comfort with friends), which has no exact English translation.

By analogy, think about cupcakes and muffins. These two types of baked goods have the same shape and are based on the same set of ingredients: flour, sugar, shortening, and salt. Both have similar accompanying ingredients such as raisins, nuts, chocolate, carrots, and bananas. You cannot distinguish a muffin from a cupcake by its chemistry, in the way you can easily distinguish flour from salt, or a bee from a bird. And yet, one is a breakfast food while the other is a dessert. Their major distinguishing feature is the time of day at which they are eaten. This difference is entirely cultural and learned, not physical. The muffin-cupcake distinction is *social reality*: when objects in the physical world, like baked goods, take on additional functions by social agreement. Likewise, emotions are social reality. A physical event like a change in heart rate, blood pressure, or respiration becomes an emotional experience only when we, with emotion concepts that we have learned from our culture, imbue the sensations with additional functions by social agreement. From the widened eyes of a friend we may perceive fear or surprise, again depending on which concepts we use. We must not confuse physical reality, such as changes in heart rate or widened eyes, with the social reality of emotion concepts.[23]

Social reality is not just about words — it gets under your skin. If you perceive the same baked good as a decadent "cupcake" or a healthful "muffin," research suggests that your body metabolizes it differently. Likewise, the words and concepts of your culture help to shape your brain wiring and your physical changes during emotion.[24]

Now that we've discarded so many assumptions of the classical view, we need a new vocabulary to discuss emotion. Familiar phrases like "facial expression" seem like common sense but tacitly assume that emotion fingerprints exist and that the face broadcasts emotion. You may have noticed in chapter 1 that I coined a more neutral term, *facial configuration,* because the English language has no word for "the set of facial muscle movements that the classical view treats as a coordinated unit." I've also disambiguated the word "emotion," because it could refer to a single instance of (say) feeling happy, or it could mean the whole category of happiness. When you construct an emotional experience of your own, I call it an *instance of emotion.* I refer to fear, anger, happiness, sadness, and so on, in general as *emotion categories,* because each word names a population of diverse instances, just like the word "cookie" names a population of diverse instances. If I were very strict, I would banish the phrase "an emotion" from our vocabulary so we don't imply its objective existence in nature, and always speak of instances and categories. But that's a bit too Orwellian, so I'll just take care to indicate when I mean an instance versus the category.

Likewise, we do not "recognize" or "detect" emotions in others. These terms imply that an emotion category has a fingerprint that exists in nature, independent of any perceiver, waiting to be found. Any scientific question about "detecting" emotion automatically presumes a certain kind of answer. In the construction mindset, I speak of *perceiving* an instance of emotion. Perception is a complex mental process that does not imply a neural fingerprint behind the emotion, merely that an instance of emotion occurred somehow. I also avoid verbs like "triggering" emotion, and phrases like "emotional reaction" and emotions "happening to you." Such wording implies that emotions are objective entities. Even when you feel no sense of agency when experiencing emotion, which is most of the time, you are an active participant in that experience.

I also do not speak of perceiving someone's emotion "accurately." Instances of emotion have no objective fingerprints in the face, body, and brain, so "accuracy" has no scientific meaning. It has a social meaning — we certainly can ask whether two people agree in their perceptions of emotion, or whether a perception is consistent with some norm. But perceptions exist within the perceiver.[25]

These linguistic guidelines might seem picky at first, but I hope you will come to see their importance. This new vocabulary is critical for understanding emotions and how they are made.

• • •

At the beginning of this chapter, you looked upon a bunch of blobs, applied a collection of concepts, and the image of a bee materialized. This was no trick of your brain but a demonstration of how your brain works all the time — you actively participate in determining what you see, and most of the time you have no awareness you are doing so. The same processes that construct meaning from mere visual input provide a solution to the puzzle of human emotion. After conducting hundreds of experiments in my lab, and reviewing thousands more by other researchers, I've come to a profoundly unintuitive conclusion shared by a growing number of scientists. Emotions do not shine forth from the face nor from the maelstrom of your body's inner core. They don't issue from a specific part of the brain. No scientific innovation will miraculously reveal a biological fingerprint of any emotion. That's because our emotions aren't built-in, waiting to be revealed. They are *made*. By *us*. We don't *recognize* emotions or *identify* emotions: we *construct* our own emotional experiences, and our perceptions of others' emotions, on the spot, as needed, through a complex interplay of systems. Human beings are not at the mercy of mythical emotion circuits buried deep within animalistic parts of our highly evolved brain: we are architects of our own experience.

These ideas do not match our experiences in daily life, where emotions seem to emerge like little bombs to disrupt whatever we were thinking or doing a moment before. Likewise, when we look at other people's faces and bodies, they seem to announce what their owners are feeling, without input or effort on our part, even when the owners themselves might be unaware. And when we look at our growling dogs and purring cats, we seem to detect their emotions too. But these personal experiences, no matter how compelling they may seem, do not reveal how the brain creates emotion, any more than our experience watching the sun move across the sky means that it revolves around the Earth.

If you're a newcomer to construction, then ideas like "emotion concepts" and "emotion perceptions" and "facial configurations" are probably not second nature for you yet. To really understand emotions — in a way that is consistent with contemporary knowledge of evolution and neuroscience — you have to give up some deeply ingrained ways of thinking. To help you along that path, in the next chapter I give you some practice with construction. We'll take a close look at a famous scientific finding about emotion that many people consider a fact, and which propelled the classical view into a dominant position in psychology for five decades. We'll unpack it from the perspective of construction and watch certainty transform into doubt. Strap on your seatbelt.

3

The Myth of Universal Emotions

Take a look at the woman in figure 3-1, who is screaming in terror. Most people who were born and raised in a Western culture can effortlessly see this emotion in her face, even with no other context in the photograph.

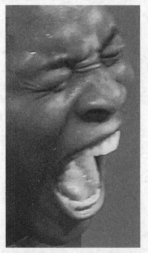

Figure 3-1: Perceiving terror
in a woman's face

Except . . . she isn't feeling terror. This photograph actually shows Serena Williams immediately after she beat her sister Venus in the 2008 U.S. Open tennis finals. Turn to page 310 (appendix C) to see the full photograph. In context, the facial configuration takes on new meaning.[1]

If Williams's face subtly transformed before your eyes once you knew the context, you are not alone. This is a common experience. How did your brain accomplish this shift? The first emotion word I used, "terror," caused your brain to simulate past facial configurations that you have seen of people feeling fear. You were almost certainly not aware of these simulations, but they shaped your perception of Williams's face. When I explained the photo's context — winning a crucial tennis match — your brain applied its conceptual knowledge of tennis and winning to simulate facial configurations that you've seen of people experiencing exultation. These simulations again influenced how you perceived Williams's face. In each case, your emotion concepts helped you make meaning from the image.[2]

In real life, we usually encounter faces in context, attached to bodies and associated with voices, smells, and other surrounding details. These details cue your brain to use particular concepts to simulate and construct your perception of emotion. That's why, in the full photo of Serena Williams, you perceive triumph, not terror. In fact, you *depend* on emotion concepts each time you experience another person as emotional. Knowledge of the concept "Sadness" is required to see a pout as sadness, knowledge of "Fear" to see widened eyes as fearful, and so on.[3]

According to the classical view, you shouldn't need concepts to perceive emotion, because emotions are supposed to have universal fingerprints that everyone around the world can recognize from birth. You're about to learn otherwise. By applying the theory of constructed emotion, combined with a little reverse engineering, you'll see that concepts are a key ingredient for perceiving emotions. We'll begin with the best experimental technique for demonstrating that certain emotions are universal: the basic emotion method used by Silvan Tomkins, Carroll Izard, and Paul Ekman (chapter 1). Then we'll systematically reduce the amount of emotion concept knowledge available to our test subjects. If their emotion perception becomes more and more impaired, then we've revealed that concepts are a critical ingredient to constructing emotion perceptions. We'll also learn how emotions can *appear* to be universally recognized under certain conditions, opening the door to a new, better understanding of how emotions are made.[4]

• • •

The basic emotion method, you may recall, was designed to study "emotion recognition." On each trial of an experiment, a test subject views the photograph of a face, carefully posed by a trained actor, to represent the so-called expressions of certain emotions: smiling for happiness, scowling for anger, pouting for sadness, and so on. Accompanying the photo is a small set of English emotion words, depicted in figure 3-2, and the subject chooses the word that best matches the face. The same words appear trial after trial. In another version of the basic emotion method, a test subject selects the best of two or three photos to match a brief story or descriptive phrase, such as "Her mother died, and she feels very sad."

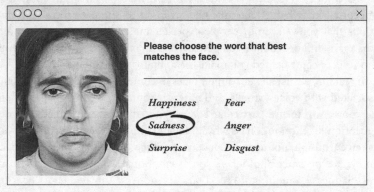

Figure 3-2: Basic emotion method: picking a word to match the face

Test subjects from all around the world (Germany, France, Italy, United Kingdom, Scotland, Switzerland, Sweden, Greece, Estonia, Argentina, Brazil, and Chile) choose the expected word or face about 85 percent of the time on average. In cultures that are less like the United States, such as Japan, Malaysia, Ethiopia, China, Sumatra, and Turkey, subjects match faces and words slightly less well, responding as expected about 72 percent of the time. Hundreds of scientific studies have used these findings to conclude that facial expressions are universally recognized and therefore universally produced, even by people in faraway cultures that had little contact with Western civilization. Ultimately, these emotion "recognition" findings have been so well replicated over the last several decades that universal emotions seem to qualify as one of those rare bulletproof scientific facts, like the law of gravity.[5]

The thing is, universal laws have this annoying habit of losing their universality. Newton's law of universal gravitation was only universal until the theory of relativity showed that it wasn't.

Watch what happens when we change the basic emotion method very slightly. Simply remove the list of emotion words. Test subjects must now *freely label* the same posed photographs from the dozens (or even hundreds) of emotion words that they know, as depicted in figure 3-3, instead of choosing a response from a short list of possibilities, as depicted in figure 3-2. When we do this, the subjects' success rate plummets. In one of the first free labeling studies ever conducted, subjects named the faces with the expected emotion words (or synonyms) only 58 percent of the time, and in subsequent studies the results were even lower. In fact, if you ask a more neutral question without referring to emotion at all — "What word best describes what's going on inside this person?" — the performance is even worse.[6]

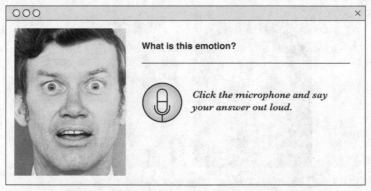

Figure 3-3: Basic emotion method with the emotion words removed

Why does such a small change make such a large difference? Because the short list of emotion words in the basic emotion method — a technique called a *forced choice* — is an unintentional cheat sheet for the test subjects. The words not only limit the available choices but also prompt the subjects to simulate facial configurations for the corresponding emotion concepts, preparing them to see certain emotions and not others. This process is called *priming*. When you first looked at Serena Williams's face, I primed you in a similar way by telling you the woman was "screaming in terror." Your simulation influenced how you categorized the sensory input from her face to see a meaningful expression. Likewise, test subjects who see a list of emotion words are primed with (i.e., they simulate) the corresponding emotion concepts to categorize the posed faces they see. Your knowledge of concepts is a key ingredient for experiencing other people as emotional, and emotion words invoke this ingredient. And they could be largely responsible for producing what looks like universal emotion perception in the hundreds of studies that use the basic emotion method.[7]

Free labeling reduced the ingredient of concept knowledge, but only somewhat. In my own lab, we went a step further and removed all emotion words, printed or spoken. If the theory of constructed emotion is correct, then this small change should impair emotion perception even more. On each trial of an experiment, we presented subjects with two *wordless* photographs side by side (figure 3-4) and asked, "Do these people feel the same emotion?" The expected answer was merely yes or no. The results of this face-matching task were telling: subjects identified the expected matches only 42 percent of the time.[8]

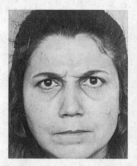 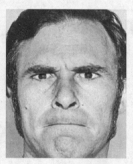

Figure 3-4: Basic emotion method with no words at all. Do these faces show the same emotion?

Next, our team reduced the ingredients even further. We actively interfered with our test subjects' access to their own emotion concepts, using a simple experimental technique. We had them repeat an emotion word like "anger" over and over. Eventually, the word becomes just a sound to the subject ("ang-gurr") that's mentally disconnected from its meaning. This technique has the same effect as creating a temporary brain lesion, but it's completely safe and lasts less than one second. Then we immediately showed subjects two wordless faces side by side as before. Their performance dropped to a dismal 36 percent: nearly *two-thirds* of their yes/no decisions were incorrect![9]

We also tested subjects with permanent brain lesions who suffer from a neurodegenerative illness called semantic dementia. These patients have trouble remembering words and concepts, including those for emotion.

We gave them thirty-six photographs: six actors each posing six different basic emotion facial configurations (smiles depicting happiness, pouts depicting sadness, scowls depicting anger, wide-eyed gasping depicting fear, nose-wrinkling depicting disgust, and neutral). The patients then sorted the photos into piles in any way that was meaningful to them. They were unable to group all scowling faces into an anger pile, all pouting faces into a sadness pile, and so on. Instead, the patients produced only positive, negative, and neutral piles, an arrangement that merely reflects pleasant versus unpleasant feeling. We now had solid evidence that emotion concepts are necessary for seeing emotion in faces.[10]

Our findings are reinforced by research on young children and infants, whose emotion concepts aren't fully developed yet. A series of experiments by psychologists James A. Russell and Sherri C. Widen showed that two- and three-year-old children, when shown basic emotion facial configurations, are not able to freely label them until they possess clearly differentiated concepts for "Anger," "Sadness," "Fear," and so on. Such young children use words like "sad," "mad," and "scared" interchangeably, like adults who exhibit low emotional granularity. It's not an issue of understanding the emotion words; even when these kids learn the meanings, they struggle to match up two pouting faces, whereas they find it easy to match a pouting face to the word "sad." Results for infants are similarly telling. Infants who are four to eight months old, for example, can distinguish smiling faces from scowling faces. This ability, however, turned out not to be related to emotion per se. In those experiments, the posed faces for happiness showed teeth while those for anger did not, and that's the cue that infants picked up on.[11]

From this sequence of experiments — removing the list of emotion words, then using wordless photographs, then temporarily disabling emotion concepts, then testing lesion patients who can no longer process emotion concepts, and finally testing infants who don't yet possess clearly defined emotion concepts — a theme emerges. As emotion concepts become more remote, people do worse and worse at recognizing the emotions that the posed stereotypes are supposedly displaying. This progression is strong evidence that people see an emotion in a face only if they possess the corresponding emotion concept, because they require that knowledge to construct perceptions in the moment.[12]

To really see the power of emotion concepts, my lab visited a remote culture in Africa with little or no knowledge of Western practices and norms.

With the fast pace of globalization, very few such isolated cultures exist any-more. My doctoral student Maria Gendron traveled to Namibia, Africa, to study emotion perception in a tribe known as the Himba, along with the cognitive psychologist Debi Roberson. Visiting the Himba was no simple task. Maria and Debi flew to South Africa and then drove for about twelve hours to their base camp in Opuwo, northern Namibia. From there, Debi, Maria, and their translator traveled many hours to reach individual villages near the Angola border, following tracks through the bush in an all-terrain vehicle, using the mountains and sun as landmarks. At night, they slept in a tent mounted on top of the car to avoid snakes and scorpions, which were numerous. I unfortunately could not join them, so they were equipped with a satellite phone and a generator so we could speak whenever a signal was available.[13]

Life among the Himba is decidedly non-Western. The people live mainly outdoors and in communal compounds made from saplings, mud, and dung. The men tend cattle day and night, while the women prepare food and care for the children. The children tend goats near the compound. The Himba speak a dialect of Otji-Herero, and they use no written language.

The Himba's reaction to the research team was fairly low-key. The chil-dren were curious and would hang around in the early morning before their chores. Some of the women were initially unsure if Maria was female since she was wearing (from their perspective) boyish clothing, which led to some finger pointing and laughter. The men must have figured it out, how-ever, because at one point, one proposed marriage. Maria's Namibian trans-lator took the simple approach by explaining politely, in Otji-Herero, that Maria was "already married to another man with a very big gun."

Maria used the face-sorting experiment with the thirty-six posed faces. It doesn't depend on words at all, let alone emotion words, so it worked nicely across the language and culture barriers. We'd created a set of photos using dark-skinned actors, because our originals featured Western faces that didn't look like Himba tribespeople. Our Himba subjects understood the task im-mediately, as we had hoped, and were able to sort the faces spontaneously by actor. When asked to sort the faces by emotion, the Himba clearly di-verged from Westerners. They placed all the smiling faces into a single pile, and most of the wide-eyed faces into a second pile, but then made many dif-ferent piles with mixtures of the remaining faces. If emotion perception is universal, then the Himba subjects should have sorted the photographs into six piles. When we asked our Himba subjects to freely label their piles, smil-

Figure 3-5: Maria Gendron (right) working with a Himba
subject in Namibia, beneath a tent attached to Maria's truck

ing faces were not "happy" (*ohange*) but "laughing" (*ondjora*). Wide-eyed
faces were not "fearful" (*okutira*) but "looking" (*tarera*). In other words, the
Himba participants categorized facial movements as behaviors rather than
inferring mental states or feelings. Overall, our Himba subjects showed no
evidence of universal emotion perception. And since we omitted all reference to English emotion concepts in our experiments, those concepts are a
prime suspect for why the basic emotion method appears to give evidence
of universality.[14]

There was still one mystery remaining, however: another group of researchers, led by psychologist Disa A. Sauter, had visited the Himba a few
years earlier and reported evidence of universal emotion "recognition." Sauter and her colleagues brought the basic emotion method to the Himba using vocal sounds (laughs, grunts, snorts, sighs, etc.) instead of photos of
posed faces. In their experiment, they offered brief emotion stories (translated into Otji-Herero) and asked their Himba participants to select which
of two vocalizations matched each story. The Himba did this well enough
that Sauter and her colleagues concluded that emotion perception was universal. We were unable to replicate these results with a different group of
Himba participants, even using the published method and the same trans-
lator as Sauter did. Maria also asked another group of Himba subjects to

freely label the vocal sounds, without accompanying stories, and again, only the laughing sounds were categorized as expected (although they labeled the sounds as "laughing" rather than "happy"). So why did Sauter and her team observe universality when we did not?[15]

In late 2014, Sauter and her colleagues inadvertently solved the mystery. They revealed that their experiment included an extra step not reported in their original publication: a step that's rich in conceptual knowledge. After the Himba participants heard an emotion story but before they listened to any sound pairs, they were asked to describe how the target person in the story was feeling. To help them in this task, Sauter and colleagues "allowed participants to listen several times to a given recorded story (if needed), *until they could explain the intended emotion in their own words.*" Whenever Himba participants described something other than the English emotion concept, they received negative feedback and were told to try again. Test subjects who were unable to provide the expected description were disqualified from the experiment. In effect, Himba participants were not permitted to listen to any sounds, let alone pick the ones that matched the story, until they had *learned* the corresponding English emotion concepts. When we attempted to replicate Sauter and colleagues' experiment, we used only the methods in their published paper, without the extra, unreported step, so our Himba test subjects did not have the opportunity to learn English emotion concepts before listening to the vocalizations.[16]

There was one other difference between our experimental method and the one used by Sauter and her colleagues. Once a Himba participant had explained the emotion concept satisfactorily—let's say it was sadness— Sauter's team played a pair of sounds, such as a cry and a laugh, and the subject chose the better match for sadness. The participant then heard more pairs of sounds, *each one containing a cry:* perhaps a cry and a sigh, then a cry and a scream, and so on. From each pair, the participant selected one sound as the better match for sadness. If the Himba participants were not confident of the link between cries and sadness at the beginning of these trials, they certainly were by the end. Our experiments avoided this problem. In each trial, Maria would read a story (through the translator), then presented a pair of sounds, and then have the participant choose the best match. Trials were in random order (e.g., a sadness trial, followed by an anger trial, followed by a happiness trial, and so on), which is a standard way to avoid learning within this type of experiment. We saw no evidence of universality.[17]

There is one emotion category that people seem able to perceive without the influence of emotion concepts: happiness. Regardless of the experimental method used, people in numerous cultures agree that smiling faces and laughing voices express happiness. So "Happy" might be the closest thing we have to a universal emotion category with a universal expression. Or it might not. For one thing, "Happiness" is usually the only pleasant emotion category that is tested using the basic emotion method, so it's trivial for subjects to distinguish it from the negative categories. And consider this fun fact: the historical record implies that ancient Romans did not smile spontaneously when they were happy. The word "smile" doesn't even exist in Latin. Smiling was an invention of the Middle Ages, and broad, toothy-mouthed smiles (with crinkling at the eyes, named the Duchenne smile by Ekman) became popular only in the eighteenth century as dentistry became more accessible and affordable. The classics scholar Mary Beard summarizes the nuances of the point:

> This is not to say that Romans never curled up the edges of their mouths in a formation that would look to us much like a smile; of course they did. But such curling did not mean very much in the range of significant social and cultural gestures in Rome. Conversely, other gestures, which would mean little to us, were much more heavily freighted with significance.

Perhaps sometime in the last few hundred years, smiling became a universal, stereotyped gesture symbolizing happiness.* Or . . . perhaps smiling in happiness is simply not universal.[18]

• • •

Emotion concepts are the secret ingredient behind the success of the basic emotion method. These concepts make certain facial configurations appear universally recognizable as emotional expressions when, in fact, they're not. Instead, we all construct perceptions of each other's emotions. We perceive others as happy, sad, or angry by applying our own emotion concepts to their moving faces and bodies. We likewise apply emotion concepts to voices and construct the experience of hearing emotional sounds. We simulate with such speed that emotion concepts work in stealth, and it seems to us as if emotions are broadcast from the face, voice, or any other body part, and we merely detect them.

---

* A proponent of the classical view might suggest that people suppressed their inborn smiles of happiness as socially inappropriate until the advent of dentistry.

A perfectly reasonable question for you to ask at this point is: how can my colleagues and I have the audacity to claim that our handful of experiments disconfirm hundreds of others that found evidence that emotions are universally recognized in expressions? The psychologist Dacher Keltner, for example, estimates that "there are a zillion data points on a perspective that conforms to Ekman."[19]

The answer is that most of these zillion experiments use the basic emotion method, which you have just seen contains a secret stash of concept knowledge about emotion. If humans actually had an inborn ability to recognize emotional expressions, then removing the emotion words from the method should not matter . . . but it did, every single time. There is very little doubt that emotion words have a powerful influence in experiments, instantly casting into doubt the conclusions of *every study ever performed* that used the basic emotion method.[20]

To date, my lab has made two expeditions to Namibia and one to Tanzania (visiting a hunter-gatherer group called the Hadza) with consistent results. The social psychologist José-Miguel Fernández-Dols has also replicated our results in an isolated culture on the Trobriand Islands in New Guinea. So, science now has a reasonable, alternative explanation for those "zillions of data points." The basic emotion method guides people to construct perceptions of Western-style emotions. That is, emotion perception is not innate but constructed.[21]

If you look closely at the original cross-cultural experiments from the 1960s, you can see clues that the conceptual elements within the basic emotion method pushed the results toward the appearance of universality. Of the seven samples using test subjects from remote cultures, the four that used the basic emotion method provided strong evidence for universality, but the remaining three used free labeling and did not show evidence of universality. These three contrary samples were not published in peer-reviewed journals but only as book chapters — a lesser form of publishing in the world of academia — and are rarely cited. As a result, the four samples supporting universality were lauded as a major breakthrough in research on our underlying human nature and set the stage for the research avalanche to come. Hundreds of subsequent studies employed the basic emotion method with forced choice, largely in cultures that had exposure to Western cultural practices and norms, taking a key condition for universality out of the experimental design but still claiming it as fact. This explains why today, many scientists and the public fundamentally misunderstand what is known about "emotional expressions" and "emotion recognition" from a scientific point of view.[22]

What might the science of emotion look like today had someone drawn different conclusions from those original studies? Consider Ekman's account of his first visit to the Fore tribe in New Guinea:

I asked them to make up a story about each facial expression [photograph]. "Tell me what is happening now, what happened before to make the person show this expression, and what is going to happen next." It was like pulling teeth. I am not certain whether it was the translation process, or the fact that they have no idea what it was I wanted to hear or why I wanted them to do this. Perhaps making up stories about strangers was just something the Fore didn't do.

Ekman might be right, but it is also possible that the Fore did not understand or accept the concept of a facial "expression," which implies an internal feeling that seeks release in a set of facial movements. Not all cultures understand emotions as internal mental states. Himba and Hadza emotion concepts, for example, appear to be more focused on actions. This is also true of certain Japanese emotion concepts. The Ifaluk of Micronesia consider emotions as transactions between people. To them, anger is not a feeling of rage, a scowl, a pounding fist, or a loud yelling voice, all within the skin of one person, but a situation in which two people are engaged in a script — a dance, if you will — around a common goal. In the Ifaluk view, anger does not "live" inside either participant.[23]

When you look at the development and history of the basic emotion method, there's a surprising amount to criticize from a scientific standpoint. Over twenty years ago, the psychologist James A. Russell catalogued many of the concerns. And remember that the "six basic facial expressions" were not a scientific discovery; the Western architects of the basic emotion method stipulated them, actors posed them, and a science was built around them. There is no known validity to these particular facial poses, and studies that use more objective methods like facial EMG and facial coding do not find evidence that people routinely make these movements in real life during episodes of emotion. Yet scientists continue to use the basic emotion method regardless. After all, it produces very consistent results.[24]

Each time a scientific "fact" is overturned it leads to new avenues for discovery. The physicist Albert Michelson won a Nobel Prize in 1907 for disproving a conjecture made by Aristotle, that light travels through empty space via a hypothetical substance called luminiferous ether. His detective work set the stage for Albert Einstein's theory of relativity. In our case, we've cast substantial doubt on the evidence for universal emotions. They only *appear* to be universal *under certain conditions*—when you give people a tiny bit of information about Western emotion concepts, intentionally or not. These observations, and others like them, set the stage for the new theory of emotion that you are about to learn. So Tomkins, Ekman, and their colleagues did contribute to a remarkable discovery. It just wasn't the discovery that they expected.[25]

The many cross-cultural studies employing the basic emotion method suggest something else exciting: it may be easy to teach emotion concepts across cultural boundaries, even unintentionally. Such a worldwide understanding would be hugely beneficial. If Saddam Hussein's half-brother had only understood the American emotion concept of anger, he might have perceived anger in Secretary of State James Baker, which might have averted the first Gulf War with the United States, saving thousands of lives.

Given how easy it is to teach emotion concepts by accident, there is also a danger in using Western stereotypes of emotion in cultural research. For instance, an ongoing series of studies called the Universal Expressions Project is attempting to document what is universal about emotional expressions in the face, body, and voice. So far, they've identified "about 30 facial expressions and 20 vocal expressions that are very similar around the world." The catch is that the project uses only the basic emotion method, so it's investigating universality with a tool that cannot provide such evidence. (Also, they're asking people to pose what they *believe* are their cultural expressions, which is not the same thing as observing actual body movements during emotion.) More importantly, if the project reaches its goal, everyone in the world might learn the Western stereotypes for emotions.[26]

In the long run, scientists who still subscribe to the basic emotion method are very likely helping to create the universality that they believe they are discovering.

Closer to home, if people believe that a face alone displays emotion, it can lead to serious mistakes with damaging repercussions. In one case, this belief changed the course of a U.S. presidential election. In 2003–2004, Governor Howard Dean of Vermont was seeking the Democratic nomination for president of the United States, an honor that ultimately went to Senator John Kerry of Massachusetts. Voters saw a lot of negative campaigning that season, and one of the most misleading examples was a video of Dean taken during a speech. In a snippet of video that went viral, Dean's face was shown alone, without context, and he looked furious. But if you watched the entire video in context, it becomes obvious that Dean was not enraged but excited, firing up the crowd with his enthusiasm. The snippet circulated on the news, spread widely, and, ultimately, Dean dropped out of the race. We can only wonder what might have happened if viewers had understood how emotions are made when they saw those misleading images.

• • •

Guided by a constructionist approach, scientists continue to replicate my lab's findings in other cultures (data from China, East Africa, Melanesia, and other regions are looking promising at press time). As they do, we are speeding the paradigm shift to a new understanding of emotion that goes beyond Western stereotypes. We can cast aside questions like "How accurately can you recognize fear?" and instead study the variety of facial movements that people actually make in fear. We can also try to understand why people hold stereotypes about facial configurations in the first place, and what their value might be.

The basic emotion method has shaped the scientific landscape and influenced public understanding of emotion. Thousands of scientific studies claim that emotions are universal. Popular books, magazine articles, radio broadcasts, and TV shows casually assume that everyone makes and recognizes the same facial configurations as expressions of emotion. Games and books teach preschool children these allegedly universal expressions. International political and business negotiation strategies are likewise based on this assumption. Psychologists assess and treat emotion deficits in people suffering from mental illness using similar methods. The growing economy of emotion-reading gadgets and apps also assumes universality, as if emotions can be read in the face or in patterns of bodily changes in the absence of context, as easily as reading words on a page. The sheer amount of time, effort, and money going into these efforts is mind-boggling. But what if the fact of universal emotions isn't a fact at all?

What if it's evidence for something else entirely . . . namely, our ability to use concepts to shape perception? This is the crux of the theory of constructed emotion: a full-fledged, alternative explanation for the mystery of human emotion that does not rely on universal emotion fingerprints. The next four chapters dive into the details of this theory and the scientific evidence that supports it.

4

The Origin of Feeling

Think about the last time you were awash in pleasure. I don't necessarily mean sexual pleasure but everyday delights: gazing at a vivid sunrise, sipping a cold glass of water when you are hot and sweaty, or enjoying a brief moment of peace at the end of a troubling day.

Now contrast this with feeling unpleasant, like the last time you were sick with a cold, or just after an argument with a close friend. Pleasure and displeasure feel qualitatively different. You and I might not agree that a specific object or event produces pleasure or displeasure — I find walnuts delicious whereas my husband calls them an offense against nature — but each of us can, in principle, distinguish one from the other. These feelings are universal, even as emotions like happiness and anger are not, and they flow like a current through every waking moment of your life.[1]

Simple pleasant and unpleasant feelings come from an ongoing process inside you called *interoception*. Interoception is your brain's representation of all sensations from your internal organs and tissues, the hormones in your blood, and your immune system. Think about what's happening within your body right this second. Your insides are in motion. Your heart sends blood rushing through your veins and arteries. Your lungs fill and empty. Your stomach digests food. This interoceptive activity produces the spectrum of basic feeling from pleasant to unpleasant, from calm to jittery, and even completely neutral.[2]

Interoception is in fact one of the core ingredients of emotion, just as flour and water are core ingredients of bread, but these feelings that come from interoception are much simpler than full-blown emotional experiences like joy and sadness. In this chapter, you'll learn how interoception works, and how it contributes to emotional experiences and perceptions. We'll need a little background first about the brain in general and how it budgets the energy in your body to keep you alive and well. That will prepare you to understand the gist of interoception, which is the origin of feeling. After that, we'll discover the unexpected and frankly astonishing influence that interoception has over your thoughts, decisions, and actions every day.

Whether you're a generally calm person, floating unperturbed in a stream of tranquility, unaffected by the vicissitudes of life; a more reactive person awash in a river of agony and ecstasy, easily moved by every little change in your surroundings; or somewhere in between, the science behind interoception, grounded in the wiring of your brain, will help you see yourself in a new light. It also demonstrates that you're not at the mercy of emotions that arise unbidden to control your behavior. You are an architect of these experiences. Your river of feelings might feel like it's flowing over you, but actually you're the river's source.

• • •

For the bulk of human history, the most learned members of our species have wildly underestimated the human brain's capabilities. This is understandable, since your brain occupies only about 2 percent of your body mass, and it looks like a blob of gray gelatin. Ancient Egyptians deemed it a useless organ and tugged it out of dead pharaohs through the nose.

The brain eventually earned its due as the seat of the mind, but it still received insufficient credit for its remarkable abilities. Brain regions were thought to be primarily "reactive," spending most of their time dormant and awakening to fire only when a stimulus arrives from the outside world. This stimulus-response view is simple and intuitive, and, in fact, neurons in your muscles work this way, lying still until stimulated, then firing to make a muscle cell respond. So scientists assumed that neurons in the brain operated similarly. When a gigantic snake slithers across your path, this stimulus was thought to launch a chain reaction in your brain. Neurons would fire in sensory regions, causing neurons in cognitive or emotional regions to fire, causing neurons in motor regions to fire, and then you'd react. The classical view typifies this mindset: when the snake appears, a "fear circuit" in your brain, which is usually in the "off" position, supposedly flips into the "on" position, causing preset changes in your face and body. Your eyes widen, you scream, and you run away.[3]

The stimulus-response view, while intuitive, is misguided. Your brain's 86 billion neurons, which are connected into massive networks, never lie dormant awaiting a jump-start. Your neurons are always stimulating each other, sometimes millions at a time. Given enough oxygen and nutrients, these huge cascades of stimulation, known as *intrinsic brain activity*, continue from birth until death. This activity is nothing like a reaction triggered by the outside world. It's more like breathing, a process that requires no external catalyst.[4]

The intrinsic activity in your brain is not random; it is structured by collections of neurons that consistently fire together, called *intrinsic networks*. These networks operate somewhat like sports teams. A team has a pool of players; at any given moment, some players are in the game and others sit on the bench, ready to jump in when needed. Likewise, an intrinsic network has a pool of available neurons. Each time the network does its job, different groupings of its neurons play (fire) in synchrony to fill all the necessary positions on the team. You might recognize this behavior as degeneracy, because different sets of neurons in the network are producing the same basic function. Intrinsic networks are considered one of neuroscience's great discoveries of the past decade.[5]

You might wonder what this hotbed of continuous, intrinsic activity is accomplishing, besides keeping your heart beating, your lungs breathing, and your other internal functions working smoothly. In fact, intrinsic brain activity is the origin of dreams, daydreams, imagination, mind wandering, and reveries, which we collectively called simulation in chapter 2. It also ultimately produces every sensation you experience, including your interoceptive sensations, which are the origins of your most basic pleasant, unpleasant, calm, and jittery feelings.[6]

To understand why this is the case, let's take your brain's perspective for a moment. Like those ancient, mummified Egyptian pharaohs, the brain spends eternity entombed in a dark, silent box. It cannot get out and enjoy the world's marvels directly; it learns what is going on in the world only indirectly via scraps of information from the light, vibrations, and chemicals that become sights, sounds, smells, and so on. Your brain must figure out the meaning of those flashes and vibrations, and its main clues are your past experiences, which it constructs as simulations within its vast network of neural connections. Your brain has learned that a single sensory cue, such as a loud bang, can have many different causes — a door being slammed, a bursting balloon, a hand clap, a gunshot. It distinguishes which of these different causes is most relevant only by their probability in different contexts. It asks, Which combination of my past experiences provides the closest match to this sound, given *this particular situation* with its accompanying sights, smells, and other sensations?[7]

And so, trapped within the skull, with only past experiences as a guide, your brain makes *predictions*. We usually think of predictions as statements about the future, like "It's going to rain tomorrow" or "The Red Sox will win the World Series" or "You will meet a tall, dark stranger." But here, I'm focusing on predictions at a microscopic scale as millions of neurons talk to one another. These neural conversations try to anticipate every fragment of sight, sound, smell, taste, and touch that you will experience, and every action that you will take. These predictions are your brain's best guesses of what's going on in the world around you, and how to deal with it to keep you alive and well.[8]

At the level of brain cells, prediction means that the neurons over here, in this part of your brain, tweak the neurons over there, in that part of your brain, without any need for a stimulus from the outside world. Intrinsic brain activity is millions and millions of nonstop predictions.

Through prediction, your brain constructs the world you experience. It combines bits and pieces of your past and estimates how likely each bit applies in your current situation. This happened when you simulated the bee in chapter 2; once you'd seen the full photograph, your brain had a new experience to draw on, so it could instantly construct a bee from the blobs. And right now, with each word that you read, your brain is predicting what the next word will be, based on probabilities from your lifetime of reading experience. In short, your experience right now was predicted by your brain a moment ago. Prediction is such a fundamental activity of the human brain that some scientists consider it the brain's primary mode of operation.[9]

Predictions not only anticipate sensory input from outside the skull but *explain* it. Let's do a quick thought experiment to see how this works. Keep your eyes open and imagine a red apple, just like you did in chapter 2. If you are like most people, you will have no problem conjuring some ghostly image of a round, red object in your mind's eye. You see this image because neurons in your visual cortex have changed their firing patterns to simulate an apple. If you were in the fruit section of a supermarket right now, these same firing neurons would be a visual prediction. Your past experience in that context (a supermarket aisle) leads your brain to predict that you would see an apple rather than a red ball or the red nose of a clown. Once the prediction is confirmed by an actual apple, the prediction has, in effect, explained the visual sensations as being an apple.[10]

If your brain predicts perfectly — say, you predicted a McIntosh apple as you came upon a display of them — then the actual visual input of the apple, captured by your retina, carries *no new information* beyond the prediction. The visual input merely confirms the prediction is correct, so the input needn't travel any further in the brain. The neurons in your visual cortex are already firing as they should be. This efficient, predictive process is your brain's default way of navigating the world and making sense of it. It generates predictions to perceive and explain everything you see, hear, taste, smell, and touch.

Your brain also uses prediction to initiate your body's movements, like reaching your arm out to pick up an apple or dashing away from a snake. These predictions occur before you have any conscious awareness or intent about moving your body. Neuroscientists and psychologists call this phenomenon "the illusion of free will." The word "illusion" is a bit of a misnomer; your brain isn't acting behind your back. You *are* your brain, and the whole cascade of events is caused by your brain's predictive powers. It's called an illusion because movement *feels* like a two-step process — decide, then move — when in fact your brain issues motor predictions to move your body well before you become aware of your intent to move. And even before you actually encounter the apple (or the snake)![11]

If your brain were merely reactive, it would be too inefficient to keep you alive. You are always being bombarded by sensory input. One human retina transmits as much visual data as a fully loaded computer network connection in every waking moment; now multiply that by every sensory pathway you have. A reactive brain would bog down like your Internet connection does when too many of your neighbors are streaming movies from Netflix. A reactive brain would also be too expensive, metabolically speaking, because it would require more interconnections than it could maintain.[12]

Evolution *literally wired* your brain for efficient prediction. As an example of this wiring in your visual system, have a look at figure 4-1, which shows how your brain predicts far more visual input than it receives.

Consider what this means: events in the world, such as a snake slithering at your feet, merely *tune* your predictions, roughly the way that your breathing is tuned by exercise. Right now, as you read these words and understand what they mean, each word barely perturbs your massive intrinsic activity, like a small stone skipping on a rolling ocean wave. In brain-imaging exper-

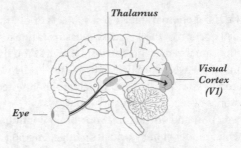

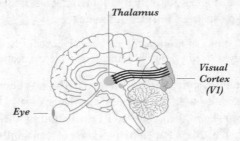

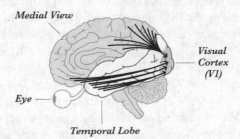

Figure 4-1: Your brain contains complete maps of your visual field. One map is located in your primary visual cortex, known as V1. If your brain merely reacted to the light waves that hit your retina and traveled to primary visual cortex (V1) via your thalamus, then it would have many neurons to carry that visual information to V1. But it has far fewer than one would expect (top image), and ten times as many projections going in the other direction, carrying visual *predictions* from V1 to the thalamus (center image). Likewise, 90 percent of all connections coming into V1 (lower image) carry predictions from neurons in other parts of cortex. Only a small fraction carries visual input from the world.[13]

iments, when we show photographs to test subjects or ask them to perform tasks, only a small portion of the signal we measure is due to the photos and tasks; most of the signal represents intrinsic activity. You might think that your perceptions of the world are driven by events in the world, but really, they are anchored in your predictions, which are then tested against those little skipping stones of incoming sensory input.[14]

Through prediction and correction, your brain continually creates and revises your mental model of the world. It's a huge, ongoing simulation that constructs everything you perceive while determining how you act. But predictions aren't always correct, when compared to actual sensory input, and the brain must make adjustments. Sometimes a skipping stone is large enough to make a splash. Consider this sentence:

> Once upon a time, in a magical kingdom far beyond the most distant mountains, there lived a beautiful princess who bled to death.

Did you find the last three words unexpected? That's because your brain predicted incorrectly based on its stored knowledge of fairy tales — it made a *prediction error* — and then adjusted its prediction in the blink of an eye based on the final words: a few skipping stones of visual information.

The same process happens when you mistake a stranger's face for someone you know, or step off a moving walkway in an airport and feel surprised by the change in your pace. Your brain computes prediction errors speedily by comparing the prediction to actual sensory input, and then it reduces the prediction error quickly and efficiently. For example, your brain can change the prediction: the stranger looks different from your friend; the moving walkway came to its end.

Prediction errors aren't problems. They're a normal part of the operating instructions of your brain as it takes in sensory input. Without prediction error, life would be a yawning bore. Nothing would be surprising or novel, and therefore your brain would never learn anything new. Most of the time, at least when you are an adult, your predictions aren't too far off-base. If they were, you would go through life feeling constantly startled, uncertain . . . or hallucinating.

Your brain's colossal, ongoing storm of predictions and corrections can be thought of as billions of tiny droplets. Each little drop represents a certain wiring arrangement that I'll call a *prediction loop,* shown in figure 4-2. This arrangement holds at many levels throughout your entire brain. Neurons participate in prediction loops with other neurons. Brain regions participate in prediction loops with other regions. Your multitudes of prediction loops run in a massive parallel process that continues nonstop for your whole life, creating the sights, sounds, smells, tastes, and touches that make up your experiences and dictate your actions.

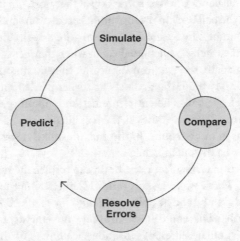

Figure 4-2: Structure of a prediction loop. Predictions become simulations of sensations and movement. These simulations are compared to actual sensory input from the world. If they match, the predictions are correct and the simulation becomes your experience. If they don't match, your brain must resolve the errors.

Suppose you are playing baseball. Someone throws the ball in your direction, and you reach out and catch it. Most likely, you'd experience this as two events: seeing a ball and then catching it. If your brain actually reacted like this, however, baseball couldn't exist as a sport. Your brain has about half a second to prepare to catch a baseball in a typical game. This isn't enough time to process the visual input, calculate where the ball will land, make the decision to move, coordinate all the muscle movements, and send the motor commands to move you into position for the catch.[15]

Prediction makes the game possible. Your brain launches predictions well before you consciously see the ball, just like it predicts a red apple in the grocery store, using your past experience. As each prediction propagates through millions of prediction loops, your brain simulates the sights, sounds, and other sensations that the predictions represent, as well as the actions you will take to catch the ball. Your brain then compares the simulations to actual sensory input. If they match . . . success! The prediction is correct, and the sensory input proceeds no further into your brain. Your body is now prepared to catch the ball, and your movement is based on your prediction. Finally, you consciously see the ball, and you catch it.[16]

That's what happens when the prediction is correct, like when I throw a baseball to my husband, who has some skill at the sport. On the other hand, when he tosses the ball back to me, my brain's predictions aren't particularly good, since I cannot play baseball to save my life. My predictions become simulations of the catch I hope to make, but when they get compared to the information I actually receive from the world, they do not match. This is a prediction error. My brain then adjusts its earlier predictions so that I can (in theory) catch the ball. The entire prediction loop process repeats, predicting and correcting many times as the ball hurtles toward me. All of this activity happens in milliseconds. In the end, most likely, I become aware of the ball sailing past my outstretched arm.

When prediction errors occur, the brain can resolve them in two general ways. The first, which we've just seen in my lame attempt to catch a baseball, is that the brain can be flexible and *change the prediction*. In this situation, my motor neurons would adjust my body movements, and my sensory neurons would simulate different sensations, leading to further predictions involving prediction loops. I could dive for the ball, for example, when it is in a different place than I expected it to be.

The brain's second alternative is to be stubborn and stick with the original prediction. It *filters the sensory input* so it's consistent with the prediction. In this situation, I could be standing in a baseball field but daydreaming (predicting and simulating) as the ball sails toward me. Even though the ball is fully within my visual field, I don't notice it until it thumps at my feet. Another example would be the food-filled diapers at my daughter's disgusting foods birthday party: our guests' prediction of a baby poo aroma dominated their actual sensory input of mashed carrots.[17]

In short, the brain is not a simple machine reacting to stimuli in the outside world. It's structured as billions of prediction loops creating intrinsic brain activity. Visual predictions, auditory predictions, gustatory (taste) predictions, somatosensory (touch) predictions, olfactory (smell) predictions, and motor predictions travel throughout the brain, influencing and constraining each other. These predictions are held in check by sensory inputs from the outside world, which your brain may prioritize or ignore.[18]

If this talk of prediction and correction seems unintuitive, think about it this way: your brain works like a scientist. It's always making a slew of predictions, just as a scientist makes competing hypotheses. Like a scientist, your brain uses knowledge (past experience) to estimate how confident you can be that each prediction is true. Your brain then tests its predictions by comparing them to incoming sensory input from the world, much as a scientist compares hypotheses against data in an experiment. If your brain is predicting well, then input from the world confirms your predictions. Usually, however, there is some prediction error, and your brain, like a scientist, has some options. It can be a responsible scientist and change its predictions to respond to the data. Your brain can also be a biased scientist and selectively choose data that fits the hypotheses, ignoring everything else. Your brain can also be an unscrupulous scientist and ignore the data altogether, maintaining that its predictions are reality. Or, in moments of learning or discovery, your brain can be a curious scientist and focus on input. And like the quintessential scientist, your brain can run armchair experiments to imagine the world: pure simulation without sensory input or prediction error.

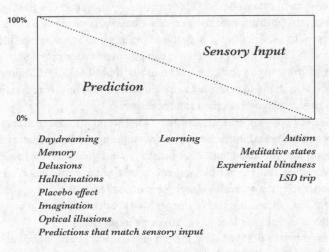

Figure 4-3: A variety of mental phenomena can be understood
as a combination of prediction and sensory input.[19]

The balance between prediction and prediction error, shown in figure 4-3, determines how much of your experience is rooted in the outside world versus inside your head. As you can see, in many cases, the outside world is irrelevant to your experience. In a sense, your brain is wired for delusion:

through continual prediction, you experience a world of your own creation that is held in check by the sensory world. Once your predictions are correct enough, they not only create your perception and action but also explain the meaning of your sensations. This is your brain's default mode. And marvelously, your brain does not just predict the future: it can imagine the future at will. As far as we know, no other animal brain can do that.

• • •

Your brain is always predicting, and its most important mission is predicting your body's energy needs, so you can stay alive and well. These crucial predictions, and their associated prediction error, turn out to be a key ingredient for making emotions. For hundreds of years, scholars believed that emotional "reactions" were caused by certain brain regions. As you'll now discover, those brain regions do the opposite of what everyone expected, helping to make emotion in a way that overturns centuries of scientific belief. And once again, the story begins with movement — not the large-scale movements of a baseball game, but the inner motion of your body.

Any movement *of* your body is accompanied by movement *in* your body. When you shift position quickly to catch a baseball, you have to breathe more deeply. To escape from a poisonous snake, your heart pumps blood faster through dilated blood vessels to rush glucose to your muscles, which increases your heart rate and changes your blood pressure. Your brain represents the sensations that result from this inner motion; this representation, you may remember, is called interoception.[20]

Your inner-body movements, and their interoceptive consequences, occur every moment of your life. Your brain must keep your heart beating and your blood pumping and your lungs breathing and your glucose metabolizing even when you are not playing sports or fleeing from a snake, even when you are sleeping or resting. Interoception is therefore continuous, just as the mechanics of hearing and vision are always operating, even when you aren't actively listening or looking at anything in particular.

From your brain's point of view, locked inside the skull, your body is just another part of the world that it must explain. Your pumping heart, your expanding lungs, and your changing temperature and metabolism send sensory input to your brain that is noisy and ambiguous. A single interoceptive cue, such as a dull ache in your abdomen, could mean a stomachache, hunger, tension, an overly tight belt, or a hundred other causes. Your brain must explain bodily sensations to make them meaningful, and its major tool for doing so is prediction. So, your brain models the world from the perspective of someone with *your body*. Just as your brain predicts the sights, smells, sounds, touches, and tastes from the world in relation to the movements of your head and limbs, it also predicts the sensory consequences of movements inside your body.[21]

Most of the time, you're unaware of the miniature maelstrom of movement inside you. (When's the last time you thought, "Hmm, my liver seems to be producing a lot of bile today"?) Of course, there are times when you directly feel a headache, a full stomach, or your heart pounding in your chest. But your nervous system isn't built for you to experience these sensations with precision, which is fortunate, because otherwise they'd overwhelm your attention.[22]

Usually, you experience interoception only in general terms: those simple feelings of pleasure, displeasure, arousal, or calmness that I mentioned earlier. Sometimes, however, you experience moments of intense interoceptive sensations as emotions. That is a key element of the theory of constructed emotion. In every waking moment, your brain gives your sensations meaning. Some of those sensations are interoceptive sensations, and the resulting meaning can be an instance of emotion.[23]

In order to understand how emotions are made, you'll need to understand a bit about some key brain regions. Interoception is actually a whole-brain process, but several regions work together in a special way that is critical for interoception. My lab has discovered that these regions form an *interoceptive network* that is intrinsic in your brain, analogous to your networks for vision, hearing, and other senses. The interoceptive network issues predictions about your body, tests the resulting simulations against sensory input from your body, and updates your brain's model of your body in the world.[24]

To simplify our discussion drastically, I'll describe this network as having two general parts with distinct roles. One part is a set of brain regions that send predictions to the body to control its internal environment: speed up the heart, slow down breathing, release more cortisol, metabolize more glucose, and so on. We'll call them your *body-budgeting regions*.* The sec-

---

* Also known as "limbic" or "visceromotor" regions. To keep things manageable — because the brain is a complicated structure — we'll focus only on body-budgeting regions in the cerebral cortex. Others can be found outside of cerebral cortex, such as the central nucleus of the amygdala. I also use "cortex" to mean "cerebral cortex."

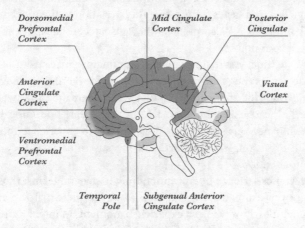

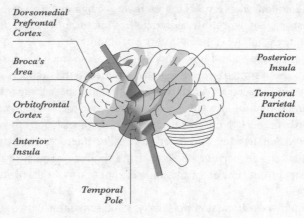

Figure 4-4: The cortical regions of the interoceptive network. Body-budgeting regions are dark gray, and primary interoceptive cortex is given its technical name, the posterior insula. Subcortical regions of this network are not shown. The interoceptive network encompasses two networks commonly known as the salience network and the default mode network. Visual cortex is shown for reference.[25]

ond part is a region that represents sensations inside your body, called your *primary interoceptive cortex.*[26]

The two parts of your interoceptive network participate in a prediction loop. Each time your body-budgeting regions predict a motor change, like speeding up the heart, they also predict the sensory consequences of that change, like a pounding feeling in your chest. These sensory predictions are called *interoceptive predictions,* and they flow to your primary interoceptive cortex, where they are simulated in the usual way. Primary interoceptive cortex also receives sensory inputs from the heart, lungs, kidneys, skin, muscles, blood vessels, and other organs and tissue as they perform their usual duties. The neurons in your primary interoceptive cortex compare the simulation to the incoming sensory input, computing any relevant prediction error, completing the loop, and ultimately creating interoceptive sensations.[27]

Your body-budgeting regions play a vital role in keeping you alive. Each time your brain moves any part of your body, inside or out, it spends some of its energy resources: the stuff it uses to run your organs, your metabolism, and your immune system. You replenish your body's resources by eating, drinking, and sleeping, and you reduce your body's spending by relaxing with loved ones, even having sex. To manage all of this spending and replenishing, your brain must constantly predict your body's energy needs, like a budget for your body. Just as a company has a finance department that tracks deposits and withdrawals and moves money between accounts, so its overall budget stays in balance, your brain has circuitry that is largely responsible for your body budget. That circuitry is within your interoceptive network. Your body-budgeting regions make predictions to estimate the resources to keep you alive and flourishing, using past experience as a guide.[28]

Why is this relevant to emotion? Because every brain region that's claimed to be a home of emotion in humans is a body-budgeting region within the interoceptive network. These regions, however, don't react in emotion. They don't react at all. They predict, intrinsically, to regulate your body budget. They issue predictions for sights, sounds, thoughts, memories, imagination, and, yes, emotions. The idea of an emotional brain region is an illusion caused by the outdated belief in a reactive brain. Neuroscientists understand this today, but the message hasn't trickled down to many psychologists, psychiatrists, sociologists, economists, and others who study emotion.[29]

Whenever your brain predicts a movement, whether it's getting out of bed in the morning or taking a sip of coffee, your body-budgeting regions adjust your budget. When your brain predicts that your body will need a quick burst of energy, these regions instruct the adrenal gland in your kidneys to release the hormone cortisol. People call cortisol a "stress hormone," but this is a mistake. Cortisol is released whenever you need a surge of energy, which happens to include the times when you are stressed. Its main purpose is to flood the bloodstream with glucose to provide immediate energy to cells, allowing, for example, muscle cells to stretch and contract so you can run. Your body-budgeting regions also make you breathe more deeply to get more oxygen into your bloodstream and dilate your arteries to get that oxygen to your muscles more quickly so your body can move. All of this internal motion is accompanied by interoceptive sensations, though you are not wired to experience them precisely. So, your interoceptive network controls your body, budgets your energy resources, and represents your internal sensations, all at the same time.[30]

Withdrawals from your body's budget don't require actual physical movement. Suppose you see your boss, teacher, or baseball coach walking toward you. You believe that she judges everything you say and do. Even though no physical movement seems called for, your brain predicts that your body needs energy and makes a budget withdrawal, releasing cortisol and flooding glucose into your bloodstream. You also have a surge in interoceptive sensations. Stop and think about this for a minute. Someone merely walks toward you while you are standing still, and your brain predicts that you need fuel! In this manner, any event that significantly impacts your body budget becomes *personally meaningful* to you.

Not long ago, my lab was evaluating a portable device for monitoring the heart. Whenever the wearer's heart rate sped up 15 percent above normal, the device would beep. One of my graduate students, Erika Siegel, was wearing the device as she worked quietly at her desk, and it remained silent for some time. At one point, I walked into the room. When Erika turned and saw me (her Ph.D. advisor), the device beeped loudly, to her embarrassed surprise and to the amusement of everyone else around us. Later in the day, I spent time wearing the device, and during a meeting with Erika, it beeped several times as I received emails from a granting agency. (So Erika had the last laugh that day.)[31]

My lab has experimentally demonstrated the brain's budgeting efforts hundreds of times (as have other labs), observing as people's body-budgeting circuitry shifts resources around, and sometimes as their body budgets fluctuate in and out of balance. We ask volunteers to sit completely motionless in front of a computer screen and view pictures of animals, flowers, babies, food, money, guns, surfers, skydivers, car crashes, and other objects and scenes. These pictures impact their body budget; heart rates go up, blood pressures change, blood vessels dilate. These budgetary changes, which prepare the body to fight or flee, occur even though the volunteers are *not moving* and *have no conscious plan to move.* When our volunteers view these pictures during an fMRI experiment, we observe their body-budgeting regions controlling these inner-body movements. And even though our subjects are lying down, completely motionless, they simulate motor movements like running and surfing, as well as the sensations from moving muscles, joints, and tendons. The pictures also change our volunteers' feelings as interoceptive changes in their bodies are being simulated and corrected. Based on these and hundreds of other studies, we now have good evidence that your brain predicts your body's responses by drawing on prior experiences with similar situations and objects, even when you're not physically active. And the consequence is interoceptive sensation.[32]

To perturb your budget, you don't even require another person or object to be present. You can just *imagine* your boss, teacher, coach, or anything else relevant to you. Every simulation, whether it becomes an emotion or not, impacts your body budget. As it turns out, people spend at least half their waking hours simulating rather than paying attention to the world around them, and this pure simulation strongly drives their feelings.[33]

When it comes to managing your body budget, your brain does not have to go it alone. Other people regulate your body budget too. When you interact with your friends, parents, children, lovers, teammates, therapist, or other close companions, you and they synchronize breathing, heart beats, and other physical signals, leading to tangible benefits. Holding hands with loved ones, or even keeping their photo on your desk at work, reduces activation in your body-budgeting regions and makes you less bothered by pain. If you're standing at the bottom of a hill with friends, it will appear less steep and easier to climb than if you are alone. If you grow up in poverty, a situation that leads to chronic body-budget imbalance and an overactive immune system, these body-budgeting problems are reduced if you have a supportive person in your life. In contrast, when you lose a close, loving relationship and feel physically ill about it, part of the reason is that your loved one is no longer helping to regulate your budget. You feel like you've lost a part of yourself because, in a sense, you have.[34]

Every person you encounter, every prediction you make, every idea you imagine, and every sight, sound, taste, touch, and smell that you fail to anticipate all have budgetary consequences and corresponding interoceptive predictions. Your brain must contend with this continuous, ever-changing flow of interoceptive sensations from the predictions that keep you alive.

Sometimes you're aware of them, and other times you're not, but they are always part of your brain's model of the world. They are, as I've said, the scientific basis for simple feelings of pleasure, displeasure, arousal, and calmness that you experience every day. For some, the flow is like the trickle of a tranquil brook. For others, it's like a raging river. Sometimes the sensations are transformed into emotions, but as you will now learn, even when they're only in the background, they influence what you do, what you think, and what you perceive.[35]

• • •

When you wake up in the morning, do you feel refreshed or crabby? In the middle of the day, do you feel dragged out or full of energy? Consider how you feel right now. Calm? Interested? Energetic? Bored? Tired? Cranky? These are the simple feelings we discussed at the beginning of the chapter. Scientists call them *affect.**

Affect is the general sense of feeling that you experience throughout each day. It is not emotion but a much simpler feeling with two features. The first is how pleasant or unpleasant you feel, which scientists call *valence*. The pleasantness of the sun on your skin, the deliciousness of your favorite food, and the discomfort of a stomachache or a pinch are all examples of affective valence. The second feature of affect is how calm or agitated you feel, which is called *arousal*. The energized feeling of anticipating good news, the jittery feeling after drinking too much coffee, the fatigue after a long run, and the weariness from lack of sleep are examples of high and low arousal. Anytime you have an intuition that an investment is risky or profitable, or a gut feeling that someone is trustworthy or an asshole, that's also affect. Even a completely neutral feeling is affect.[36]

Philosophers from the West and the East describe valence and arousal as basic features of human experience. Scientists largely agree that affect is present from birth and that babies can feel and perceive pleasure and displeasure, even as they disagree whether newborns emerge into the world with fully formed emotions.[37]

Affect, you may recall, depends on interoception. That means affect is a constant current throughout your life, even when you are completely still or asleep. It does not turn on and off in response to events you experience as emotional. In this sense, affect is a fundamental aspect of consciousness,

---

* The noun "affect" is pronounced like "apple," with its accent on the first syllable and a short "a": \'A-fekt\.

like brightness and loudness. When your brain represents wavelengths of light reflected from objects, you experience brightness and darkness. When your brain represents air pressure changes, you experience loudness and softness. And when your brain represents interoceptive changes, you experience pleasantness and unpleasantness, and agitation and calmness. Affect, brightness, and loudness all accompany you from birth until death.[38]

Let's be clear on one thing: interoception is not a mechanism dedicated to manufacturing affect. Interoception is a fundamental feature of the human nervous system, and why you experience these sensations as affect is one of the great mysteries of science. Interoception did not evolve for you to have feelings but to regulate your body budget. It helps your brain track your temperature, how much glucose you are using, whether you have any tissue damage, whether your heart is pounding, whether your muscles are stretching, and other bodily conditions, all at the same time. Your affective feelings of pleasure and displeasure, and calmness and agitation, are simple summaries of your budgetary state. Are you flush? Are you overdrawn? Do you need a deposit, and if so, how desperately?[39]

When your budget is unbalanced, your affect doesn't instruct you how to act in any specific way, but it prompts your brain to search for explanations. Your brain constantly uses past experience to predict which objects and events will impact your body budget, changing your affect. These objects and events are collectively your *affective niche*. Intuitively, your affective niche includes everything that has any relevance to your body budget in the present moment. Right now, this book is within your affective niche, as are the letters of the alphabet, the ideas you're reading about, any memories that my words bring to mind, the air temperature around you, and any objects, people, and events from your past that impacted your body budget in a similar situation. Anything outside your affective niche is just noise: your brain issues no predictions about it, and you do not notice it. The feel of your clothing against your skin is usually not in your affective niche (though it is now, since I just mentioned it), unless it happens to be relevant, say, to your physical comfort.[40]

The psychologist James A. Russell developed a way of tracking affect, and it's become popular among clinicians, teachers, and scientists. He showed that you can describe your affect in the moment as a single point on a two-dimensional space called a *circumplex,* a circular structure with two dimensions, as in figure 4-5. Russell's two dimensions represent valence and arousal, with distance from the origin representing intensity.[41]

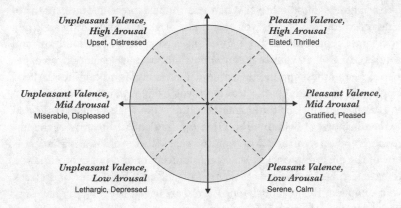

Unpleasant Valence,
High Arousal
Upset, Distressed

Pleasant Valence,
High Arousal
Elated, Thrilled

Unpleasant Valence,
Mid Arousal
Miserable, Displeased

Pleasant Valence,
Mid Arousal
Gratified, Pleased

Unpleasant Valence,
Low Arousal
Lethargic, Depressed

Pleasant Valence,
Low Arousal
Serene, Calm

Figure 4-5: An affective circumplex

Your affect is always some combination of valence and arousal, repre-sented by one point on the affective circumplex. When you sit quietly, your affect is at a central point of "neutral valence, neutral arousal" on the circum-plex. If you're having fun at a lively party, your affect might be in the "pleas-ant, high arousal" quadrant. If the party turns boring, your affect might be "unpleasant, low arousal." Younger American adults tend to prefer the upper right quadrant: pleasant, high arousal. Middle-aged and older Americans tend to prefer the lower right quadrant (pleasant, low arousal), as do people from Eastern cultures like China and Japan. Hollywood is a $500 billion in-dustry because people are willing to pay to see movies so that, for a few hours, they can travel within this affective map. You don't even have to open your eyes to have an affective adventure. When you daydream and have a large change in interoception, your brain will swirl with affect.[42]

Affect has far-reaching consequences beyond simple feeling. Imagine you are a judge presiding over a prisoner's parole case. You are listening to the inmate's story, hearing about his behavior in prison, and you have a bad feeling. If you agree to parole, he could hurt someone else. Your hunch is that you should keep him locked up. So you deny parole. Your bad feel-ing, which is unpleasant affect, seems like evidence that your judgment was correct. But could your affect have misled you? This exact situation was the subject of a 2011 study of judges. Scientists in Israel found that judges were significantly more likely to deny parole to a prisoner if the hearing was just before lunchtime. The judges experienced their interoceptive sensations not as hunger but as evidence for their parole decision. Immediately after lunch, the judges began granting paroles with their customary frequency.[43]

When you experience affect without knowing the cause, you are more likely to treat affect as information about the world, rather than your experience of the world. The psychologist Gerald L. Clore has spent decades performing clever experiments to better understand how people make decisions every day based on gut feelings. This phenomenon is called *affective realism*, because we experience supposed facts about the world that are created in part by our feelings. For example, people report more happiness and life satisfaction on sunny days, but only when they are *not* explicitly asked about the weather. When you apply for a job or college or medical school, make sure you interview on a sunny day, because interviewers tend to rate applicants more negatively when it is rainy. And the next time a good friend snaps at you, remember affective realism. Maybe your friend is irritated with you, but perhaps she didn't sleep well last night, or maybe it's just lunchtime. The change in her body budget, which she's experiencing as affect, might not have anything to do with you.[44]

Affect leads us to believe that objects and people in the world are inherently negative or positive.* Photographs of kittens are deemed pleasant. Photographs of rotting human corpses are deemed unpleasant. But these images do not have affective properties inside them. The phrase "an unpleasant image" is really shorthand for "an image that impacts my body budget, producing sensations that I experience as unpleasant." In these moments of affective realism, we experience affect as a property of an object or event in the outside world, rather than as our own experience. "I feel bad, therefore you must have done something bad. You are a bad person." In my lab, when we manipulate people's affect without their knowing, it influences whether they experience a stranger as trustworthy, competent, attractive, or likable, and they even see the person's face differently.[45]

People employ affect as information, creating affective realism, throughout daily life. Food is "delicious" or "bland." Paintings are "beautiful" or "ugly." People are "nice" or "mean." Women in certain cultures must wear scarves and wigs so as not to "tempt men" by showing a bit of hair. Sometimes affective realism is helpful, but it also shapes some of humanity's most troubling problems. Enemies are "evil." Women who are raped are perceived

---

* Affective realism is a common but powerful form of naive realism, the belief that one's senses provide an accurate and objective representation of the world.

as "asking for it." Victims of domestic violence are said to "bring it on them-selves."[46]

The thing is, a bad feeling doesn't always mean something is wrong. It just means you're taxing your body budget. When people exercise to the point of labored breathing, for example, they feel tired and crappy well be-fore they run out of energy. When people solve math problems and perform difficult feats of memory, they can feel hopeless and miserable, even when they are performing well. Any graduate student of mine who never feels dis-tress is clearly doing something wrong.[47]

Affective realism can also lead to tragic consequences. In July 2007, an American gunner aboard an Apache helicopter in Iraq mistakenly killed a group of eleven unarmed people, including several Reuters photojournal-ists. The soldier had misjudged a journalist's camera to be a gun. One ex-planation for this incident is that affective realism caused the soldier, in the heat of the moment, to imbue a neutral object (a camera) with unpleasant valence. Every day, soldiers must make quick decisions about other people, whether they are embedded in a unit during wartime, on a peacekeeping mission, negotiating in a cross-cultural setting, or collaborating with unit members on a stateside base. These quick judgments are extremely difficult to negotiate, especially in such high-stakes, high-arousal settings where er-rors are often made at the expense of someone's life.[48]

A little closer to home, affective realism may also play a role in police shootings of unarmed civilians. The U.S. Department of Justice analyzed shootings by Philadelphia police officers between 2007 and 2013 and found that 15 percent of the victims were unarmed. In half of these cases, an offi-cer reportedly misidentified "a nonthreatening object (e.g., a cell phone) or movement (e.g., tugging at the waistband)" as a weapon. Many factors may contribute to these tragedies, ranging from carelessness to racial bias, but it is also possible that some of the shooters actually perceive a weapon when none is present due to affective realism in a high-pressure and dangerous context.* The human brain is wired for this sort of delusion, in part because moment-to-moment interoception infuses us with affect, which we then use as evidence about the world.[49]

---

* I am absolutely not saying that affective realism is the primary cause of police shootings. I'm just making the scientific point that the brain is wired for prediction. All of us literally see what we believe based on our past experiences, unless our predictions are corrected by sensory inputs from the world.

People like to say that seeing is believing, but affective realism demonstrates that believing is seeing. The world often takes a backseat to your predictions. (It's still in the car, so to speak, but is mostly a passenger.) And as you're about to learn right now, this arrangement is not limited to vision.

• • •

Suppose you're walking alone in the forest, and you hear a rustle in the leaves and see a vague movement on the ground. As always, your body-budgeting regions initiate predictions — say, that there's a snake nearby. These predictions prepare you to see and hear a snake. At the same time, these regions predict that your heart rate should increase and your blood vessels should dilate, for instance, in preparation to run. A pounding heart and surging blood would cause interoceptive sensations, so your brain must predict those sensations as well. As a result, your brain simulates the snake, the bodily changes, and the bodily sensations. These predictions translate into feeling; in this case, you'll begin to feel agitated.[50]

What happens next? Maybe a snake slithers out from the brush. In this case, the sensory input matches your predictions and you run. Or perhaps no snake is present — the leaves were just rustled by the wind — but you see a snake anyway. That's affective realism. Now consider the third possibility: there is no snake, and you don't see a snake. In this case, your visual predictions of a snake are corrected quickly; however, your interoceptive predictions are not. Your body-budgeting regions keep predicting adjustments to your budget long after the predicted need is over. You therefore may take a long time to calm down, even if you know there is nothing wrong. Remember when I compared your brain to a scientist who makes and tests hypotheses? Your body-budgeting regions are like a mostly deaf scientist: they make predictions but have a hard time listening to the incoming evidence.[51]

Some of the time, your body-budgeting regions are sluggish to correct their predictions. Think about the last time you ate too much and felt bloated. You might be able to blame your body-budgeting regions. One of their jobs is to predict your level of circulating glucose, which determines how much food you need, but they don't receive the message "I'm full" from your body in a timely manner, so you keep eating. If you've ever heard the advice, "Wait 20 minutes before you take a second helping, to see if you're really still hungry," now you know why it works. Whenever you make a big deposit or withdrawal from your body budget — eating, exercising, injuring yourself — you might have to wait for your brain to catch up. Marathon runners learn this; they feel fatigue early in the race when their body budget is still solvent, so they keep running until the unpleasant feeling goes away. They ignore the affective realism that insists they're out of energy.[52]

Take a moment and consider what this means for your day-to-day life. You've just learned that the sensations you feel from your body don't always reflect the actual state of your body. That's because familiar sensations like your heart beating in your chest, your lungs filling with air, and, most of all, the general pleasant, unpleasant, aroused, and quiescent sensations of affect *are not really coming from inside your body.* They are driven by simulations in your interoceptive network.[53]

In short, you feel what your brain believes. Affect primarily comes from prediction.

You've already learned that you see what your brain believes — that's affective realism. Now you know the same is true for most feelings you've experienced in your life. Even the feeling of the pulse in your wrist is a simulation, constructed in sensory regions of your brain and corrected by sensory input (your actual pulse). Everything you feel is based on prediction from your knowledge and past experience. You are truly an architect of your experience. Believing is feeling.

These ideas are not just speculation. Scientists with the right equipment can change people's affect by directly manipulating body-budgeting regions that issue predictions. Helen S. Mayberg, a pioneering neurologist, has developed a deep brain stimulation therapy for people suffering from treatment-resistant depression. These people don't just experience the anguish of a major depressive episode — they are in agony, trapped in a pit of self-loathing and unending torment. Some of them can barely move. During surgery, Mayberg works with a team of neurosurgeons who drill small holes in the skull and sink electrodes into a key predictive area in the patient's interoceptive network. When the neurosurgeons turn on the electrodes, Mayberg's patients report *immediate* relief from their agony. As the electrical current is turned off and on, the patients' crippling wave of dread approaches and recedes in synchrony with the stimulation. Mayberg's remarkable work might represent the first time in scientific history that direct stimulation of the human brain has consistently changed people's affective feelings, potentially leading to new treatments for mental illness.[54]

While predictive brain circuitry is important for affect, it likely is not necessary. Consider the case of Roger, a fifty-six-year-old patient whose relevant circuitry was destroyed by a rare illness. He has an above-normal IQ and a college education but also plenty of mental difficulties, such as se-

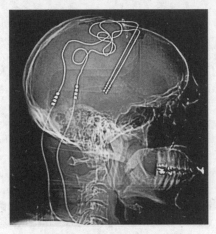

Figure 4-6: Deep brain stimulation

vere amnesia and difficulty with smell and taste. Nevertheless, Roger experiences affect. Most likely, his affect is driven by actual sensory inputs from his body; other brain regions could be supplying the predictions, an example of degeneracy (different sets of neurons producing the same outcome). The opposite situation can also occur. Patients with spinal cord damage or Pure Autonomic Failure, a degenerative disease of the autonomic nervous system, have interoceptive predictions but don't receive sensory inputs from their organs and tissue. These patients likely experience affect based primarily on uncorrected predictions.[55]

• • •

Your interoceptive network doesn't just help determine how you feel. Its body-budgeting regions are some of the most powerful and well-connected predictors in your entire brain. These regions are loud and bossy, like a mostly deaf scientist with a big megaphone. They launch predictions for vision, hearing, and your other senses; your primary sensory regions, which don't issue predictions of their own, are wired to listen.[56]

Let me show you what this means. You might think that in everyday life, the things you see and hear influence what you feel, but it's mostly the other way around: that what you feel alters your sight and hearing. Interoception in the moment is more influential to perception, and how you act, than the outside world is.

You might believe that you are a rational creature, weighing the pros and cons before deciding how to act, but the structure of your cortex makes this an implausible fiction. Your brain is wired to listen to your body budget. Affect is in the driver's seat and rationality is a passenger. It doesn't matter whether you're choosing between two snacks, two job offers, two investments, or two heart surgeons — your everyday decisions are driven by a loudmouthed, mostly deaf scientist who views the world through affect-colored glasses.[57]

Antonio Damasio, in his bestseller *Descartes' Error*, observes that a mind requires passion (what we would call affect) for wisdom. He documents that people with damage to their interoceptive network, particularly in one key body-budgeting region, have impaired decision-making. Robbed of the capacity to generate interoceptive predictions, Damasio's patients were rudderless. Our new knowledge of brain anatomy now compels us to go one step further. Affect is not just necessary for wisdom; it's also irrevocably woven into the fabric of every decision.[58]

The shouting power of body-budgeting circuitry has serious implications for the financial world. It helped to precipitate the greatest economic disasters of our time, most recently the global financial meltdown of 2008 that cast countless families into economic ruin.

The science of economics used to employ a concept called the rational economic person (*homo economicus*), who controls his or her emotions to make reasoned economic judgments. This concept was a foundation of Western economic theory, and though it has fallen out of favor among academic economists, it has continued to guide economic practice. However, if body-budgeting regions drive predictions to every other brain network, then the model of the rational economic person is based on a biological fallacy. You cannot be a rational actor if your brain runs on interoceptively infused predictions. An economic model at the foundation of the U.S. economy — some might say the global economy — is rooted in a neural fairy tale.[59]

Every economic crisis in the last thirty years has been related, at least in some part, to the rational economic person model. According to journalist Jeff Madrick, author of *Seven Bad Ideas: How Mainstream Economists Have Damaged America and the World*, several of economists' most fundamental ideas caused a series of financial crises leading up to the Great Recession. A common theme running through these ideas is that unregulated free-market economies work well. In these economies, decisions regarding investments, production, and distribution are based on supply and demand with no government regulation or oversight. Mathematical models indicate that under certain conditions, unregulated free-market economies do work well. But one of those "certain conditions" is that people are rational decision makers. I have lost count of the number of experiments published over the past fifty years showing that people are not rational actors. You cannot overcome emotion through rational thinking, because the state of your body budget is the basis for every thought and perception you have, so interoception and affect are built into every moment. Even when you experience yourself as rational, your body budget and its links to affect are there, lurking beneath the surface.[60]

If the idea of the rational human mind is so toxic to the economy, and it's not backed up by neuroscience, why does it persist? Because we humans have long believed that rationality makes us special in the animal kingdom. This origin myth reflects one of the most cherished narratives in Western thought, that the human mind is a battlefield where cognition and emotion struggle for control of behavior. Even the adjective we use to describe ourselves as insensitive or stupid in the heat of the moment — "thoughtless" — connotes a lack of cognitive control, of failing to channel our inner Mr. Spock.

This origin myth is so strongly held that scientists even created a model of the brain based on it. The model begins with ancient subcortical circuits for basic survival, which we allegedly inherited from reptiles. Sitting atop those circuits is an alleged emotion system, known as the "limbic system," that we supposedly inherited from early mammals. And wrapped around the so-called limbic system, like icing on an already-baked cake, is our allegedly rational and uniquely human cortex. This illusory arrangement of layers, which is sometimes called the "triune brain," remains one of the most successful misconceptions in human biology. Carl Sagan popularized it in *The Dragons of Eden,* his bestselling (some would say largely fictional) account of how human intelligence evolved. Daniel Goleman employed it in his bestseller *Emotional Intelligence.* Nevertheless, humans don't have an animal brain gift-wrapped in cognition, as any expert in brain evolution knows. "Mapping emotion onto just the middle part of the brain, and reason and logic onto the cortex, is just plain silly," says neuroscientist Barbara L. Finlay, editor of the journal *Behavior and Brain Sciences.* "All brain divisions are present in all vertebrates." So how do brains evolve? They reorganize as they expand, like companies do, to keep themselves efficient and nimble.[61]

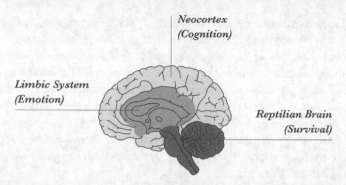

*Neocortex
(Cognition)*

*Limbic System
(Emotion)*

*Reptilian Brain
(Survival)*

Figure 4-7: The "triune brain" idea, with so-called cognitive circuitry layered on top of so-called emotion circuitry. This illusory arrangement depicts how thinking supposedly regulates feeling.

The bottom line is this: the human brain is anatomically structured so that no decision or action can be free of interoception and affect, no matter what fiction people tell themselves about how rational they are. Your bodily feeling right now will project forward to influence what you will feel and do in the future. It is an elegantly orchestrated, self-fulfilling prophecy, embodied within the architecture of your brain.

• • •

Your brain, with its billions of neurons, has much more going on than I've sketched out in this chapter. Most neuroscientists agree that we are decades away from knowing the intricacies of how a brain works, let alone how it creates consciousness. Still, we can be fairly sure of some things.

Right now, as your brain makes meaning from these words, it is predicting changes in your body budget. Every thought, memory, perception, or emotion that you construct includes something about the state of your body: a little piece of interoception. A visual prediction, for example, doesn't just answer the question, "What did I see last time I was in this situation?" It answers, "What did I see last time I was in this situation *when my body was in this state*?" Any change in affect you feel while reading these words — more or less pleasant, or more or less calm — is a result of those interoceptive predictions. Affect is your brain's best guess about the state of your body budget.

Interoception is also one of the most important ingredients in what you experience as reality. If you didn't have interoception, the physical world would be meaningless noise to you. Consider this: Your interceptive predictions, which produce your feelings of affect, determine what you care about in the moment — your affective niche. From the perspective of your brain, anything in your affective niche could potentially influence your body budget, and nothing else in the universe matters. That means, in effect, that *you construct the environment in which you live.* You might think about your environment as existing in the outside world, separate from yourself, but that's a myth. You (and other creatures) do not simply find yourself in an environment and either adapt or die. You construct your environment — your reality — by virtue of what sensory input from the physical environment your brain selects; it admits some as information and ignores some as noise. And this selection is intimately linked to interoception. Your brain expands its predictive repertoire to include anything that might impact your body budget, in order to meet your body's metabolic demands. This is why affect is a property of consciousness.

Interoception, as a fundamental part of the predictive process, is a key ingredient of emotion. However, interoception alone cannot explain emotion. An emotion category like anger or sadness is far more complex than a simple feeling of unpleasantness and arousal.

When Connecticut Governor Dannel Malloy's voice wavered during his speech after the Sandy Hook Elementary School massacre, he didn't cry, he didn't pout, and at one point he actually smiled. And yet, somehow, viewers inferred that he was experiencing intense sadness. Sensation and simple feeling are not sufficient to explain how an audience of thousands perceived the depth of Malloy's anguish.

Affect alone also doesn't explain how we construct our own experiences of sadness, nor how one instance of sadness differs from another. Nor does affect tell you what sensations mean or what to do about them. That's why people eat when they are tired or find a defendant guilty when they are hungry. You must make the affect *meaningful* so your brain can execute a more specific action. One way to make meaning is to construct an instance of emotion.

So, how do interoceptive sensations become emotions? And why do we experience these sensations (really predictions) in such diverse ways: as physical symptoms, as perceptions of the world, as simple affective feeling, and sometimes as emotion? That is the next mystery we'll address.

5

Concepts, Goals, and Words

When you look at a rainbow, you see discrete stripes of color, roughly like the drawing on the left side of figure 5-1. But in nature, a rainbow has no stripes — it's a continuous spectrum of light, with wavelengths that range from approximately 400 to 750 nanometers. This spectrum has no borders or bands of any kind.

Why do you and I see stripes? Because we have mental *concepts* for colors like "Red," "Orange," and "Yellow." Your brain automatically uses these concepts to group together the wavelengths in certain ranges of the spectrum, *categorizing* them as the same color. Your brain downplays the variations within each color category and magnifies the differences between the categories, causing you to perceive bands of color.[1]

Figure 5-1: Rainbows, drawn with stripes (left)
and continuous as in nature (right)

Human speech also is continuous — a stream of sound — yet when you listen to your native language, you hear discrete words. How does that happen? Once again, you use concepts to categorize the continuous input. Beginning in infancy, you learn regularities in the stream of speech that reveal the boundaries between phonemes, the smallest bits of sound that you can distinguish in a language (for example, the sound of "D" or "P" in English). These regularities become concepts that your brain later uses to categorize the stream of sound into syllables and words.[2]

This remarkable process is filled with challenges because the audio stream is ambiguous and highly variable. Consonant sounds vary with context: the sound of "D" is acoustically different in the words "Dad" and "Death," yet somehow we hear both as a "D." Vowel sounds vary with the age, sex, and size of the speaker, as well as by context within the same speaker. An incredible 50 percent of the words we hear cannot be understood out of context (when presented in isolation). But using your concepts, your brain learns to categorize, constructing phonemes in tens of milliseconds within all this variable, noisy information, ultimately permitting you to communicate with others.[3]

Everything you perceive around you is represented by concepts in your brain. Take a look at any object near you. Then, look slightly to the left of the object. You just accomplished something remarkable without even knowing it. Your head and eye movements seemed inconsequential but caused a *gigantic* change in the visual input reaching your brain. If you think of your field of vision as a big TV screen, then your slight eye movement just changed millions of pixels on that screen. And yet, you did not experience blurry streaks across your visual field. That's because you don't see the world in terms of pixels: you see objects, and they changed very little as you moved your eyes. You perceive low-level regularities like lines, contours, streaks, and blurs, as well as higher-level regularities like complex objects and scenes. Your brain learned these regularities as concepts long ago, and it uses those concepts now to categorize your continually changing visual input.[4]

Without concepts, you'd experience a world of ever-fluctuating noise. Everything you ever encountered would be unlike everything else. You'd be experientially blind, like when you first saw the blobby picture in chapter 2, but permanently so. You'd be incapable of learning.[5]

All sensory information is a massive, constantly changing puzzle for your brain to solve. The objects you see, the sounds you hear, the odors you smell, the touches you feel, the flavors you taste, and the interoceptive sensations you experience as aches and pains and affect . . . they all involve continuous sensory signals that are highly variable and ambiguous as they reach your brain. Your brain's job is to predict them before they arrive, fill in missing details, and find regularities where possible, so that you experience a world of objects, people, music, and events, not the "blooming, buzzing confusion" that is really out there.[6]

To achieve this magnificent feat, your brain employs concepts to make the sensory signals meaningful, creating an explanation for where they came from, what they refer to in the world, and how to act on them. Your perceptions are so vivid and immediate that they compel you to believe that you experience the world *as it is,* when you actually experience a world *of your own construction.* Much of what you experience as the outside world begins inside your head. When you categorize using concepts, you go beyond the information available, just as you did when you perceived a bee within blobs.

In this chapter, I explain that each time you experience emotion or perceive it in others, you are again categorizing with concepts, making meaning of sensations from interoception and the five senses. This is a key theme of the theory of constructed emotion.

My point is not to say, "You construct instances of emotion by categorization: isn't that unique?" Rather, it's to show that categorization constructs *every* perception, thought, memory, and other mental event that you experience, so *of course* you construct instances of emotion in the same manner. This is not effortful, conscious categorization, as when an entomologist pores over some new specimen of weevil, deciding whether it's a member of the *anthribidae* or *nemonychidae* family. I'm speaking of the rapid, automatic categorization performed constantly by your brain, in every waking moment, in milliseconds, to predict and explain the sensory input that you encounter. Categorization is business as usual for your brain, and it explains how emotions are made without needing fingerprints.

We'll be informal for now about the inner workings (i.e., the neuroscience) of categorization and just deal with some of the more basic questions. What are concepts? How are they formed? What sort of concepts are emotion concepts? And in particular, what superpower must a human mind possess to create meaning from scratch? Many of these questions are still active areas of research. When solid evidence exists, I present it. When there is less evidence, I make educated guesses. The answers not only explain how emotions are made but reveal a glimpse at the core of what it means to be human.[7]

• • •

Philosophers and scientists define a *category* as a collection of objects, events, or actions that are grouped together as equivalent for some purpose. They define a *concept* as a mental representation of a category. Traditionally, categories are supposed to exist in the world, while concepts exist in your head. For example, you have a concept of the color "Red." When you apply this concept to wavelengths of light to perceive a red rose in a park, that red color is an instance of the category "Red."* Your brain downplays the differences between the members of a category, such as the diverse shades of red roses in a botanical garden, to consider those members equivalent as "red." Your brain also magnifies differences between members and nonmembers (say, red versus pink roses) so that you perceive firm boundaries between them.

Imagine walking down the street in your city or town with a brain full of concepts. You see many objects all at once: flowers, trees, cars, houses, dogs, birds, bees. You see people walking, moving their bodies and faces. You hear sounds and smell diverse scents. Your brain puts this information together to perceive events like children playing in a park, a person gardening, an old couple holding hands on a bench. You create your experience of these objects, actions, and events by categorizing using concepts. Your ever-predicting brain swiftly anticipates sensory input, asking "Which of my concepts is this like?" For example, if you view a car head-on and then from the side, and you have a concept for that car, you can know it's the same one even

---

* I apologize on behalf of the world's philosophers, sages, luminaries, and other professional thinking persons for the muddled state of affairs regarding the distinction between categories and concepts. Categories like cars and birds are said to exist in the world, whereas concepts are said to exist in your brain, but if you think about it for a moment, who is creating the category? Who is grouping its members together to treat them as equivalent? You are. Your brain is doing it. So categories, like concepts, exist in your brain. (Their separation is rooted in a problem called "essentialism" that you'll learn about in chapter 8.) In this book, I refer to a "concept" when talking about knowledge, like knowledge of redness. I refer to a "category" when we talk about the instances that we construct with knowledge, like the red roses we perceive. (Tip of the hat to Douglas Adams for the phrase "philosophers, sages, luminaries, and other professional thinking persons.")

though the visual information hitting your retina from these two angles is entirely different.[8]

When your brain instantly categorizes sensory input as (say) a car, it's utilizing a concept of "Car." The deceptively simple phrase "concept of Car" stands for something more complex than you might expect. So, what exactly is a concept? That depends on which scientists you ask, which is business as usual in science. We must expect a certain amount of controversy around a topic as fundamental as "how knowledge is organized and represented in the human mind." And the answer is crucial to understanding how emotions are made.

If I asked you to describe the concept "Car," you might say a method of transportation that typically has four wheels, is made of metal, has an engine, and runs on some kind of fuel. Early scientific approaches assumed that a concept works exactly like this: a dictionary definition stored in your brain, describing necessary and sufficient features. "A car is a vehicle with an engine, four wheels, seats, doors, and a roof." "A bird is an egg-laying, flying animal with wings." This classical view of concepts assumes that their corresponding categories have firm boundaries. Instances of the category "Bee" are never in the category "Bird." Also in this view, every instance is an equivalently good representative of the category. Any bee is representative, so it goes, because all bees have something in common, either the way they look or what they do, or an underlying fingerprint that makes them bees. Any variation from bee to bee is considered irrelevant to the fact that they are bees. You might notice a parallel here to the classical view of emotion, in which every instance of the category "Fear" is similar, and instances of "Fear" are distinct from instances of "Anger."[9]

Classical concepts dominated philosophy, biology, and psychology from antiquity until the 1970s. In real life, the instances of a category vary tremendously from one another. There exist cars with no doors, such as a golf cart, or with six wheels like the Covini C6W. And some instances of a category really are more representative than others: nobody would call an ostrich a representative bird. In the 1970s, the classical view of concepts finally collapsed. Well, except in the science of emotion.[10]

From the ashes of classical concepts, a new view arose. It said that a concept is represented in the brain as the best example of its category, known as the *prototype*. For example, the prototypical bird has feathers and wings and can fly. Not all instances of "Bird" have these features, such as ostriches and emus, but they are still birds. Variation from the prototype is perfectly fine, but not too much variation: a bee is still not a bird, even though it has wings and can fly. In this view, as you learn about a category, your brain supposedly represents the concept as a single prototype. It might be the most frequent example of the category, or the most typical example, meaning the instance that is the closest match or has a majority of the category's features.[11]

Where emotion is concerned, people seem to have an easy time describing prototypical features of a given emotion category. Ask an American to describe prototypical sadness and he'll say it features a frowning or pouting face, a slumped posture, crying, moping around, a monotonous tone of voice, and that it begins with a loss of some sort and ends with an overall feeling of fatigue or powerlessness. Not every instance of sadness has every feature, but the description should be typical of sadness.[12]

So, prototypes might seem to be a good model for emotion concepts, if not for one paradoxical detail. When we measure actual instances of sadness using scientific tools, this frowning/pouting prototype of loss is not the most frequently or typically observed pattern. Everybody seems to know the prototype, but it's rarely found in real life. Instead, as you learned throughout chapter 1, we find great variability in sadness and every other emotion category.[13]

If there are no emotion prototypes stored in the brain, how do people list their features so easily? Most likely, your brain constructs prototypes as you need them, on the spot. You have experienced a diverse population of instances of the concept "Sadness," which reside in bits and pieces in your head, and in the blink of an eye, your brain constructs a summary of sadness that best fits the situation. (An example of population thinking in the brain.)[14]

Scientists have shown that people can construct similar prototypes in the lab. Print a random pattern of dots on a sheet of paper, then create a dozen variations of that pattern, and show people only the dozen variations. People can produce the original prototype pattern even though they've never seen it, simply by finding similarities in the variations. This means a prototype need not be found in nature, yet the brain can construct one when needed. Emotion prototypes, if that's what they indeed are, could be constructed in the same manner.[15]

Thus, concepts aren't fixed definitions in your brain, and they're not prototypes of the most typical or frequent instances. Instead, your brain has

Figure 5-2: Inferring a "prototype" pattern (step 5) from examples (steps 1–4). Test subjects first saw a variety of 9-dot patterns on a 30×30 grid. They classified each pattern into one of two categories, A and B. This was called the "learning phase" of the experiment. Next, they classified more patterns, some old and some new, including the prototypes of categories A and B, which the subjects had never seen. Subjects easily categorized the prototypes but had a more difficult time with the other, new variants. That meant each subject's brain must have constructed the prototypes despite not having seen them during the learning phase.

many instances — of cars, of dot patterns, of sadness, or anything else — and it imposes similarities between them, in the moment, according to your *goal* in a given situation. For example, your usual goal for a vehicle is to use it for transportation, so if an object meets that goal for you, then it's a vehicle, whether it's a car, a helicopter, or a sheet of plywood with four wheels nailed on. This explanation of concepts comes from Lawrence W. Barsalou, one of the world's leading cognitive scientists studying concepts and categories.[16]

Goal-based concepts are super flexible and adaptable to the situation. If you're in a pet shop to replenish your home aquarium and the salesperson asks, "What kind of fish would you like?" you might say "a goldfish" or "a black molly" but probably not "a poached salmon." Your concept "Fish" in this situation serves a goal to purchase a pet, not to order dinner, so you'll construct instances of the concept "Fish" that best suit your fish tank. If you're on a snorkeling expedition, you will use "Fish" in service of a goal to find exciting wildlife, so the best instance might be a huge nurse shark or a colorful spotted boxfish. Concepts are not static but remarkably malleable and context-dependent, because your goals can change to fit the situation.

A single object can also be part of different concepts. For example, a car does not always serve the goal of transportation. Sometimes a car is an instance of the concept "Status Symbol." In the right circumstances, a car can be a "Bed" for a homeless person, or even a "Murder Weapon." Drive a car into the ocean and it becomes an "Artificial Reef."

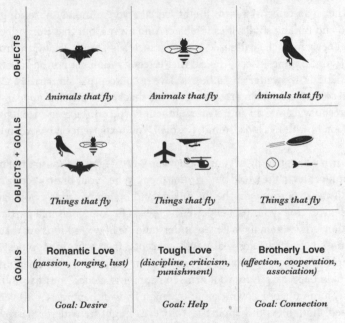

Figure 5-3: Concepts and goals. Row 1 illustrates concepts centered on a perceptual similarity, such as wings. Row 2 demonstrates that categories of objects can be goal-based. Bats, helicopters, and Frisbees share no perceptual features but can be described by a mental similarity: a common goal to move through the air. Row 3 illustrates similarity that is purely mental. The concept "Love" can be associated with different goals depending on context.

To see the real power of goal-based concepts, consider a purely mental concept such as "Things That Can Protect You from Stinging Insects."

Instances of the category are remarkably diverse: a flyswatter, a beekeeper's suit, a house, a Maserati, a large trash can, a vacation in Antarctica, a calm demeanor, even a university degree in entomology. They share no perceptual features. This category is clearly and entirely a construction of the human mind. Not all instances work in every context: for example, when you're gardening, whacking away at a bed of overgrown iris, and you accidentally disturb a bees' nest and unleash a swarm in your direction, a nearby house would be far better protection than a flyswatter. Yet your brain lumps all these instances into the same category because they can achieve the same goal, safety from stings. In fact, the goal is the *only* thing that holds together the category.

When you categorize, you might feel like you're merely observing the world and finding similarities in objects and events, but that cannot be the case. Purely mental, goal-based concepts such as "Things That Can Protect You from Stinging Insects" reveal that categorization cannot be so simple and static. A flyswatter and a house have no perceptual similarities. Goal-based concepts therefore free you from the shackles of physical appearance. When you walk into an entirely new situation, you don't experience it based solely on how things look, sound, or smell. Your experience it based on your goal.

So, what's happening in your brain when you categorize? You are not finding similarities in the world but *creating* them. When your brain needs a concept, it constructs one on the fly, mixing and matching from a population of instances from your past experience, to best fit your goals in a particular situation. And herein lies a key to understanding how emotions are made.[17]

Emotion concepts are goal-based concepts. Instances of happiness, for example, are highly variable. You can smile in happiness, sob in happiness, scream in happiness, raise your arms in happiness, clench your fists in happiness, jump up and down doling out high fives in happiness, or even be stunned motionless in happiness. Your eyes might be wide or narrowed; your breathing rapid or slow. You can have the heart-pounding, exciting happiness of winning the lottery or the calm, relaxed happiness of lying on a picnic blanket with your lover. You've also perceived many other people as happy in various ways. Altogether, this motley assortment of experiences and perceptions can involve different actions and inner-body changes, they may feel affectively different, and they can include different sights, sounds, and smells. To you, in the moment, however, these sets of physical changes are equivalent for some goal. Perhaps your goal is to feel accepted, to feel pleasure, to achieve an ambition, or to find meaning in life. Your concept of "Happiness" in the moment is centered on such a goal, binding together the diverse instances from your past.

Let's unpack an example. Suppose that you are in an airport waiting for your close friend to arrive for a visit, her first one in a long time. As you stare at the exit gates and await her imminent arrival, your brain is busily issuing thousands of predictions based on your concepts, in milliseconds, all outside of your awareness. After all, there are a host of different emotions you might experience in such a situation. You could experience the happiness of seeing your friend, the anticipation that she's about to appear, the fear that she won't arrive, or worry that you might no longer have anything in common. You could also have a non-emotional experience, like the exhaustion of your long drive to the airport, or the perception of tightness in your chest as a symptom that you're coming down with a cold.

Using this storm of predictions, your brain makes meaning of sensations based on your past experiences with airports and friends and illnesses and related situations. Your brain weighs its predictions based on probabilities; they compete to explain what caused your sensations, and they determine what you perceive, how you act, and what you feel in this situation. Ultimately, the most probable predictions become your perception: say, you are happy and your friend is walking through the gates right now. Not every instance of "Happiness" from your past matches the present situation, because "Happiness" is a goal-based concept composed of wildly diverse instances, but some of them had bits and pieces that matched well enough to win the competition. Do these predictions match the actual sensory input from the world and your body? Or is there prediction error that must be resolved? That's a matter for your prediction loops to work out and, if necessary, to correct.

Let's suppose your friend arrived safely, and later over coffee, she describes her turbulent plane flight that scared her out of her wits. She constructs an instance of "Fear" with the goal of communicating what it feels like to be strapped into the airplane seat, eyes closed, hot and queasy as the plane bumped up and down, her mind racing about her safety. When she says the word "frightened," you also construct an instance of "Fear," but it needn't have exactly the same physical features as hers; you probably won't squeeze your eyes shut, for example. Yet you can still perceive her fear and feel empathy for her. As long as your instances concern the same goal (detecting danger) in the same situation (a turbulent airplane ride), you and your friend are communicating clearly enough. On the other hand, if you constructed some other instance of "Fear," such as the exuberant fear of riding a rollercoaster, you might have trouble understanding why your friend was so upset by the flight. Successful communication requires that you and your friend are using synchronized concepts.

Think back to Darwin's ideas about the importance of variation within a species (chapter 1). Each animal species is a population of unique individuals who vary from one another. No feature or set of features is necessary, sufficient, or even frequent or typical of every individual in the population. Any summary of the population is a statistical fiction that applies to no individual. And most importantly, variation within a species is meaningfully related to the environment in which individuals live. Some individuals are more fit than others to pass their genetic material to the next generation. In a similar manner, some instances of concepts are more effective in a particular context to achieve a particular goal. Their competition in your brain is like Darwin's theory of natural selection but carried out in milliseconds; the most suitable instances outlive all rivals to fit your goal in the moment. That is categorization.[18]

• • •

Where do emotion concepts come from? How can a concept like "Awe" have such diversity: awe of the vastness of the universe; awe of Erik Weihenmayer, who scaled Mount Everest while blind; and awe that a tiny worker ant can carry five thousand times its body weight? The classical view proposes that you are born with these concepts, or that your brain finds emotion fingerprints in people's expressions and internalizes them as concepts. But we know that scientists haven't found such fingerprints, and infants show no evidence of being born knowing "Awe."

The human brain, it turns out, bootstraps a conceptual system into its wiring within the first year of life. This system is responsible for the wealth of emotion concepts that you now employ to experience and perceive emotions.

The newborn brain has the ability to learn patterns, a process called *statistical learning*. The moment that you burst into this strange new world as a baby, you were bombarded with noisy, ambiguous signals from the world and from your body. This barrage of sensory input was not random: it had some structure. Regularities. Your little brain began computing probabilities of which sights, sounds, smells, touches, tastes, and interoceptive sensations go together and which don't. "Those edges form a boundary. Those two blobs are part of a bigger blob. That brief silence was a separator." Little by little, but with surprising speed, your brain learned to resolve this ocean of vague sensation into patterns: sights and sounds, smells and tastes, touches and interoceptive sensations, and combinations thereof.[19]

Scientists have debated for hundreds of years over what you're born with versus what you learn, and I won't enter that debate. Let's just say that one thing you're born with is a fundamental ability to learn from regularities and probabilities around you. (In fact, you learn statistically even in utero, which makes it complicated to determine whether certain concepts are innate or learned.) Your prodigious capacity for statistical learning set you on the path toward the particular kind of mind, with the particular system of concepts, that you have today.[20]

Statistical learning in humans was first discovered in studies of language development. Babies have a natural interest in listening to speech, perhaps because the sounds occurred alongside body budgeting from birth, and even in utero. As they hear the sounds streaming along, they gradually infer the boundaries between phonemes, syllables, and words. From blobs of sound like *itstimefordinner, areyouhungryfordinnernow,* and *dinnertimeyummyyummycarrots,* infants learn which syllables are paired together more frequently ("din-ner," "yum-my") and therefore likely to be part of a single word. Syllables that co-occur relatively rarely are more likely to be part of different words. Babies learn these regularities extremely quickly, even within a few minutes of exposure. This learning process is so powerful that it changes the wiring in a baby's brain. Babies are born able to hear the differences between all sounds in all languages, but by the time they reach one year of age, statistical learning has reduced this ability to the sounds contained only in the languages they have heard spoken by live humans. Babies become wired for their native languages by statistical learning.[21]

Statistical learning is not the only way that humans acquire knowledge, but this learning begins very early in life and goes well beyond language. Studies show that babies easily learn statistical regularities in sound and vision, and it's reasonable to assume the same for the rest of the senses plus interoceptive sensations. What's more, babies can learn complex regularities that span multiple senses. If you fill a box with blue and yellow balls, and the yellow balls make a squeaking sound while the blue ones are silent, infants can generalize the association between color and sound.[22]

Babies use statistical learning to make predictions about the world, guiding their actions. Like little statisticians, they form hypotheses, assess probabilities based on their knowledge, integrate new evidence from the environment, and perform tests. In one creative study by the developmental psychologist Fei Xu, ten- to fourteen-month-old children first expressed a preference for pink or black lollipops, then were shown two candy jars: one containing more black lollipops than pink, and one with more pink than black. The experimenter then closed her eyes and drew one lollipop from each jar so infants could see only the stick, not the color. Each lollipop was placed into a separate, opaque cup with only the stick showing. Infants crawled to the cup that was statistically more likely to contain their preferred color, because it came from a jar where that color was in the majority. Experiments like this demonstrate that infants are not merely reactive to the world. Even from a very young age, they actively estimate probabilities based on patterns that they observe and learn, to maximize the outcomes they desire.[23]

Humans are not the only animals that learn statistically: non-human primates, dogs, and rats can do it, among others. Even single-celled animals engage in statistical learning and then prediction: they not only respond to changes in their environment but anticipate them. Human infants, however, do more than statistically learn simple concepts. They also quickly learn that some of the information they need about the world *resides in the minds of the people around them.*[24]

You might have noticed that young children assume that other people share their preferences. A one-year-old who likes crackers better than broccoli believes that everyone else in the world does too. She cannot infer mental states in others the way that Governor Malloy's audience inferred that he was filled with sorrow during his speech about the Sandy Hook massacre. Even so, Xu and her students have successfully observed the rudiments of mental inference even in young children as they learn statistically. Sixteen-month-old children were shown two bowls, one containing boring white cubes and the other full of more interesting, colorful Slinky toys. When these toddlers were allowed to choose an object from either bowl, sure enough they chose a favorite Slinky for themselves and for the experimenter. But then the experimenter revealed a third bowl containing many Slinkys and only a few cubes, and in full view of the children, he chose five white cubes for himself. When the children were asked to pick from that bowl, they gave the experimenter a cube! In other words, the children were able to learn a subjective preference of the experimenter that was different from their own. This realization, that an object has positive value for someone else, is an example of mental inference.[25]

Going beyond preferences, babies can even infer other people's goals statistically. They can tell the difference when an experimenter chooses a pattern of colored balls randomly versus with intent. In the latter case, they can infer that the experimenter's goal is to choose particular colors, and they'll expect that the experimenter will continue following it.* It seems as if infants automatically try to guess the goal behind another person's actions; they form a hypothesis (based on past experience in similar situations) and predict the outcome that will occur several minutes later.[26]

Statistical learning alone, however, does not equip humans to learn purely mental, goal-based concepts whose instances share no perceptual similarities. Take the concept "Money," for example. You can't learn it simply by viewing a piece of colored paper, a gold nugget, a seashell, and a pile of barley or salt, each of which has been deemed currency by some society in history. Likewise, instances of an emotion category such as "Fear" don't have enough statistical regularity — as demonstrated in chapter 1 — to allow a human brain to build a concept based on perceptual similarities. To build a purely mental concept, you need another secret ingredient: words.

From infancy, little human brains have an affinity for processing speech signals and quickly realize that speech is one way to access the information inside other people's minds. They're particularly attuned to adult "baby talk" with a higher and more variable pitch, shorter sentences, and strong eye contact.[27]

Even before infants understand what words mean in a conventional sense, the sounds of the words introduce statistical regularity that speeds concept learning. The developmental psychologists Sandra R. Waxman and Susan A. Gelman, leaders in this area of research, hypothesize that words invite an infant to form a concept, but only when adults speak with intent to communicate: "Look, sweetie: a *flower!*"[28]

---

* In case you're wondering how scientists can know what an infant is "expecting," here's the trick. Babies pay more attention to the unexpected. If the experimenter does something predictable, like selecting colored balls that conform to his goal, the baby will barely pay attention. However, if the experimenter selects a different set of balls, the baby will pay close attention and look for a longer time, indicating the pattern was unexpected. In psychology, this is called the habituation paradigm.

Waxman demonstrated this power of words in infants as young as three months. The infants first viewed pictures of different dinosaurs. As each image was shown, infants heard an experimenter speak a made-up word, "toma." When these infants were later shown pictures of a new dinosaur and a non-dinosaur such as a fish, those who had heard the word could distinguish more reliably which pictures depicted a "toma," implying that they had formed a simple concept. When the same experiment was performed with audio tones instead of human speech, the effect never materialized.[29]

Spoken words give the infant brain access to information that can't be found by observing the world and resides only in the minds of other people, namely, *mental similarities:* goals, intentions, preferences. Words allow infants to begin growing goal-based concepts, including emotion concepts.

A little human brain, bathed in the words of others around it, accumulates simple concepts. Some concepts are learned without words, but words confer distinct advantages to a developing conceptual system. A word might begin as a mere stream of sounds to the infant, just one part of the whole statistical learning package, but it quickly becomes more than that. It becomes an invitation for the infant to *create* similarities among diverse instances. A word tells the infant, "Do you see all these objects that look different physically? They have an equivalence that is mental." That equivalence is the basis for a goal-based concept.[30]

Fei Xu and her students have demonstrated this experimentally by showing objects to ten-month-old infants, giving the objects nonsense names like "wug" or "dak." The objects were wildly dissimilar, including dog-like and fish-like toys, cylinders with multicolored beads, and rectangles covered in foam flowers. Each one also made a ringing or rattling noise. Nevertheless, the infants learned patterns. Infants who heard the same nonsense name across several objects, regardless of their appearance, expected those objects to make the same noise. Likewise, if two objects had different names, the infants expected them to make different noises. This is a remarkable feat for infants because they used the sounds of a word to predict whether objects made the same noise or not, learning a pattern that transcended mere physical appearance. Words encourage infants to form goal-based concepts by inspiring them to represent things as equivalent. In fact, studies show that infants can more easily learn a goal-based concept, given a word, than a concept defined by physical similarity without a word.[31]

I don't know about you, but every time I think about this, I find it bloody amazing. Any animal can view a bunch of similar-looking objects and form a concept of them. But you can show human infants a bunch of objects that look different, sound different, and feel different, and merely add a word — a *WORD* — and these little babies form a concept that overcomes the physical differences. They understand that the objects have some kind of psychological similarity that can't be immediately perceived through the five senses. This similarity is what we called the goal of the concept. The infant creates *a new piece of reality*, a thing called a "wug" with the goal "to make a ringing noise."

From an infant's perspective, the concept "Wug" did not exist in the world before an adult taught it to her. This sort of social reality, in which two or more people agree that something purely mental is real, is a foundation of human culture and civilization. Infants thereby learn to categorize the world in ways that are consistent, meaningful, and predictable to us (the speakers), and eventually to themselves. Their mental model of the world becomes similar to ours, so we can communicate, share experiences, and perceive the same world.

When my daughter, Sophia, was a toddler and I bought her a toy car, I didn't realize I was helping to extend her goal-based categories, honing her conceptual system for creating social reality. She'd hold that car close to a toy truck, and they'd transform into "mama" and "baby" as she made them "kiss." Sometimes our goddaughter, Olivia, would visit, who is the same age, and the two girls would climb into the bathtub and engage in elaborate, imaginary dramas for hours, imposing new functions on toys, bars of soap, towels, and various bathroom items as the props in their water opera. A defining moment of humanity occurs when one child becomes an all-powerful being by draping a washcloth over her head and brandishing a toothbrush, and the second child kneels before her in supplication.

When we, as adults, speak a word to a child, an act of great significance takes place without fanfare. In that moment, we offer the child a tool to expand reality — a similarity that is purely mental — and she incorporates it into the patterns that are being laid down inside her own brain for future use. In particular, as we shall now see, we hand her the tools to make and perceive emotions.

• • •

Infants are born unable to see faces. They have no perceptual concept of "Face" and so are experientially blind. They quickly learn to see human faces, however, from the perceptual regularities alone: two eyes up top, a nose in the middle, and a mouth.[32]

If we observe this through the lens of the classical view of emotion, we could tell a story that infants statistically learn emotion concepts the same way, from perceptual regularities in the instances of happiness, sadness, surprise, anger, and other emotion categories that exist in the body or in other people's so-called emotional expressions. Many researchers, inspired by the classical view, have simply assumed that children's emotion concepts are scaffolded onto an inborn or early-to-develop understanding of facial expressions. This supposedly explains how children learn emotion words and also the causes and consequences of emotions.[33]

The stumbling block for this whole idea, we have learned, is that consistent emotion fingerprints don't exist in the face and body. Children must be gaining emotion concepts in some other way.

We've also just seen that words invite infants to equate wildly dissimilar objects. Words encourage infants to search for similarities beyond the physical, similarities that act like a mental glue for concepts. Babies could reasonably learn emotion concepts in this manner. Instances of "Anger" might share no perceptual similarities, but the word "angry" could be grouping them into a single concept, just as infants grouped "wugs" and "daks." I'm speculating for the moment, but the idea fits the data we've discussed.

I try to imagine how my daughter, Sophia, might have learned emotion concepts when she was an infant, guided by the emotion words that my husband and I spoke to her intentionally. In our culture, one goal in "Anger" is to overcome an obstacle that someone blameworthy has put in your path. So, when a little friend would smack Sophia, sometimes she would cry and other times she'd swat back. When she didn't like her food, sometimes she'd spit it out and other times she'd smile and tip the bowl onto the floor. These physical actions were accompanied by different facial movements, different changes in her body budget (to match her physical actions), and different interoceptive patterns. Within this ongoing stream of activity, her father and I would utter streams of sounds: "Sophie, sweetie, are you angry?" "Don't be angry, honey." "Sophie, you're feeling angry."[34]

At first, these noises must have been novel to Sophia, but over time, if my hypothesis is correct, she learned statistically to associate these diverse body patterns and contexts with the sounds "an-gry," just like associating a squeaking toy with the sound "wug." Eventually, the word "angry" invited my daughter to search for a way in which these instances were the same, even if on the surface they looked and felt different. In effect, Sophia formed a rudimentary concept whose instances were characterized by a common goal: overcoming an obstacle. And most importantly, Sophia learned which actions and feelings most effectively achieved this goal in each situation.

In this way, Sophia's brain would have bootstrapped the concept "Anger" into its neural architecture. When we first used the word "angry" with Sophia, we constructed her experiences of anger with her. We focused her attention, guiding her brain to store each instance in all its sensory detail. The word helped her to create commonalities with all the other instances of "Anger" already in her brain. Her brain also captured what preceded and followed those experiences. All of this became her concept of "Anger."[35]

In our earlier encounter with Connecticut Governor Malloy, I described how viewers inferred his emotional state — intense sadness — by observing his movements and voice in a certain context. I think children learn to do the same thing. As they learn a concept such as "Anger," they can predict and give meaning to other people's movements and vocalizations — smiles, shrugs, shouts, whispers, tightened jaws, widened eyes, even motionlessness — as well as their own bodily sensations, to construct perceptions of anger. Or, they can focus on predicting and giving meaning to their own interoceptive sensations, along with sensations from the world, to construct an emotional experience. As Sophia grew older, she extended her concept of "Anger" to people who slam doors, adding to her population of instances. And when she encountered a sneezing person and said, "Mama, that man is angry," and I corrected her, she honed her concept of "Anger" yet again. Her brain gave sensations meaning, using concepts that fit the situation, to construct an instance of emotion.[36]

If I am correct, then, as children continue to develop their concept of "Anger," they learn that not all instances of "Anger" are constructed for the same goal in every situation. "Anger" can also be for protecting oneself against an offense, dealing with someone who acted unfairly, desiring aggression toward another person, wanting to win a competition or to enhance performance in some way, or wishing to appear powerful.[37]

Following this line of reasoning, Sophia eventually would learn that anger-related words like "irritation," "scorn," and "vengeance" each referred to distinct goals that glued together variable populations of instances. And with this, Sophia developed an expert vocabulary of anger-related concepts that prepared her for the life of a typical American teenager. (For the record, she's not much for experiencing scorn or vengeance on a regular basis, but the concepts come in handy with other adolescents.)

My guiding hypothesis, as you can see from my story of Sophia's development, is that emotion words hold the key to understanding how children learn emotion concepts in the absence of biological fingerprints and in the presence of tremendous variation. Not the words in isolation, mind you, but words spoken by other humans in the child's affective niche who use emotion concepts. These words invite a child to form goal-based concepts for "Happiness," "Sadness," "Fear," and every other emotion concept in the child's culture.

So far, my hypothesis about emotion words is only reasoned speculation because the science of emotion is missing a systematic exploration of this question. Certainly nothing like the creative studies of Waxman, Xu, Gelman, and other developmental psychologists has yet been conducted for emotion concepts and categories. But we have some compelling evidence that is consistent with this hypothesis.

Some of the evidence comes from careful testing of children in the lab, which suggests that they don't develop adult-like emotion concepts like "Anger," "Sadness," and "Fear" until around age three. Younger children in Western cultures use words like "sad," "scared," and "mad" interchangeably to mean "bad"; they exhibit low emotional granularity, just like my graduate school test subjects for whom "depressed" and "anxious" meant nothing more than "unpleasant." As parents, we may look at our infants and perceive emotions in their cries, wriggles, and smiles. Certainly infants feel pleasure and distress from birth, and affect-related concepts (pleasant/unpleasant) show up by three to four months of age. But there's a lot of research to indicate that adult-like emotion concepts develop later. Just how much later is an open question.[38]

Other evidence for my hypothesis about emotion words comes from a surprising source: people who work with chimpanzees. Jennifer Fugate, a former postdoctoral fellow in my lab, collected photographs of chimpanzee facial configurations that some scientists treat as emotional expressions, including "play" faces, "scream" faces, "bared teeth" faces, and "hoot" faces. She tested chimp experts and novices to see if they could recognize these configurations, and at first, none of them could do it. So we performed an experiment similar to those used with infants: half of our experts and novices viewed pictures of chimp facial configurations alone, and half viewed them labeled with made-up words, such as "peant" for the play face and "sahne" for the scream face. In the end, only our subjects who learned the words could correctly categorize new chimp facial configurations, demonstrating that they had acquired the concepts for the face categories.[39]

As children grow up, they definitely form a whole conceptual system for emotion. This includes all the emotion concepts they've learned in their lives, anchored by the words that name those concepts. They categorize different facial and bodily configurations as the same emotion, and a single configuration as many different emotions. Variation is the norm. So where is the statistical regularity that holds together a concept like "Happiness" or "Anger"? In the words themselves. The most visible commonality that all instances of "Anger" share is that they're all called "anger."

Once children have the initial emotion concept, other factors besides words become important to their developing conceptual system for emotion. They come to realize that emotions are events that develop over time. An emotion has a beginning or cause that precedes it ("My mommy walked into the room"). Then there's a middle, the goal itself that is happening now ("I am happy to see my mommy"). Then there's an end, the consequence of meeting the goal, which happens later ("I'll smile and my mommy will smile back and give me a hug"). This means that an instance of an emotion concept helps to make sense of longer continuous streams of sensory input, dividing them into distinct events.[40]

You see emotions in blinks, furrowed brows, and other muscle twitches; you hear emotions in the pitch and lilt of voices; you feel emotions in your own body, but the emotional information is not in the signal itself. Your brain was not programmed by nature to recognize facial expressions and other so-called emotional displays and then to reflexively act on them. The emotional information is in your perception. Nature provided your brain with the raw materials to wire itself with a conceptual system, with input from a chorus of helpful adults who spoke emotion words to you in a deliberate and intentional way.

Concept learning does not stop in childhood — it continues throughout life. Sometimes a new emotion word appears in your primary language, engendering a new concept. For example, *schadenfreude,* a German emotion word meaning "pleasure from someone else's misfortune," has now been incorporated into English. Personally, I'd like to add the Greek word *stenahoria* to English, which refers to a feeling of doom, hopelessness, suffocation, and constriction. I can think of a few romantic relationships where this emotion concept would have come in handy.[41]

Other languages commonly have emotion words whose associated concepts have no equivalent in English. For example, Russian has two distinct concepts for what Americans call "Anger." German has three distinct "Angers" and Mandarin has five. If you were to learn any of these languages, you'd need to acquire these new emotion concepts to construct perceptions and experiences with them. You'll develop these concepts faster if you live with native speakers of the new language. The new concepts are affected by the older ones from your primary language. Native speakers of English who learn Russian, for example, must learn to distinguish between anger at a person, called *serdit'sia,* and anger for more abstract reasons such as the political situation, known as *zlit'sia.* The latter concept is more similar to the English concept of "Anger," but Russian speakers use the former more frequently; as a result, English speakers use *serdit'sia* more frequently as well and wind up misapplying it. This is not an error in a biological sense, since neither concept has a biological fingerprint, but in a cultural sense.[42]

New emotion concepts from a second language can also modify those of your primary language. A research scientist in my lab, Alexandra Touroutoglou, came from Greece to learn neuroscience. As she became more proficient at speaking English, her Greek and English emotion concepts began to blend. For example, Greek has two concepts for "Guilt," one for minor infractions and another for serious transgressions. English covers both situations with the single word "guilty." When Alex would speak with her sister who was still in Greece, Alex would use the "major" guilt word (*enohi*) when describing, say, that she ate too much pie at our lab's beach party. To her sister, Alex came across as overly dramatic. In this case, Alex constructed her dessert experience using the English concept for guilt.[43]

I hope by now you appreciate the drama that is going on here. Emotion words are not about emotional facts in the world that are stored like static files in your brain. They reflect the varied emotional meanings you construct from mere physical signals in the world using your emotion knowledge. You acquired that knowledge, in part, from the collective knowledge contained in the brains of those who cared for you, talked to you, and helped you to create your social world.

Emotions are not reactions *to* the world; they are your constructions *of* the world.

• • •

Once your conceptual system is established in your brain, you need not explicitly recall or speak an emotion word to construct an instance of an emotion. In fact, you can experience and perceive an emotion even if you don't have a word for it. Most of us who speak English were able to enjoy someone else's misfortune long before the word *schadenfreude* entered our language. All you need is a concept. How do you get a concept without a word? Well, your brain's conceptual system has a special power called *conceptual combination*. It combines existing concepts to create your very first instance of a novel concept of emotion.[44]

My friend Batja Mesquita is a Dutch cultural psychologist, and the first time I traveled to visit her in Belgium, she told me that we were sharing the emotion *gezellig*. Curled up in her living room, sharing wine and chocolates, she explained that this emotion means the comfort, coziness, and togetherness of being at home, with friends and loved ones. *Gezellig* is not an internal feeling that one person has for another but a way of experiencing oneself in the world. No single word in English describes the experience of *gezellig*, but once Batja explained it to me, I immediately experienced it. Her use of the word invited me to form a concept as infants do, but through conceptual combination — I automatically employed my concepts of "Close Friend," "Love," and "Delight," with a touch of "Comfort" and "Well-Being." This translation was not perfect, though, because in my American way of experiencing *gezellig*, I used emotion concepts that focus more on internal feelings than those that describe the situation.[45]

Conceptual combination is a potent capability of the brain. Scientists still debate on the mechanisms responsible for it, but they pretty much agree that it's a basic function of the conceptual system. It allows you to construct a potentially limitless number of novel concepts from your existing ones. This includes goal-based concepts like "Things That Can Protect You from Stinging Insects," in which the goal is short-lived.[46]

Conceptual combination is powerful, but it is far less efficient than having a word. If you asked me what I had for dinner this evening, I could say "baked dough with tomato sauce and cheese," but this is much less efficient than saying "pizza." Strictly speaking, you don't need an emotion word to construct an instance of that emotion, but it's easier when you have a word. If you want the concept to be efficient, and you want to transmit the concept to others, then a word is pretty handy.

Infants can benefit from this "pizza effect" before they can speak. For example, prelinguistic infants generally can hold about three objects in mind at a time. If you hide toys in a box while an infant watches, she can remember up to three hiding places. However, if you label several toys with a nonsense word like "dax" and several more with "blicket" before hiding them — assigning the toys to categories — the infant can hold up to six objects in mind! This happens even if all six toys are physically identical, strongly suggesting that infants gain the same efficiency benefits from conceptual knowledge that adults do. Conceptual combination plus words equals the power to create reality.[47]

In many cultures, you will find people who have hundreds, perhaps thousands of emotion concepts, that is, they exhibit high emotional granularity. In English, for example, they might have concepts for anger, sadness, fear, happiness, surprise, guilt, wonder, shame, compassion, disgust, awe, excitement, pride, embarrassment, gratitude, contempt, longing, delight, lust, exuberance, and love, to name a few. They'll also have distinct concepts for interrelated words like "aggravation," "irritation," "frustration," "hostility," "rage," and "disgruntlement." This person is an emotion expert. A sommelier of emotion. Each word corresponds to its own emotion concept, and each concept can be used in the service of at least one goal, but usually many different goals. If an emotion concept is a tool, then this person has a gigantic toolbox fit for a skilled craftsperson.

People who exhibit moderate emotional granularity might have dozens of emotion concepts rather than hundreds. In English, they might have concepts for anger, sadness, fear, disgust, happiness, surprise, guilt, shame, pride, and contempt; perhaps not many more than the so-called basic emotions. For these folks, words like "aggravation," "irritation," "frustration," "hostility," "rage," "disgruntlement," and so on would all belong to the concept "Anger." This person has your run-of-the-mill little red toolbox, filled with some pretty handy tools. Nothing fancy, but they get the job done.

People who exhibit low emotional granularity will have only a few emotion concepts. In English, they might have words in their vocabulary like "sadness," "fear," "guilt," "shame," "embarrassment," "irritation," "anger," and "contempt," but those words all correspond to the same concept whose goal is something like "feeling unpleasant." This person has a few tools — a hammer and Swiss Army knife. Maybe this person gets along fine, but a few new tools wouldn't hurt, at least if he or she lives in a Western cultural setting. (My husband jokes that before we met, he knew only three emotions: happy, sad, and hungry.)

When a mind has an impoverished conceptual system for emotion, can it perceive emotion? From scientific experiments in our own lab, we know that the answer is generally no. As you learned in chapter 3, we can easily interfere with people's ability to perceive anger in a scowl, sadness in a pout, and happiness in a smile by impairing their access to their emotion concepts.

If people lack a well-developed conceptual system for emotion, what is their emotional life like? Will they feel only affect? These questions are difficult to test scientifically. Emotional experiences have no objective fingerprints in the face, body, or brain that would enable us to compute an answer. The best we can do is ask people how they feel, but they'd have to use emotion concepts to answer the question, defeating the purpose of the experiment!

The way around this conundrum is to study people who have a naturally impoverished conceptual system for emotion, a condition called alexithymia, which by one estimate affects about 10 percent of the world's population. Its sufferers do have difficulty experiencing emotion, as the theory of constructed emotion would predict. In a situation where a person with a working conceptual system might experience anger, people with alexithymia are more likely to experience a stomachache. They complain of physical symptoms and report feelings of affect but fail to experience them as emotional. People with alexithymia have difficulty perceiving emotion in others as well. If a person with a working conceptual system saw two men shouting at each other, she might make a mental inference and perceive anger, whereas a person with alexithymia would report perceiving only shouting. People with alexithymia also have a restricted emotion vocabulary and have difficulty remembering emotion words. These clues provide further evidence that concepts are critical for experiencing and perceiving emotion.[48]

• • •

Concepts are linked to everything you do and perceive. And as you learned in the previous chapter, everything you do and perceive is linked to your body budget. Therefore, concepts must be linked to your body budget. And, in fact, they are.

When you were born, you couldn't regulate your budget, so your caregivers did it for you. Each time your mother picked you up to feed you was a multisensory event with regularities: the sight of your mother's face, the sound of her voice, her motherly aroma, her touch, the taste of her milk (or formula), and your interoceptive sensations associated with being held and cuddled and fed. Your brain captured the *entire sensory context* in the moment, as a pattern of sights, sounds, smells, tastes, touches, and interoceptive sensations. This is how concepts begin to form. You learn in a multisensory way. Your inner-body changes and their interoceptive consequences are part of every concept that you learn, whether you're aware of it or not.[49]

When you categorize with your multisensory concepts, you're also regulating your body budget. When you played with a ball as a baby, you categorized it not just by its color and shape and texture (and by the smell of the room, the feel of the floor against your hands and knees, the lingering taste of whatever you last ate, and so on), but also by your interoceptive sensations in the moment. This allowed you to predict your actions, like swatting the ball or putting it into your mouth, which influenced your body budget.

As an adult, when you learn that an event is an instance of some emotion, such as "Embarrassment," you likewise capture the event's sights, sounds, smells, tastes, touches, and interoceptive sensations together as your concept. And when you make meaning using that concept, your brain again takes into account your entire situation. For example, if you surface from under the ocean waves onto the beach and notice that your swimsuit has fallen off, your brain might construct an instance of "Embarrassment." Your conceptual system samples instances of embarrassed nakedness from your past, which is more taxing on your body budget than the refreshed nakedness after stepping out of a sauna, or the comfortable nakedness after a passionate afternoon with your lover. Depending on the immediate circumstances, your brain might also sample fully clothed instances of "Embarrassment" where you felt exposed, like answering a question wrongly in class, but not more private embarrassment like forgetting your best friend's birthday. Your brain samples from your larger conceptual system, as you've seen, according to your goal in a given situation. The winning instance guides you to regulate your body budget appropriately.[50]

All categorizations are based on probabilities. For example, if you are on vacation in Paris and you perceive a stranger frowning at you in a subway car, you might not have any past experience with that stranger or that subway, and might not have visited Paris before, but your brain does have past experiences of other frowning people in unfamiliar places. Your brain can then construct a sample of concepts, based on past experience and probability, to use as predictions. Each added piece of context (Are you alone or is the car crowded? Is it a man or a woman? With raised or furrowed eyebrows?) allows your brain to hone the probabilities until it settles on the best-fitting concept that will minimize prediction error. This is categorization with emotion concepts. You aren't detecting or recognizing emotion in someone's face. You aren't recognizing a physiological pattern in your own body. You are predicting and explaining the meaning of those sensations based on probability and experience. This happens each time you hear an emotion word or are faced with an array of sensations.[51]

All of this categorization, context, and probability may seem remarkably counterintuitive. When I'm walking through the woods and see a monstrous snake in my path, I certainly don't say to myself, "Well, I actively predicted that snake from a population of competing concepts, which were constructed from the past and have some degree of similarity to this current set of sensations, thereby creating my perception." I just "saw a snake." And when I gingerly turn on my heels and run, I don't think, "I honed my many predictions down to one winning instance of the emotion category 'Fear,' causing me to run away." No, I just feel terrified with an urge to flee. The fear comes on suddenly and uncontrollably, as if a stimulus (the snake) triggered a little bomb (a neural fingerprint) causing the response (fear and running).

When I relate the snake story to my friends later, over coffee, I don't tell them, "Having constructed an instance of the concept 'Fear' to fit my surroundings using my past experience, my brain changed the firing of my visual neurons before the snake appeared on the path, preparing me to see the snake and to run in the other direction, and once my prediction was confirmed, my sensations were categorized, and I constructed an experience of fear that explained my sensations in terms of a goal, and I made a mental inference to perceive the snake as the cause of my feelings, and the running away as their consequence." No, my story is much simpler: "I saw a snake. I screamed and fled."

Nothing about my encounter with the snake tells me that I was an architect of the whole experience. Nevertheless, I was that architect, whether I felt it or not, just as you were with the blobby bee. Even before I was aware of the snake, my brain was busy constructing an instance of fear. Or, if I am an eight-year-old girl hoping for a pet snake someday, I might construct an instance of excitement. If I am her parent who will allow a snake into my house over my dead body, I might construct an instance of irritation. The stimulus-response brain is a myth, brain activity is prediction and correction, and we construct emotional experiences outside of awareness. This explanation fits the architecture and operation of the brain.[52]

Simply put: I did not see a snake and categorize it. I did not feel the urge to run and categorize it. I did not feel my heart pounding and categorize it. I categorized sensations in order to see the snake, to feel my heart pounding, and to run. I correctly predicted these sensations, and in doing so, explained them with an instance of the concept "Fear." This is how emotions are made.

Right now, as you read these words, your brain is wired with a powerful conceptual system for emotion. It began purely as an information-gaining system, acquiring knowledge about your world through statistical learning. But words allowed your brain to go beyond the physical regularities that you learned, to invent part of your world, in a collective with other brains. You created powerful, purely mental regularities that helped you control your body budget in order to survive. Some of these mental regularities are emotion concepts, and they function as mental explanations for why your heart thumps in your chest, why your face flushes, and why you feel and act the way you do in certain circumstances. When we share those abstractions with each other, by synchronizing our concepts during categorization, we can perceive each other's emotions and communicate.

That, in a nutshell, is the theory of constructed emotion — an explanation for how you experience and perceive emotion effortlessly without the need for emotion fingerprints. The seeds of emotion are planted in infancy, as you hear an emotion word (say, "annoyed") over and over in highly varied situations. The word "annoyed" holds this population of diverse instances together as a concept, "Annoyance." The word invites you to search for the features that the instances have in common, even if those similarities exist only in other people's minds. Once you have this concept established in your conceptual system, you can construct instances of "Annoyance" in the presence of highly variable sensory input. If the focus of your attention is on yourself during categorization, then you construct an experience of annoyance. If your attention is on another person, you construct a perception of annoyance. And in each case, your concepts regulate your body budget.

When another driver cuts you off in traffic and your blood pressure rises, your hands become sweaty, and you shout as you slam on the brakes and feel annoyed . . . this is an act of categorization. When your young child picks up a sharp knife and your breathing slows, your hands are dry, you smile, and you calmly ask her to put it down as you feel annoyed inside . . . this is an act of categorization. When you see another person staring at you oddly with wide eyes and perceive him as annoyed, this is also an act of categorization. In all these instances, your conceptual knowledge of "Annoyance" drives the categorization, and your brain makes meaning that is tied to context. My story in chapter 2 about the guy in graduate school who asked me to lunch, when I thought I felt attraction but in fact I had the flu, is another example of categorization. My body budget was disrupted by a virus, but I experienced the resulting change in affect as attraction to my lunch partner because I'd constructed an instance of infatuation. If I'd categorized my symptoms in a different context, I might have understood them as something that a few Tylenol and a couple of days' rest could cure.

Your genes gave you a brain that can wire itself to its physical and social environment. The people around you, in your culture, maintain that environment with their concepts and help you live in that environment by transmitting those concepts from their brains to yours. And later, you transmit your concepts to the brains of the next generation. It takes more than one human brain to create a human mind.

What I have not yet explained, however, is how this all works inside the brain: the biology of categorization. What brain networks are involved? How is this process related to your brain's intrinsic, predictive powers, and how does it affect your all-important body budget? That is what we'll discuss next as you learn the final piece of the puzzle for how emotions are made in the brain.

6

How the Brain Makes Emotions

Have you ever wanted to punch your boss? I would never advocate workplace violence, of course, and many bosses are terrific work partners. But sometimes we are blessed with supervisors who personify the German emotion word *Backpfeifengesicht,* meaning "a face in need of a fist."

Suppose you have such a boss, and he's been handing you extra projects for almost a year. You've been expecting a promotion for all your good work, but he has just informed you that the promotion went to someone else. How would you feel?

If you live in a Western culture, you'd likely feel angry. Your brain would issue numerous predictions of "Anger" simultaneously. One prediction might be to pound your fist on the desk and yell at your boss. Another is to stand up and walk slowly across the room toward your boss, leaning in menacingly to whisper, "You will regret this." Or you could sit quietly in your chair as you scheme to undermine your boss's career.[1]

These diverse predictions of "Anger" have similarities, such as the boss, the lost promotion, and the common goal to exact vengeance. They also have plenty of differences, because yelling, whispering, and silence require different sensory and motor predictions. Your action also is different in each case (pounding, leaning, sitting), so your inner-body changes are different, as are the consequences for your body budget, and therefore the interoceptive and affective consequences are different as well. Ultimately, through a process we'll discuss shortly, your brain selects a *winning instance* of "Anger" that best fits your goal in this particular situation. The winning instance determines how you behave and what you experience. This process is categorization.

The scenario with your boss could play out differently, however. You could be angry with a different goal, like changing your boss's mind, or maintaining social relations with the coworker who got the promotion in your place. Or you could construct an instance of a different emotion such as "Regret" or "Fear," or a non-emotion like "Emancipation," or a physical symptom like a "Headache," or a perception that your boss is an "Idiot." In each case, your brain follows a similar process, categorizing to best fit the entire situation and your internal sensations, based on past experience. Categorization means selecting a winning instance that becomes your perception and guides your action.[2]

It takes a rich set of concepts to construct emotion, as you read in the preceding chapter. Now you'll learn *how* your brain acquires and uses your conceptual system from your earliest moments as an infant. Along the way, you'll also learn the neural basis for several important topics you've seen previously: emotional granularity, population thinking, why emotions feel triggered rather than constructed, and why your body-budgeting regions can affect every decision and action you make.* When taken as a whole, these explanations hint at a unifying framework for *how the brain makes meaning*: one of the most extraordinary mysteries of the human mind.

• • •

The infant brain is missing most of the concepts that we have as adults. Babies don't know what telescopes are, or sea cucumbers, or picnics, let alone purely mental concepts like "Whimsy" or "*Schadenfreude.*" A newborn is experientially blind to a great extent. Not surprisingly, the infant brain does not predict well. A grown-up brain is dominated by prediction, but an infant brain is awash in prediction error. So babies must learn about the world from sensory input before their brains can model the world. This learning is a primary task of the infant brain.

At first, much of the onslaught of sensory input is new to an infant's brain, and its significance is undetermined, so little will be ignored. If sensory input is like a skipping stone on an ocean wave of brain activity, for infants the stone is a boulder. Infants absorb the sensory input around them and learn, learn, learn. The developmental psychologist Alison Gopnik de-

---

* More detailed scientific evidence supporting this chapter can be found in appendix D.

scribes babies as having a "lantern" of attention that is exquisitely bright but diffuse. In contrast, your adult brain has a network to shut out information that might sidetrack your predictions, allowing you to do things like read this book without distraction. You have a built-in "spotlight" of attention that illuminates some things, such as these words, while leaving other things in the dark. The infant brain's "lantern" cannot focus in this manner.[3]

As the months pass, if everything is working properly, the infant brain begins to predict more effectively. Sensations from the outside world have become concepts in the infant's model of the world; what was outside is now inside. These sensory experiences, over time, create the opportunity for the infant brain to make *coordinated* predictions that span the senses. A rumbling tummy in a bright room after awakening means that it's morning, whereas a warm wetness with bright overhead light means that it's evening bath time. When my daughter, Sophia, was only a few weeks old, we capitalized on such multisensory predictions to help her develop sleep patterns that would not reduce us to sleep-deprived zombies. We exposed her to distinct songs, stories, colored blankets, and other rituals to help her distinguish statistically between naptime and bedtime, so she would sleep for shorter or longer stretches.[4]

How does the infant brain, equipped with a smattering of concrete concepts and dominated by prediction error, eventually encompass thousands of complex, purely mental concepts like "Awe" and "Despair," each of which is a population of diverse instances? This is actually a question of engineering, and its solution can be found in the architecture of the human cerebral cortex. It all comes down to some basic problems of efficiency and energy. An infant brain must continually learn and update its concepts in a changing environment. This task requires a mighty powerful, efficient brain. But this brain has practical constraints. Its networks of neurons can grow only so big and still fit inside a skull that can be birthed through a human pelvis. Neurons are also expensive little cells to keep alive (they require a lot of energy), and so a brain has a limit on how many connections it can support metabolically and still run. So the infant brain must transfer information *efficiently* by passing it to as few neurons as possible.

The solution to this engineering challenge is a cortex that represents concepts so that *similarities are separated from differences.* This separation, as you will now see, leads to a tremendous optimization.

Whenever you watch a video on YouTube, you're witnessing efficient information transfer of a similar kind. A video is a sequence of still images or "frames" displayed in rapid succession. There is great redundancy from one frame to the next, however, so when YouTube's server sends the stream of video information over the Internet to your computer or phone, it needn't send every single pixel from every frame. It's more efficient to communicate only what has *changed* from one frame to the next, because any static areas of the previous frame have already been transmitted. YouTube separates the video's similarities from its differences to speed up transmission, and software on your computer or phone assembles the pieces into a cohesive video.

The human brain does much the same thing when it processes prediction error. The sensory information from sight is highly redundant like a video, and the same is true for sound, smell, and the other senses. The brain represents this information as patterns of firing neurons, and it's advantageous (and efficient) to represent it with as few neurons as possible.

For example, the visual system represents a straight line as a pattern of neurons firing in primary visual cortex. Suppose that a second group of neurons fires to represent a second line at a ninety-degree angle to the first line. A third group of neurons could *summarize* this statistical relationship between the two lines efficiently as a simple concept of "Angle." The infant brain might encounter a hundred different pairs of intersecting line segments of varying length, thickness, and color, but conceptually they are all instances of "Angle," each of which gets efficiently summarized by some smaller group of neurons. These summaries eliminate redundancy. In this manner, the brain separates statistical similarities from sensory differences.

In the same manner, the instances of the concept "Angle" are themselves part of other concepts. For example, an infant receives visual input about her mother's face from many different vantage points: while nursing, while sitting face to face, in the morning and the evening. Her concept of "Angle" will be part of her concept "Eye" that summarizes the continuously changing lines and contours of her mother's eyes seen at different angles and in different luminances. Different groups of neurons fire to represent the various instances of the concept "Eye," allowing the infant to recognize those eyes as her mother's eyes each time, regardless of the sensory differences.[5]

As we go from very specific to increasingly general concepts (in this example, from line to angle to eye), the brain creates similarities that are progressively more efficient summaries of the information. For example, "Angle" is an efficient summary with respect to lines but is a sensory detail with respect to eyes. The same logic works for the concepts "Nose" and "Ear" and so on. Together, these concepts are part of the concept "Face," whose instances are yet more efficient summaries of the sensory regularities in facial features. Eventually, the infant's brain forms summary representations for enough visual concepts that she can see one stable object, despite incredible variation in low-level sensory details. Think about it: each of your eyes transmits millions of tiny pieces of information to your brain in a moment, and you simply see "a book."

This principle — finding similarities in the service of efficiency — doesn't just describe the visual system; it also operates within each sensory system (sounds, smells, interoceptive sensations, and so on), and for patterns of different senses in combination. Consider a purely mental concept like "Mother." As a baby nurses one morning, groups of neurons fire in her various sensory systems, in statistically related patterns, to represent the mother's visual image, the sound of her voice, her scent, the tactile sensations of being held, an increase in energy from being fed, the sensations of a full tummy, plus the pleasure of feeding and being cuddled. All of these representations are interrelated, and their summary is represented elsewhere, in the pattern of firing within a smaller group of neurons, as a rudimentary, multisensory instance of "Mother." During nursing again later in the day, other summaries of the concept "Mother" will be similarly created, using similar, but not identical, groupings of neurons. And as the infant swats at a hanging toy above the crib, watches the toy swing through the air, and feels any associated tactile and interoceptive sensations, all of which are linked with a decrease in energy due to her movement, her brain summarizes these statistically related events as a rudimentary, multisensory instance of the concept "Self."[6]

In this manner, an infant's brain distills widely dispersed firing patterns for individual senses into one multisensory summary. This process reduces redundancy and represents the information in a minimal, efficient form for future use. It's like dehydrated food that takes up less space but needs to be reconstituted before eating. This efficiency makes it practical for the brain to form rudimentary concepts such as "Mother" and "Self" that result from learning.

As a child gets older, her brain begins to predict more effectively using her concepts — but of course she still makes mistakes. When Sophia was three years old, for instance, we were in a shopping mall when she spotted a man ahead of us, with his hair in dreadlocks. She knew three people with dreadlocks at that time: her beloved Uncle Kevin, who is medium height and dark-skinned; an acquaintance who also has dark skin but is quite tall and broad-shouldered; and one of our neighbors, who is female and short with light skin. In that moment, Sophia's brain was furiously launching multiple, competing predictions that could potentially become her experience. For the sake of argument, let's say this included 100 predictions of Uncle Kevin from Sophia's past experience, from different places and times and angles, along with 14 predictions of her acquaintance, and 60 predictions of her female neighbor. Each prediction was assembled from bits and pieces of patterns in her brain, all mixed and matched. These 174 predictions were also accompanied by many other predictions of people, places, and things from Sophia's prior experiences—anything at all that was statistically related to the scene in front of her.

In total, Sophia's population of 174 predictions is what we've been calling a "concept" (in this case, the concept "People with Dreadlocks"). When we say these instances are "grouped" as a concept, be aware that there is no "grouping" stored anywhere in Sophia's brain. Any given concept is not represented in the information flow among one single set of neurons; each concept is itself a population of instances, and these instances are represented in different patterns of neurons on each occasion. (This is degeneracy.) The concept is constructed in the moment, ad hoc. And among these myriad instances, one of them will be the most similar (by pattern matching) to Sophia's current situation. That's what we've been calling the "winning instance."[7]

On that particular day, Sophia leaped out of her stroller, ran across the mall, and wrapped her little arms around the man's leg, shouting, "Uncle KEVIN!" Her delight was short-lived, however, as Uncle Kevin was at home six hundred miles away. She looked up into a total stranger's face and shrieked.*

The same general process occurs for purely mental concepts such as "Sadness." A child hears the word "sad" spoken in three different situations. These three instances are represented in the child's brain in bits and pieces. They are not "grouped together" in any concrete way. On a fourth occasion, the child sees a boy in her classroom crying, and a teacher uses the word "sad." The child's brain constructs the three prior instances as predictions,

---

* And as luck would have it, his name was Kevin.

along with other predictions that are statistically similar in any way to the current situation. This collection of predictions is a concept created in the moment, by virtue of some purely mental similarity among the instances of "Sadness." Once again, the prediction that is most similar to the current situation becomes her experience — an instance of emotion.[8]

* * *

It's time for me to explain something directly what so far I have only implied. Two of the phenomena I've been discussing are actually one and the same. I'm speaking of concepts and predictions.

When your brain "constructs an instance of a concept," such as an instance of "Happiness," that is equivalent to saying your brain "issues a prediction" of happiness. When Sophia's brain issued 100 predictions about Uncle Kevin, each one was an instance of the momentary concept "Uncle Kevin" that she formed before grabbing the stranger's leg.[9]

I separated the ideas of predictions and concepts earlier to simplify some explanations. I could have used the word "prediction" throughout the book and never mentioned the word "concept," or vice versa, but information transmission is easier to understand in terms of predictions flying across the brain, whereas knowledge is more readily understood in terms of concepts. Now that we're discussing how concepts work in the brain, we must acknowledge that concepts are predictions.

Early in life, you build up concepts from detailed sensory input (as prediction error) from your body and the world. Your brain efficiently compresses the sensory input it receives, just like YouTube compresses video, extracting similarities out of differences, eventually creating an efficient, multisensory summary. Once your brain has learned a concept in this manner, it can run this process in reverse, expanding the similarities into differences to construct an instance of the concept, much as your computer or phone expands the incoming YouTube video for display. This is a prediction. Think of prediction as "applying" a concept, modifying the activity in your primary sensory and motor regions, and correcting or refining as needed.

Imagine that you're in a shopping mall, as I was with my daughter, strolling from store to store. The mall is filled with sounds, people are bustling about, the shop windows are overflowing with tempting products for sale, and your brain is busy issuing thousands of simultaneous predictions as usual. "There is motion in front of me." "There is motion to my left." "My breathing is slowing down." "My stomach is rumbling." "I hear laughter." "I am calm." "I am lonely." "I see my neighbor." "I see that nice guy who works at the post office." "I see my Uncle Kevin." Let's say that those last three predictions about people are instances of a concept for "Happiness," having to do with feeling connected to friends. Your brain simultaneously constructs many instances of this concept, based on past experiences in similar situations when you have unexpectedly bumped into friends. Each instance has some probability of being correct at that moment.

Let's give our focus to one of those instances, your prediction that you see your beloved Uncle Kevin unexpectedly in a shopping mall. Your brain issues this prediction because, at some time in the past, you saw Uncle Kevin in a similar situation and experienced sensations that you categorized as happiness. How well will this prediction match your incoming sensory inputs right now? If it matches better than all the other predictions, then you will experience this instance of "Happiness." If not, then your brain will adjust the prediction, and you might experience an instance of "Disappointment." Or if need be, your brain will *make* the prediction match the sensory input, and you will mistakenly perceive someone else to be your Uncle Kevin, as Sophia did in the shopping mall that day.

So there you are, standing in the mall, and your brain must determine whether its prediction of Uncle Kevin ultimately becomes your perception and directs your action, or whether a course correction is required. To determine the details, the brain unpacks the summary of all the sensory input into a gigantic *cascade* of more detailed predictions, like uncompressing a YouTube video for viewing, or adding water to dehydrated food to make it edible. This process, shown in figure 6-1, is the same one that builds up a concept from details, but in reverse.

For example, when the prediction of "Happiness" reaches the upper portions of the visual system, the prediction might unpack into details of Uncle Kevin's appearance, say, whether he is facing toward you or away from you, and what clothing he is wearing. These details are themselves predictions based on probabilities (e.g., Uncle Kevin never wears plaid), so your brain can compare the simulation to actual sensory input and compute and resolve any prediction error. This resolution does not happen in a single step but in millions of bits and pieces (as the prediction loops discussed in chapter 4). Each visual detail is unpacked into even more detailed predictions in turn, for (say) colors, shirt texture, and so on, each of which involves more prediction loops and cascading and unpacking. The cascade ends in the brain's primary visual cortex, which represents your lowest-level visual concepts in a tornado of ever-changing lines and edges.

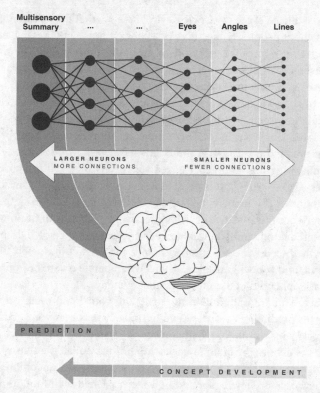

Figure 6-1: The concept cascade. When you develop a concept (right to left), sensory input is compressed into efficient, multisensory summaries. When you construct an instance of a concept by prediction (left to right), those efficient summaries unpack into ever more detailed predictions, which are checked against actual sensory input at each stage.

Cascades begin — of all places — within our old friend the interoceptive network.* That's where multisensory summaries are constructed in your brain. Cascades end in your primary sensory regions, where the tiniest details of your experience are represented, not just for vision as in our example but also for sound, touch, interoception, and the rest of your senses.

If one cascade of predictions accounts for the incoming sensory input

* Specifically, in a portion of the interoceptive network known as the default mode network. Appendix D has the details.

— Uncle Kevin is indeed in front of you, his hair pulled back in a particular way, wearing a particular shirt, his voice sounding a particular way, your body in a particular state, and so on — then you have constructed an instance of "Happiness" having to do with feeling connected to friends. That is, the *entire cascade* is that instance of the concept "Happiness" as you glimpse your uncle. You are feeling happy.

The concept cascade reveals the neural reasons for several of the claims I've made earlier in the book. First, your cascade of predictions explains why an experience like happiness feels triggered rather than constructed. You're simulating an instance of "Happiness" even before categorization is complete. Your brain is preparing to execute movements in your face and body before you feel any sense of agency for moving, and is predicting your sensory input before it arrives. So emotions seem to be "happening to" you, when in fact your brain is actively constructing the experience, held in check by the state of the world and your body.[10]

Second, the cascade explains a statement I made in chapter 4, that every thought, memory, emotion, or perception that you construct in your life includes something about the state of your body. Your interoceptive network, which regulates your body budget, is launching these cascades. Every prediction you make, and every categorization your brain completes, is always in relation to the activity of your heart and lungs, your metabolism, your immune function, and the other systems that contribute to your body budget.

Third, the cascade also highlights the neural advantages of high emotional granularity, the phenomenon (described in chapter 1) of constructing more precise emotional experiences. When your brain constructs multiple instances of "Happiness" at seeing Uncle Kevin, it must sort out which one best resembles your current sensory input, to become the winning instance. This is a big job for your brain with some metabolic cost. But imagine if the English language had a more specific word than "happiness" for feeling attachment to a close friend, such as the Korean word *jeong* (정). Your brain would require less effort to construct with this more precise concept. Even better, if you had a special word for "happiness at feeling close to my Uncle Kevin," your brain could be even more efficient at determining the winning instance. On the other hand, if you were constructing with the very broad concept "Pleasant Feeling" rather than "Happiness," your brain's job would be harder. Preciseness leads to efficiency; this is a biological payoff of higher emotional granularity.[11]

Finally, we're seeing population thinking in action in the brain, because multiple predictions make up a concept in the moment. You do not construct just one instance of "Happiness" and experience it. You construct a large population of predictions, each of which has its own cascade. That population is a concept. It doesn't represent the sum total of everything you know about happiness, just summaries that fit your goal — bumping into a friend — in a similar situation. In a different, happiness-related situation, like receiving a gift or hearing your favorite song, your interoceptive network would launch very different summaries (and cascades) representing "Happiness" in that moment. These dynamic constructions are another example of efficiency in the brain.

Scientists have known for some time that knowledge from the past, wired into brain connections, creates simulated experiences of the future, such as imagination. Other scientists focus on how this knowledge creates experiences of the present moment. The Nobel laureate and neuroscientist Gerald M. Edelman called your experiences "the remembered present." Today, thanks to advances in neuroscience, we can see that Edelman was correct. An instance of a concept, as an entire brain state, is an anticipatory guess about how you should act in the present moment and what your sensations mean.[12]

My description of the concept cascade is just a sketch of a much larger parallel process. In real life, your brain never categorizes 100 percent with one concept and 0 percent with others. Predictions are more probabilistic than that. Your brain launches thousands of predictions simultaneously in every moment, in a storm of probabilities, and never lingers on a single winning instance. When you construct one hundred varied, simultaneous predictions of Uncle Kevin in a moment, each one is a cascade. (If you're interested in more details about the neuroscience, see appendix D.)[13]

• • •

Each time you categorize with concepts, your brain creates many competing predictions while being bombarded by sensory input. Which predictions should be the winners? Which sensory input is important, and which is just noise? Your brain has a network to help resolve these uncertainties, known as your *control network*. This is the same network that transforms an infant's "lantern" of attention into the adult "spotlight" you have now.[14]

The famous optical illusion in figure 6-2 illustrates your control network in action. Depending on the context, whether you read horizontally or vertically, you'll perceive the central symbol as a "B" or a "13." Your control network helps select the winning concept — letter or number? — in each moment.[15]

Figure 6-2: The control network helps the brain select among
potential categorizations: in this case, "B" versus "13."

Your control network also helps construct instances of emotion. Suppose you've recently argued with your significant other, and now you're having chest pain. Is it a heart attack, indigestion, an experience of anxiety, or a perception that your partner's being unreasonable? Your interoceptive network will launch hundreds of competing instances of different concepts, each a brain-wide cascade, to resolve this quandary. Your control network assists in efficiently constructing and selecting among the candidate instances so your brain can pick a winner. It helps neurons to participate in certain constructions rather than others, and keeps some concept instances alive while suppressing others. The result is akin to natural selection, in which the instances most suitable to the current environment survive to shape your perception and action.[16]

The name "control network" is unfortunate because it implies a central position of authority, as if the network were making decisions and conducting the process. This is not the case. Your control network is more of an optimizer. It constantly tinkers with the information flow among neurons, ramping up the firing rate of some neurons and slowing down others, which moves sensory input in and out of your attentional spotlight, making some predictions fit while others become irrelevant. It's like a car-racing team that constantly optimizes the engine and body to make a car slightly faster and safer. This tinkering ultimately helps your brain simultaneously to regulate your body budget, produce a stable perception, and launch an action.[17]

Your control network helps select between emotion and non-emotion concepts (is this anxiety or indigestion?), between different emotion concepts (is this excitement or fear?), between different goals for an emotion concept (in fear, should I escape or attack?), and between different instances (when running to escape, should I scream or not?). When you're watching a movie, your control network might favor your visual and auditory systems, transporting you into the story. At other times it might background the traditional five senses in favor of more intense affect, resulting in an experience of emotion. Much of this tinkering happens outside your awareness.[18]

Some scientists refer to the control network as an "emotion regulation" network. They assume that emotion regulation is a cognitive process that exists separately from emotion itself, say, when you're pissed off at your boss but refrain from punching him. From the brain's perspective, however, regulation is just categorization. When you have an experience that feels like your so-called rational side is tempering your emotional side — a mythical arrangement that you've learned is not respected by brain wiring — you are constructing an instance of the concept "Emotion Regulation."[19]

Your control network and interoceptive network, as you've now seen, are critical for constructing emotion. Moreover, these two core networks together contain most of the major hubs for communication throughout the entire brain. Think about the world's largest airports that serve multiple airlines. A traveler in JFK International Airport in New York can switch between American Airlines and British Airways because the two airlines overlap there. Likewise, information can pass efficiently between different networks in your brain via the major hubs in the interoceptive and control networks.[20]

These major hubs help to synchronize so much of your brain's information flow that they might even be a prerequisite for consciousness. If any of these hubs become damaged, your brain is in big trouble: depression, panic disorder, schizophrenia, autism, dyslexia, chronic pain, dementia, Parkinson's disease, and attention deficit hyperactivity disorder are all associated with hub damage.[21]

The major hubs in your interoceptive and control networks make possible what I describe in chapter 4, that your everyday decisions are driven by your body-budgeting regions — your inner, loudmouthed, mostly deaf scientist who views the world through affect-colored glasses. You see, your brain's body-budgeting regions *are* major hubs. Through their massive connections, they broadcast predictions that alter what you see, hear, and otherwise perceive and do. That's why, at the level of brain circuitry, no decision can be free of affect.

•   •   •

I've said several times that the brain acts like a scientist. It forms hypotheses through prediction and tests them against the "data" of sensory input. It corrects its predictions by way of prediction error, like a scientist adjusts his or her hypotheses in the face of contrary evidence. When the brain's predictions match the sensory input, this constitutes a model of the world in that instant, just like a scientist judges that a correct hypothesis is the path to scientific certainty.

Several years ago, my family was eating dinner in our kitchen in Boston when suddenly, simultaneously, all of us had a sensation that was entirely new. Our chairs tipped backward for a moment, then righted themselves, but in a curvy sort of way like cresting an ocean wave. This completely novel experience left us in a state of experiential blindness, so we started forming hypotheses. Did we all simply lose our balance momentarily? No, that wasn't likely to happen to three people at once. Did a car crash outside the house? No, we hadn't heard anything. Had a building exploded far away, out of audible range, making the ground tremble? Maybe, but the feeling wasn't so much a tremble as a swoop. What about an earthquake? Maybe, but we'd never been in an earthquake before, and ours had lasted only one second, much shorter than earthquakes we'd seen in disaster movies. However, the rising and falling shape, an almost sinusoidal motion, was consistent with our understanding of earthquakes. An earthquake was the best match to our knowledge, so we settled on that hypothesis. A few hours later, we learned that a magnitude 4.5 earthquake had struck in nearby Maine and rippled throughout New England.

This same process of elimination that my family performed consciously, the brain does naturally, automatically, and extremely rapidly. Your brain has a mental model of the world as it will be in the next moment, developed from past experience. This is the phenomenon of *making meaning* from the world and the body using concepts. In every waking moment, your brain uses past experience, organized as concepts, to guide your actions and give your sensations meaning.

I've been calling this process "categorization," but it's known by many other names in science. Experience. Perception. Conceptualization. Pattern completion. Perceptual inference. Memory. Simulation. Attention. Morality. Mental Inference. In the folk psychology of daily life, these words mean different things, and scientists often study them as different phenomena, assuming each is produced by a distinct process in the brain. But really, they arise via the same neural processes.

When I feel cheerful as my nephew Jacob exuberantly wraps his little arms around my neck for a big hug, this is conventionally called "an emotional experience." When I see happiness in the big smile on his face as he is hugging me, I am no longer experiencing but "perceiving." When I recollect the hug and how warm it made me feel, I am no longer perceiving but "remembering." When I contemplate whether I was feeling happy or sentimental, I am no longer remembering but "categorizing." My view is that these terms don't mark sharp distinctions, and they can all be accounted for with the same brain ingredients for making meaning.

To make meaning is to go beyond the information given. A fast-beating heart has a physical function, such as getting enough oxygen to your limbs so you can run, but categorization allows it to become an emotional experience such as happiness or fear, giving it additional meaning and functions understood within your culture. When you experience affect with unpleasant valence and high arousal, you make meaning from it depending on how you categorize: Is it an emotional instance of fear? A physical instance of too much caffeine? A perception that the guy talking to you is a jerk? Categorization bestows new functions on biological signals, not by virtue of their physical nature but by virtue of your knowledge and the context around you in the world. If you categorize the sensations as fear, you are making meaning that says, "Fear is what caused these physical changes in my body." When the concepts involved are emotion concepts, your brain constructs instances of emotion.[22]

When you perceived the blobby picture in chapter 2 as a bee, you made meaning from the visual sensations. Your brain accomplished this feat by predicting a bee and simulating lines to connect the blobs. Prior experience — seeing the real bee photograph — encouraged your brain to leave the prediction uncorrected. As a result, you perceived a bee in the blobs. Your prior experiences shape the meaning of momentary sensations. This same miraculous process makes emotion.

Emotions are meaning. They explain your interoceptive changes and corresponding affective feelings, in relation to the situation. They are a prescription for action. The brain systems that implement concepts, such as the interoceptive network and the control network, are the biology of meaning-making.

So, now you know how emotions are made in the brain. We predict and categorize. We regulate our body budgets, as any animal does, but wrap this regulation in purely mental concepts like "Happiness" and "Fear," that we construct in the moment. We share these purely mental concepts with other adults, and we teach them to our children. We make a new kind of reality and live in it every day, mostly unaware that we are doing so. That's the topic of the next chapter.

7

Emotions as Social Reality

If a tree falls in the forest and no one is present to hear it, does it make a sound? This clichéd question has been asked to death by philosophers and grade-school teachers, but it also reveals something critical about human experience and, in particular, how we experience and perceive emotion.

The common-sense answer to this riddle is yes, of course a falling tree makes a sound. If you and I were walking in the forest at the time, we would clearly hear the cracking of the wood, the rustling of the leaves, and the monstrous thud as the trunk slammed into the forest floor. It seems obvious that this sound would be present even if you and I were not.

The scientific answer to the riddle, however, is no. A falling tree itself makes no sound. Its descent merely creates vibrations in the air and the ground. These vibrations become sound only if something special is present to receive and translate them: say, an ear connected to a brain. Any mammalian ear will do nicely. The outer ear gathers changes in air pressure and focuses them on the eardrum, producing vibrations in the middle ear. These vibrations move fluid in the inner ear over little hairs that translate the pressure changes into electrical signals that are received by the brain. Without this special machinery, there is no sound, only air movement.

Even after the brain receives these electrical signals, its task is not complete. This wave must still be interpreted as the sound of a toppling tree. For this, the brain needs the concept of "Tree" and what trees can do, such as fall in a forest. This concept can come from prior experience with trees, or from learning about trees in a book, or from another person's description. Without the concept, there is no crashing timber, only the meaningless noise of experiential blindness.

A sound, therefore, is not an event that is *detected* in the world. It is an experience *constructed* when the world interacts with a body that detects changes in air pressure, and a brain that can make those changes meaningful.[1]

Without a perceiver there is no sound, only physical reality. In this chapter, we explore another kind of reality that we humans construct, which exists only for those who are equipped to perceive it. Within this effortless ability lies an answer to the question, "What is an emotion?" It also explains how emotions are passed down through the generations without biological fingerprints.

Next, consider another question: "Is an apple red?" This is also a riddle, but less obviously so than the one about the falling tree. Again, the common-sense answer is yes, the apple is red (or yellow or green if you prefer). The scientific answer, however, is no. "Red" is not a color contained in an object. It is an experience involving reflected light, a human eye, and a human brain. We experience red only when light of a certain wavelength (say, 600 nanometers) reflects from an object (in the midst of other reflections at other wavelengths), and only while a receiver translates this contrasting array of light into visual sensations. Our receiver is the human retina, which uses its three types of photoreceptors, called cones, to convert the reflected light into electrical signals made meaningful by a brain. In a retina that's missing a medium or long cone, light at 600 nanometers is experienced as gray. And in the absence of a brain, there is no experience of color at all, only reflected light in the world.[2]

Even with the right equipment in place (the eye and the brain), the experience of a red apple is not a done deal. For the brain to convert a visual sensation into the experience of red, it must possess the concept "Red." This concept can come from prior experience with apples, roses, and other objects you perceive as red, or from learning about red from other people. (Even people who are blind since birth have a concept of "Red" that they learn from conversations and books.) Without this concept, the apple would be experienced differently. For instance, to the Berinmo people of Papua New Guinea, apples reflecting light at 600 nanometers are experienced as brownish, because Berinmo concepts for color divide up the continuous spectrum differently.[3]

These riddles about apples and trees invite us, as perceivers, to wrestle with two conflicting points of view. On one hand, common sense tells us that sounds and colors exist in the world beyond our skin, and we detect them with eyes and ears that carry the information to the brain. On the other hand, as we learned in chapters 4–6, we humans are architects of our own experiences. We do not passively detect physical changes in the world. We actively participate in constructing our experiences even though we are mostly unaware of that fact. An object might seem to transmit information about its color into your brain, but the information required for you to experience color comes mainly from your predictions, corrected by the light that your brain takes in from the world.

With prediction, you can "see" color in your mind's eye on demand. Try right now to see the green colors of a verdant forest. The colors might not be as vivid as usual and the experience may be fleeting, but you can probably do it. And as you do, neurons in your visual cortex change their firing. You are simulating green. You can also imagine a crashing tree and hear the sound in your mind. Try it, and neurons in your auditory cortex will change their firing.

Changes in air pressure and wavelengths of light exist in the world, but to us, they are sounds and colors. We perceive them by going beyond the information given to us, making meaning from them using knowledge from past experience, that is, concepts. Every perception is constructed by a perceiver, usually with sensory inputs from the world as one ingredient. Only certain changes in air pressure are heard as trees falling. Only some of the wavelengths of light striking our retinas are transformed into the experience of red or green. To believe otherwise is naive realism, as if perceptions were synonymous with reality.

A third and final riddle is, "Are emotions real?" You might think this question is ridiculous, a classic example of academic indulgence. Of course emotions are real. Think about the last time you were thrilled or sad or furious. These were clearly real feelings. But in fact, this third riddle is like the falling tree and the red apple: a dilemma about what exists in the world versus in the human brain. The riddle forces us to confront our assumptions about the nature of reality and our role in creating it. But here, the answer is a bit more complex, because it depends on what we mean by "real."

If you talk to a chemist, "real" is a molecule, an atom, a proton. To a physicist, "real" is a quark, a Higgs boson, or maybe a collection of little strings vibrating in eleven dimensions. They are supposed to exist in the natural world whether or not humans are present — that is, they are thought to be *perceiver-independent* categories. If all human life left this planet tomorrow, subatomic particles would still be here.[4]

But evolution has provided the human mind with the ability to create another kind of real, one that is completely dependent on human observers. From changes in air pressure, we construct sounds. From wavelengths of light, we construct colors. From baked goods, we construct cupcakes and muffins that are indistinguishable except by name (chapter 2). Just get a couple of people to agree that something is real and give it a name, and they create reality. All humans with a normally functioning brain have the potential for this little bit of magic, and we use it all the time.

Figure 7-1: Queen Anne's lace

If you doubt your power as a conjurer of reality, look at figure 7-1. This plant is *daucus carota,* better known as Queen Anne's lace. Usually the outer blooms are white, but in rare cases they are pink (i.e., they reflect light at a wavelength that people in my culture experience as pink). My friend Kevin ("Uncle Kevin" from the previous chapter) once went to extraordinary lengths to purchase a pink Queen Anne's lace, which he planted proudly at the center of his garden. One day, he and I were having tea in his yard when another friend stopped by. Kevin and I popped inside to get some tea for her. We returned just in time to watch the friend shake her head, stoop, and with deftness born from decades of experience, rip the Queen Anne's lace out of the ground.

Nothing in the natural world indicates whether a plant is definitively a flower or a weed. Queen Anne's lace is a flower to Kevin but a weed to his friend. The distinction depends on the perceiver. A rose is usually considered a flower, but it becomes a weed if you discover it in a field of vegetables. A dandelion is often considered a weed, but it transforms into a flower when placed in a bouquet of wildflowers or if it's a gift from your two-year-old child. Plants exist objectively in nature, but flowers and weeds require a perceiver in order to exist. They are *perceiver-dependent* categories. Albert Einstein illustrated this point nicely when he wrote, "Physical concepts are free creations of the human mind, and are not, however it may seem, uniquely determined by the external world."[5]

Common sense leads us to believe that emotions are real in nature and exist independent of any observer, in the same manner as Higgs bosons and plants. Emotions seem to be present in wiggling eyebrows and wrinkled noses, in sagging shoulders and sweaty palms, in racing hearts and squirts of cortisol, and in silence, screams, and sighs.

Science, however, tells us that emotions require a perceiver, just as colors and sounds do. When you experience or perceive emotion, sensory input is transformed into patterns of firing neurons. At the time, if you focus your attention on your body, you experience emotions as if they are happening in your body, just like you experience red color in the apple and sound in the world. If you're instead focusing attention on the world, you experience faces and voices and bodies as if they express emotion for you to decode. But as we learned in chapter 5, your brain categorizes using emotion concepts to make these sensations meaningful. The result is that you construct instances of happiness, fear, anger, or other emotion categories.

Emotions are real, but real in the same manner of the sound of a tree falling, the experience of red, and the distinctions between flowers and weeds. They are all constructed in the brain of a perceiver.

You move your facial muscles all the time. Your eyebrows scrunch. Your lips curl. Your nose wrinkles. These actions are perceiver-independent and they help you sample the sensory world. Widening your eyes enhances your peripheral vision, so you can more easily detect objects surrounding you. Narrowing your eyes improves your visual acuity for objects right in front of you. Wrinkling your nose helps to block noxious chemicals. But these movements are not intrinsically emotional.[6]

Inside your body, your heartbeat, blood pressure, breathing, temperature, and cortisol level fluctuate throughout the day. These changes have physical functions to regulate your body in the world; they are perceiver-independent. They also are not intrinsically emotional.

Your muscle movements and bodily changes become functional as instances of emotion *only when you categorize them that way,* giving them new functions as experiences and perceptions. Without emotion concepts, these new functions don't exist. There are only moving faces, beating hearts, circulating hormones, and so on, just as without color and sound concepts, "red" and the sound of a falling tree would not exist. There'd be only light and vibrations.

Historically, scientists have debated whether emotion categories like fear and anger are real in nature or illusory. We learned in chapter 1 that those who adhere to the classical view believe that emotion categories are carved in nature, with every instance of (say) "Fear" sharing a common biological fingerprint. Emotion concepts in your head, they say, exist separately from those natural categories. Critics usually counter that anger, fear, and so on, are mere words from folk psychology and should be discarded for scientific endeavors. Early in my journey, I took this latter view, but I now think there's another possibility that's more realistic.[7]

The distinction between "real in nature" versus "illusory" is a false dichotomy. Fear and anger are real to a group of people who *agree* that certain changes in the body, on the face, and so on, are meaningful as emotions. In other words, emotion concepts have *social reality.* They exist in your human mind that is conjured in your human brain, which is part of nature. The biological processes of categorization, which are rooted in physical reality and are observable in the brain and body, create socially real categories. Folk concepts like "fear" and "anger" are not mere words to be discarded from scientific thought but play a critical role in the story of how the brain creates emotion.

• • •

Social reality is not just about trivial-sounding examples like flowers, weeds, and red apples. Human civilization is literally built with social reality. Most things in your life are socially constructed: your job, your street address, your government and laws, your social status. Wars are waged and neighbor slaughters neighbor, all for the sake of social reality. When Benazir Bhutto, the late prime minister of Pakistan, said that "You can kill a man, but not an idea," she was proclaiming the power of social reality to reshape the world.

Money is a classic example of social reality. Given a rectangle of paper with a dead leader's face printed on it, or a metal disk or a shell or some barley, a group of people categorized that object as money, and it *became* money. We exchange billions of dollars every day based on social reality called a stock market. We study economies scientifically with complicated mathematical equations. The disastrous effects of the financial crisis of 2008 were a product of social reality. In a matter of moments, a collection of mortgages — themselves constructs of social reality — went from valuable to worthless, hurling people into economic ruin. Nothing objective in biology or physics caused this to happen. It was just one collective and devastating change of imagination. And consider this: what is the difference between two hundred one-dollar bills and a silk-screened painting of two hundred one-dollar bills? The answer is, "about $43.8 million." That's the price paid in 2013 for Andy Warhol's painting "200 One Dollar Bills." The painting is exactly what its title sounds like, scarcely different from the currency it depicts. The colossal difference in value is entirely social reality. The price also fluctuates — the work sold for a mere $300,000 in the 1990s, a relative bargain — which also reflects social reality. If $43.8 million seems like a high price to you, then you're a participant in this social reality.[8]

Make something up, give it a name, and you've created a concept. Teach your concept to others, and as long as they agree, you've created something real. How do we work this magic of creation? We categorize. We take things that exist in nature and impose new functions on them that go beyond their physical properties. Then we transmit these concepts to each other, wiring each other's brains for the social world. This is the core of social reality.[9]

Emotions are social reality. We construct instances of emotion in exactly the same manner as colors, falling trees, and money: using a conceptual system that is realized within the brain's wiring. We transform sensory inputs from the body and the world, which are perceiver-independent, into an instance of (say) happiness in the context of a concept, "Happiness," found in many human minds. The concept imposes new functions on these sensations, creating reality where there was none before: an experience or perception of emotion.

Instead of asking, "Are emotions real?" the better question is, "How do emotions become real?" Ideally, the answer lies in building a bridge from the perceiver-independent biology of the brain and body, like interoception, to the everyday folk concepts that we live our lives around, like "Fear" and "Happiness."

Emotions become real to us through two human capabilities that are prerequisites for social reality. First, you need a group of people to agree that a concept exists, such as "Flower" or "Cash" or "Happiness." This shared knowledge is called *collective intentionality*. Most people barely think about collective intentionality, but it nevertheless is a foundation of every society. Even your own name is made real through collective intentionality.[10]

Emotion categories, in my view, are made real through collective intentionality. To communicate to someone else that you feel angry, both of you need a shared understanding of "Anger." If people agree that a particular constellation of facial actions and cardiovascular changes is anger in a given context, then it is so. You needn't be explicitly aware of this agreement. You don't even have to agree whether a particular instance is anger or not. You just have to agree in principle that anger exists with certain functions. At that point, people can transmit information about that concept among themselves so efficiently that anger seems inborn. If you and I agree that a furrowed brow indicates anger in a given context, and I furrow my brow, I am efficiently sharing information with you. My movement itself does not carry anger to you, any more than vibrations in the air carry sound. By virtue of the fact that we share a concept, my movement initiates a prediction in your brain . . . a uniquely human brand of magic. It is categorization as a cooperative act.[11]

Collective intentionality is necessary for social reality but not sufficient. Certain non-human animals are capable of a rudimentary form of collective intentionality without social reality. Ants work together toward a common activity, as do bees. Flocks of birds and schools of fish move in synchrony. Certain troupes of chimpanzees use tools, such as sticks for retrieving and eating termites, and rocks for cracking nuts, whose uses are passed down to offspring. Chimps even appear to learn a concept of "Tool" by realizing that different-looking objects can be used for a common purpose — for instance, obtaining food with some sort of object that is held in the hands, like a wooden stick or a screw driver.

Humans are unique, however, because our collective intentionality involves mental concepts. We can look at a hammer, a chainsaw, and an ice pick and categorize them all as "Tools," then change our minds and categorize them all as "Murder Weapons." We can impose functions that would not otherwise exist, thereby inventing reality. We can work this magic because we have the second prerequisite for social reality: language.

No other animals have collective intentionality combined with words. A few other animal species do have symbolic communication of a sort. Elephants appear to communicate through low-frequency vocal rumblings that can travel over a mile. Certain great apes appear to use sign language in a limited way, on the order of a two-year-old human, usually linked in some way to securing a reward. But only human animals have both language and collective intentionality. The two abilities build on one another in complex ways, allowing a human infant to bootstrap a conceptual system into her brain, changing its wiring in the process. The combination also allows people to categorize cooperatively, which is the basis of communication and social influence.[12]

Words invite us to form concepts, as we learned in chapter 5, by grouping together physically dissimilar things for some purpose. A trumpet, a timpani, a violin, and a military cannon look nothing alike, but the phrase "musical instrument" allows us to treat them as similar to meet a goal, such as performing Pyotr Tchaikovsky's *1812 Overture*. The word "fear" groups together diverse instances that have greatly varied movements, interoceptive sensations, and events in the world. Even prelinguistic infants use words to form concepts about balls and noisemakers, as long as the words are spoken intentionally by live humans.

Words are also the most efficient shorthand we know for communicating concepts that are shared by a group. When I order a pizza, I never have a conversation like this one:

> ME: Hello, I'd like to place an order, please.
> VOICE ON TELEPHONE: Sure, what would you like?
> ME: I'd like a lump of dough that's been rolled flat and shaped into a circle or sometimes a rectangle with tomato sauce and cheese on top of it that's been baked in a very hot oven long enough for the cheese to melt and the crust to brown. For eating.
> VOICE: That'll be $9.99. It'll be ready when the big hand is on the twelve and the little hand is on the seven.

The word "pizza" would shorten this telephone call considerably because we have shared experience, and therefore shared knowledge, concerning pizza in our culture. I would describe the individual properties of a pizza only to someone who had never encountered pizza before, someone who would likewise labor to understand a pizza, feature by feature.

Words also have power. They let us place ideas directly into another person's head. If I seat you in a chair, perfectly motionless, and say the word "pizza" to you, neurons in your brain will change their firing pattern automatically, making predictions. You might even salivate as you simulate the taste of mushrooms and pepperoni. Words give us our own special form of telepathy.

Words also encourage *mental inference:* figuring out the intentions, goals, and beliefs of others. Human infants learn critical information resides in the minds of other people, as we discussed in chapter 5, and words are a vehicle for inferring this information.

Words are not the only way to communicate a concept, of course. If I am married and want to indicate this to the world, I don't have to walk around repeating, "I'm married, I'm married, *I'm married."* I can just wear a ring, preferably with some very large diamonds in it. Or in northern India, I can wear a *bindi* (red dot) on my forehead. Likewise, if I'm happy, I don't need words to communicate this. I can simply smile, and others around me understand through collective intentionality, as a torrent of predictions are unleashed in their brains. When my daughter was a preschooler, I only had to widen my eyes to warn her away from mischief. No words were required.

Nevertheless, you need a word to teach a concept efficiently. Collective intentionality requires that everyone in a group shares a similar concept, be it "Flower" or "Weed" or "Fear." The instances of each of these concepts vary widely with few statistical regularities in their physical features, but all group members must learn the concepts somehow. For all practical purposes, this learning requires a word.

Which comes first, a concept or a word? This is an ongoing scientific and philosophical debate that we won't solve here; however, it's clear that people form certain concepts before knowing the word. Within a few days after birth, infants rapidly learn the perceptual concept of a face without knowing the word "face," as we noted in chapter 5, because faces have statistical regularity: two eyes, a nose, and a mouth. Similarly, we distinguish the concepts "Plant" and "Human Being" without requiring words for them: plants photosynthesize and people do not. The difference is perceiver-independent, regardless of how the two concepts are named.[13]

On the other hand, certain concepts require words. Consider the category of "Pretend Telephones." We've all seen children hold an object to their ear and converse into it, emulating their parents' phone behavior. The choice of object varies broadly: it might be a banana, a hand, a cup, even a security blanket. These instances have no significant statistical regularities, and yet a father can hand a banana to his young son and say, "Ring, ring, ring, it's for you," and this shorthand is sufficient for a shared understanding of what to do next. On the other hand, if you did not know the concept "Pretend Telephone," and you saw a two-year-old pressing a toy car against her ear and speaking, you would see only a talking child holding a toy to the side of her head.

Similarly, emotion concepts are most easily learned with emotion words. You've learned that emotion categories have no consistent fingerprint in the face, body, or brain. That means instances of a single emotion concept, like "Surprise," need no physical similarity for your brain to group them together. And any two emotion concepts, like "Surprise" and "Fear," need no consistent fingerprints to reliably distinguish them. So we, as a culture, *introduce* mental similarity using words. From childhood we hear people say "fear" and "surprise" in particular contexts. The sound of each word (or, later in life, the written form of each word) creates enough statistical regularity within each category, and statistical differences between them, to get us started. The words quickly prompt us to infer the goals to anchor each concept. Without the words "fear" and "surprise," these two concepts would likely not spread from person to person. Nobody knows whether the concepts form before the words or vice versa, but it's clear that words are vitally linked to the way we develop and transmit purely mental concepts.

• • •

Classical view theorists debate endlessly about how many emotions there are. Is love an emotion? How about awe? Curiosity? Hunger? Do synonyms like happy, cheerful, and delighted refer to different emotions? What about lust, desire, and passion: are they distinct? Are they emotions at all? From the standpoint of social reality, these debates are nonissues. Love (or curiosity, hunger, etc.) is an emotion as long as people agree that its instances serve the functions of an emotion.[14]

We've characterized some of these functions in previous chapters. The first stems from the fact that emotion concepts, like all concepts, *make meaning*. Suppose you find yourself breathing rapidly and sweating. Are you are excited? Afraid? Physically exhausted? Different categorizations represent different meanings: that is, different likely explanations for your physical state in this situation, based on your past experience. Once you've made an instance of emotion, by categorizing with an emotion concept, your sensations and actions are explained.

The second function of emotions stems from the fact that concepts *prescribe action*: If you're breathing rapidly and sweating, what should you do? Should you grin broadly in excitement, run away in fear, or lie down for a nap? An instance of emotion, constructed from a prediction, tailors your action to meet a particular goal in a particular situation, using past experience as a guide.

The third function is related to a concept's ability to *regulate your body budget*. Depending how you categorize your sweating, panting state, your body budget may be affected differently. A categorization of excitement might lead to a moderate release of cortisol (say, to raise your arms); a categorization of fear might lead to a greater release of cortisol (as you prepare to run away); whereas napping requires no additional cortisol. Categorization literally gets under your skin. Every instance of emotion involves some body budgeting for the immediate future.

These three functions have something in common: they're about you alone. You don't need any other people involved in the experience in order to make meaning, to act, or to regulate your body budget. But emotion concepts have two other functions that draw other individuals into your circle of social reality. One function is *emotion communication,* in which two people categorize with concepts in synchrony. If you see a man taking quick breaths and sweating, it communicates one thing if he's wearing a jogging suit and something else entirely if he's wearing a groom's tuxedo. Categorization here communicates meaning and explains why the man acts as he does. The other function is *social influence*. Concepts like "Excitement," "Fear," and "Exhaustion" are tools for you to regulate other people's body budgets, not just your own. If you can get someone else to perceive your panting, sweaty state as fear, you influence their actions in a way that mere quick breaths and damp brows cannot achieve on their own. You can be an architect of other people's experiences.[15]

These latter two functions require that other people — the ones you are communicating with or influencing — agree that certain body states or physical actions serve particular functions in certain contexts. Without this collective intentionality, one person's actions, no matter how meaningful they are to him, will be perceived by others as meaningless noise.

Suppose you and a friend are walking together when you see a man stamping his foot forcefully on the pavement. You categorize the man as angry. Your friend categorizes the man as dejected. The man himself believes he is just clomping some caked mud off his shoe. Does that mean you and your friend are wrong? Could the man be unaware of his own emotion in the moment? Who is correct in this case?

If this were a question of physical reality, you could settle the matter definitively. If I say that my shirt is made of silk and you say no, it's made of polyester, we can perform a chemical test to discover the answer. With social reality, however, there is no such thing as accuracy. If I say my shirt is a thing of beauty and you say it's hideous, neither of us is objectively correct. The same is true for perceiving emotion in the stamping man. Emotions have no fingerprints, so there can be no accuracy. The best you can do is find consensus. We can ask other people if they agree with you or with me about the shirt or the stamping man, or we can compare our categorizations to the norms of our culture.[16]

You, your friend, and the stamping man each construct a perception by prediction. The stamping man himself might be feeling unpleasant arousal, and he may categorize his interoceptive sensations, together with those he predicted from the outside world, as an instance of "Removing Mud from My Shoe." You may construct a perception of anger and your friend a perception of dejection. Each construction is real, so questions of accuracy are unanswerable in a strictly objective sense. This is not a limitation of science: it is just the wrong question to be asking in the first place. There are no observer-independent measurements that can reliably and specifically adjudicate the matter. When you can't find an objective criterion to compute accuracy and are left with consensus, this is a clue that you are dealing with social, not physical, reality.[17]

This point is easily and frequently misunderstood, so let me be clear. I am not saying emotions are illusions. They are real, but socially real in the manner of flowers and weeds. I'm not saying that everything is relative. If that were true, civilization would fall apart. I am also not saying that emotions are "just in your head." That phrase trivializes the power of social reality. Money, reputation, laws, government, friendship, and all of our most fervent beliefs are also "just" in human minds, but people *live and die* for them. They are real because people agree that they're real. But they, and emotions, exist only in the presence of human perceivers.

· · ·

Imagine the feeling of reaching into a bag of potato chips and discovering that the previous chip you ate was the last one. You feel disappointed that the bag is empty, relieved that you won't be ingesting any more calories, slightly guilty that you ate the entire bag, and yet hungry for another chip. I have just invented an emotion concept, and there is surely no word for it in the English language. And yet, as you read my prolonged description of this complex feeling, you most likely simulated the whole thing, right down to the crinkle of the bag and the cheerless little crumbs at the bottom. You experienced this emotion without a word for it.

Your brain accomplished this feat by combining instances of concepts you already know, such as "Bag," "Chips," "Disappointment," "Relief," "Guilt," and "Hunger." This powerful ability of your brain's conceptual system, which we called conceptual combination in chapter 5, creates your very first instance of this new chip-related category of emotion, ready for simulation. Now if I name my new creation "Chiplessness" and teach it to our fellow citizens, it becomes every bit as real an emotion concept as "Happiness" and "Sadness." People can predict with it, categorize with it, regulate their body budgets with it, and construct diverse instances of "Chiplessness" in different situations.

This brings us to one of the most challenging ideas in this book: you *need* an emotion concept in order to experience or perceive the associated emotion. It's a requirement. Without a concept for "Fear," you cannot experience fear. Without a concept for "Sadness," you cannot perceive sadness in another person. You could learn the necessary concept, or you could construct it in the moment through conceptual combination, but your brain must be able to make that concept and predict with it. Otherwise, you will be experientially blind to that emotion.

I realize this idea might sound counterintuitive, so let's start with a few examples.

You are probably unfamiliar with an emotion called *liget*. It's a feeling of exuberant aggression experienced by a headhunting tribe from the Philippines, the Ilongot. *Liget* involves intense focus, passion, and energy while pursuing a hazardous challenge with a group of people who are competing against another group. The danger and energy instill a sense of togetherness and belonging. *Liget* is not just a mental state but a complex situation with social rules about which activities bring it on, when it is appropriate to feel, and how other people should treat you during an episode. To a member of the Ilongot tribe, *liget* is every bit as real an emotion as happiness and sadness are to you.

Westerners surely do experience pleasant aggression. Athletes feel it in the heat of competition. Videogame players cultivate it during first-person shooter games. But these people are not experiencing *liget* with all its meaning, prescribed actions, body-budget changes, communication, and social influence unless they can construct "*Liget*" using conceptual combination. *Liget* is the whole conceptual package, and if your brain cannot make this concept, then you cannot experience *liget,* although you can experience parts of it: the pleasant, high arousal affect; the aggression; the thrill of pursuing a risky challenge; or the feeling of brother- or sisterhood that comes from being part of a group.

Next, consider an emotion concept that's more recently adopted by U.S. culture. In a recent meeting with my lab members, I learned that an acquaintance (call him Robert) failed in his bid to win a Nobel prize. Robert had treated me poorly in the past (which is polite scientist-speak for "he acted like an ass"), so when I heard the news, I have to admit that I had a complex emotional experience: I felt some empathy for Robert, plus a small measure of gratification about his misfortune, plus a large wave of guilt at my pettiness, as well as embarrassment that someone might discover my uncharitable feeling.

Imagine if I'd described my conceptual combination to my lab members: "Robert probably feels horrible about his failure, and I am pleased about that." My words would have been highly inappropriate. No one else in my lab knew my history with Robert, nor my simultaneous guilt and embarrassment, so they wouldn't have understood my perspective and might have viewed me as an ass myself. So instead, I said, "I am feeling a bit of *schadenfreude,*" and everybody in the room smiled and nodded with recognition. One word efficiently communicated my emotional experience and made it socially acceptable, because everyone else in the lab had the concept and could construct a perception of *schadenfreude.* We couldn't have done that with mere pleasantly valenced affect at someone else's misfortune.

The situation is exactly the same for a more familiar Western emotion like sadness. Any healthy human can experience low-arousal, unpleasant affect. But you cannot experience sadness with all of its cultural meaning, appropriate actions, and other functions of emotion unless you have the concept "Sadness."

Some scientists argue that without an emotion concept, the emotion still exists but the affected person doesn't realize it, implying a state of emotion outside of consciousness. I suppose this is a possibility, but I doubt it. If you had no concept of "Flower" and someone showed you a rose, you'd experience only a plant, not a flower. No scientist would claim that you're seeing a flower but just "don't realize it." Similarly, the blobby image in chapter 2 does not have a hidden bee in it. You perceived the bee only because of conceptual knowledge. The same reasoning applies to emotions; without the concept "*Liget*" or "Sadness" or "Chiplessness" to categorize with, there is no emotion, only a pattern of sensory signals.

Think of how useful the concept of "*Liget*" could be in Western culture. When military cadets train in the art of war, a small percentage of them reportedly develop a feeling of pleasure in killing. They do not seek to kill to feel pleasure; they are not psychopaths. But when they do kill, they experience pleasure. Their stories of combat often depict intense feelings of pleasure from the thrill of the hunt, or from a job well-done with comrades-in-arms. In Western culture, however, killing with pleasure is considered terrible and shameful; it is difficult to empathize with or muster compassion for those who have experienced this feeling. So consider this: what if we taught the concept and the word *liget* to cadets, including a set of social rules for when *liget* is appropriate to feel? We could embed this emotion concept in our broader cultural context of values and norms, just like we did with *schadenfreude*. The concept might even allow servicepeople to flexibly cultivate the experience of *liget* when needed for their military duties. New emotion concepts like *liget* could broaden their emotional granularity, improving their unit's cohesion and their job performance, all the while protecting mental health for these members of our armed forces, both in battle and when they come home from deployment.[18]

I realize I'm saying something provocative: that each of us needs an emotion concept before we can experience or perceive that emotion. This definitely doesn't match common sense or everyday experience; emotions feel so built-in. But if emotions are constructed by prediction, and you can predict only with the concepts you possess, well . . . there you have it.

• • •

The emotions that you experience so effortlessly, and which feel built-in, most likely were also known in your parents' generation, and their parents' as well. The classical view explains this progression by proposing that emotions — separate from emotion concepts — are built into the nervous system through evolution. I have an evolutionary story to tell as well, but it's about social reality, and it doesn't require emotion fingerprints in the nervous system.

Emotion concepts like "Fear," "Anger," and "Happiness" are passed down from one generation to the next. This occurs not merely because we propagate our genes but because those genes allow each generation to wire the brains of the next one. Infants grow minds full of concepts as they learn the mores and values of their culture. This process goes by many names: Brain development. Language development. Socialization.

One of humanity's major adaptive advantages — why we've flourished as a species — is that we live in social groups. This arrangement has allowed us to expand across the globe, creating livable habitats by feeding, clothing, and learning from each other in otherwise inhospitable physical conditions. We can therefore amass information across generations — stories, recipes, traditions, anything that we can describe — that helps each generation to shape the brain wiring of the next. This trove of intergenerational knowledge lets us actively shape the physical environment, rather than just adapt to it, and to create civilizations.[19]

Living in groups has some drawbacks, of course, particularly a major dilemma that every human must face: getting along versus getting ahead. Everyday concepts like "Anger" and "Gratitude" are critical tools for dealing with these two competing concerns. They are instruments of culture. They prescribe situation-specific actions, allow you to communicate, and influence the behaviors of others, all in the service of managing your body budget.

Just because fear appears generation after generation in your culture does not prove that fear is coded into the human genome, nor that it was sculpted by natural selection in our hominin ancestors millions of years ago on the African savanna. These single-cause explanations discount the enormous power of collective intentionality (not to mention copious evidence from modern neuroscience). Evolution surely allowed humans to create culture, and part of that culture is a system of goal-based concepts to manage ourselves and each other. Our biology allows us to create goal-based concepts, but exactly which concepts may be a matter of cultural evolution.[20]

The human brain is a cultural artifact. We don't load culture into a virgin brain like software loading into a computer; rather, culture helps to *wire* the brain. Brains then become carriers of culture, helping to create and perpetuate it.

All humans who live in groups must solve common problems, so it's not surprising to find some concepts that are similar across cultures. Most human societies, for example, have myths about supernatural beings: nymphs from ancient Greece, fairies from Celtic legends, leprechauns from Ireland, little people from Native American tales, Menehune from Native Hawaiian folklore, trolls from Scandinavia, the Aziza from Africa, the Agloolik from Inuit culture, the Mimis from Aboriginal Australia, the Shin from China, the

Kami from Japan, and countless others. Tales of these magical creatures are an important part of human history and literature. They do not, however, mean that magical creatures actually exist or have ever existed in nature (no matter how much we wish we could attend Hogwarts). The category "Magical Creature" is constructed by human minds, and since it exists in so many different cultures, it probably serves some important function. In the same manner, "Fear" exists in many cultures (but not all, such as the !Kung people of the Kalahari Desert) by virtue of having important functions. As far as I know, no emotion concept is universal, but even if one were, universality itself does not automatically imply a perceiver-independent reality.[21]

Social reality is a driving force behind human culture. It's perfectly plausible for emotion concepts, as elements of social reality, to be learned from others during infancy, or even much later when someone moves from one culture to another (more on this shortly). Social reality is therefore one conduit for transmitting behaviors, preferences, and meanings from ancestors to descendants via natural selection. Concepts are not merely a social veneer on top of biology. They are a biological reality that is wired into your brain by culture. People who live in cultures with certain concepts, or more diverse concepts, may be more fit to reproduce.[22]

In chapter 5, we looked at the illusory stripes that we carve into a rainbow, as we categorize the wavelengths of light with our concepts for colors. If you visit the Russian Google (images.google.ru) and search for the Russian word for rainbow, радуга, you'll see that Russian drawings contain seven colors, not six: the Western blue stripe has been subdivided into light blue and dark blue, as in figure 7-2:[23]

Figure 7-2: Rainbow drawings are culture-specific

These pictures demonstrate that concepts of color are influenced by culture. In Russian culture, the colors синий (blue) and Голубой (sky blue to a Westerner) are different categories, as distinct as blue and green are to an American. This distinction is not due to inborn, structural differences in the visual system of Russians versus Americans, but to culture-specific, learned concepts of color. People raised in Russia are simply taught that light and dark blue are distinct colors with different names. These color concepts become wired into their brains, and so they perceive seven stripes.[24]

Words represent concepts, and concepts are tools of culture. We pass them down from parent to child, from one generation to the next, just like your great-great-grandmother's candlesticks from the Old Country. "Rainbows have six stripes." "Money is traded for goods." "Cupcakes are a dessert and muffins are a breakfast food."

Emotion concepts are also cultural tools. They come with a rich set of rules, all in the service of regulating your body budget or influencing someone else's. These rules can be specific to a culture, stipulating when it's acceptable to construct a given emotion in a given situation. In the United States, it's appropriate to feel fear when you're on a rollercoaster, or about to hear the results of a cancer screening, or if someone points a gun at you. In the United States, it's not appropriate to feel fear each time you walk out of your house in a safe neighborhood: that feeling would be considered pathological, an anxiety disorder called agoraphobia.

My friend Carmen, who was born in Bolivia, was surprised when I told her that emotion concepts vary widely from culture to culture. "I thought everybody in the world has the same emotions," she explained to me in Spanish. "Well, Bolivians do have stronger emotions than Americans. *Más fuerte.*" Most people have lived with one set of emotion concepts their whole lives, so like Carmen, they find this cultural relativity surprising. Yet, scientists have documented numerous emotion concepts around the world that don't exist in English. Norwegians have a concept for an intense joy of falling in love, calling it "*Forelsket.*" The Danes have the concept "*Hygge*" for a certain feeling of close friendship. The Russian "*Tocka*" is a spiritual anguish, and the Portuguese "*Saudade*" is a strong, spiritual longing. After a little research, I located a Spanish emotion concept that has no direct equivalent in English, called "*Pena Ajena.*" Carmen described it to me as "sadness over another person's loss," but I've also seen it characterized as discomfort or embarrassment on someone else's behalf. Here are a few more I find compelling:[25]

- *Gigil* (Filipino): The urge to hug or squeeze something that is unbearably adorable.[26]
- *Voorpret* (Dutch): Pleasure felt about an event before the event takes place.[27]
- *Age-otori* (Japanese): The feeling of looking worse after a haircut.[28]

Some emotion concepts from other cultures are incredibly complicated, perhaps impossible to translate into English, yet natives experience them as a matter of course. The concept of "*Fago*" in Ifaluk (Micronesian) culture can mean love, empathy, pity, sadness, or compassion, depending on context. In Czech culture, the concept of "*Litost*" is said to be untranslatable but roughly "torment over one's own misery combined with the desire for revenge." The Japanese emotion concept "*Arigata-meiwaku*" is felt when someone has done you a favor that you didn't want from them, and which may have caused difficulty for you, but you're required to be grateful anyway.[29]

When I speak to audiences in the United States about emotion concepts as variable and culture-specific, and then suggest that our own English-language concepts are similarly local to our culture, some people are very surprised, as my friend Carmen was. "But happiness and sadness are *real* emotions," they insist, as if the emotions of other cultures are not as real as our own. To this I usually say: you are exactly right. *Fago, litost,* and the rest are not emotions . . . to *you.* That's because you don't know these emotion concepts; the associated situations and goals are not important in middle-class American culture. Your brain cannot issue predictions based on "*Fago,*" so the concept doesn't feel automatic the way that happiness and sadness do to you. To understand *fago,* you have to combine other concepts that you do know, performing conceptual combination and expending mental effort. But the Ifaluk do have this emotion concept. Their brains automatically predict with it. When they experience *fago,* it feels just as automatic and real as happiness or sadness does to you, as if *fago* just happens to them.

Yes, *fago, litost,* and the rest are just words made up by people, but so are "happy," "sad," "fearful," "angry," "disgusted," and "surprised." Invented words are the very definition of social reality. Would you say that your local currency is real money and the currencies of other cultures are just made up? To someone who has never traveled, it might seem that way, lacking the concept for another currency. But experienced travelers have the concept "Currency from Another Culture." I'm asking you to learn the concept of

"Emotion from Another Culture," so you understand that its instances are just as real to others as your own emotions are to you.

If you've found these ideas challenging, try this one: some cherished Western emotion concepts are completely absent in other cultures. Utka Eskimos have no concept of "Anger." The Tahitians have no concept of "Sadness." This last item is very difficult for Westerners to accept . . . life without sadness? Really? When Tahitians are in a situation that a Westerner would describe as sad, they feel ill, troubled, fatigued, or unenthusiastic, all of which are covered by their broader term *pe'ape'a*. Someone who believes in the classical view of emotion would explain away this variability, saying that a frowning Tahitian really is in a biological state of sadness, whether he knows it or not. A constructionist does not have the luxury of such certainty, because people frown for many reasons such as while thinking, exerting effort, in humor, when censoring a thought, or when feeling *pe'ape'a*.[30]

Beyond individual emotion concepts, different cultures don't even agree on what "emotion" is. Westerners think of emotion as an experience inside an individual, in the body. Many other cultures, however, characterize emotions as interpersonal events that require two or more people. This includes the Ifaluk of Micronesia, the Balinese, the Fula, the Ilongot of the Philippines, the Kaluli of Papua New Guinea, the Minangkabau of Indonesia, the Pintupi Aborigines of Australia, and the Samoans. More intriguingly, some cultures don't even have a unified concept of "Emotion" for the experiences that Westerners lump together as emotional. The Tahitians, the Gidjingali Aborigines of Australia, the Fante and Dagbani of Ghana, the Chewong of Malaysia, and our friends the Himba from chapter 3 are a few well-studied examples.[31]

Most scientific research on emotion is conducted in English, using American concepts and American emotion words (and their translations). According to noted linguist Anna Wierzbicka, English has been a conceptual prison for the science of emotion. "English terms of emotion constitute a folk taxonomy, not an objective, culture-free analytic framework, so obviously we cannot assume that English words such as disgust, fear, or shame are clues to universal human concepts, or to basic psychological realities." To make matters even more imperialistic, these emotion words are from twentieth-century English, and there's evidence that some are fairly modern. The concept of "Emotion" itself is an invention of the seventeenth century. Before that, scholars wrote about passions, sentiments, and other concepts that had somewhat different meanings.[32]

Different languages describe diverse human experience in different ways — emotions and other mental events, colors, body parts, direction, time, spatial relations, and causality. The diversity from language to language is astonishing. The experiences of my friend Batja Mesquita, the cultural psychologist whom you met in chapter 5, provide an example. She was born and raised in the Netherlands and immigrated to America for her postdoctoral training. Over the next fifteen years, she married, raised a family, and was a professor at Wake Forest University in North Carolina. When living in the Netherlands, Batja felt that her emotions were, for lack of a better word, natural. After moving to the United States, however, she soon noticed her emotions were not a good fit for American culture. Americans struck her as unnaturally happy. We constantly spoke in an upbeat tone of voice. We smiled a tremendous amount. When Batja asked how people were doing, we would always answer positively ("I'm doing great!"). Batja's own emotional responses seemed inadequate in the U.S. cultural context. When asked how she was feeling, she did not respond with sufficient enthusiasm or say she was "fabulous" or "wonderful." I once heard her give a talk on her experiences, and I nodded through the entire thing, clapped vigorously at the end, and then walked up to her, gave her a hug, and said "excellent job!" It took me a moment to realize I had just confirmed every one of her observations.[33]

Batja's experience is not unique. Our colleague Yulia Chentsova Dutton from Russia says that her cheeks ached for an entire year after moving to the United States because she had never smiled so much. My neighbor Paul Harris, a transplanted emotion researcher from England, has observed how American academics are always excited by scientific puzzles — a high arousal, pleasant feeling — but never merely curious, perplexed, or confused, which are low arousal and fairly neutral experiences that are more familiar to him. In general, Americans prefer high arousal, pleasant states. We smile a lot. We praise, compliment, and encourage each other. We give each other awards for all levels of accomplishment, even "Certificates of Participation." It seems like every other week there is an awards show on television. I have lost count of how many books on happiness have been published in the United States in the last ten years. We are a culture of positivity. We like to be happy and to celebrate how great we are.[34]

The more time that Batja spent in America, the more her emotions became attuned to the American context. Her pleasant emotion concepts expanded and became more variable. She became more granular, experiencing the American style of happiness as distinct from satisfaction and contentment. Her brain bootstrapped new concepts for American norms and customs. This process is called *emotion acculturation*. From a new culture, you acquire new concepts, which translate into new predictions. Using those predictions, you become able to experience and perceive the emotions of your newly adopted home.

The scientist who discovered emotion acculturation is, in fact, Batja herself. She found that people's emotion concepts not only vary from culture to culture but also transform. For example, situations that bring about anger in Belgium, like having your goals blocked by a coworker, in Turkey will also include feelings of (what Americans experience as) guilt, shame, and respect. But for Turkish immigrants in Belgium, their emotional experiences come to look more "Belgian" the longer they live there.[35]

A brain that is bathed in the situations of a new culture is probably somewhat like an infant's brain: driven more by prediction error than prediction. Lacking the emotion concepts of the new culture, the immigrant brain soaks up sensory input and builds new concepts. The new emotional patterns don't replace the old ones, though they may cause interference, as was the case for my research associate Alexandra from Greece whom you met in chapter 5. You can't predict efficiently when you don't know the local concepts. You must get by with conceptual combination, which can be effortful and yields only an approximate meaning. Or you will be awash in prediction error much of the time. The process of acculturation therefore taxes your body budget. In fact, people who are less emotionally acculturated report more physical illness. Once again, categorization gets under your skin.[36]

• • •

In this book, I am trying to acculturate you into a new way of thinking about emotion. Whether you realize it or not, you have a set of concepts about emotions: what they are, where they come from, and what they mean. Perhaps you began this book with classical view concepts such as "Emotional Reaction" and "Facial Expression" and "Emotion Circuit in the Brain." If so, I've been slowly replacing them with a new set, including "Interoception," "Prediction," "Body Budget," and "Social Reality." In a sense, I am attempting to draw you into a new culture called the theory of constructed emotion. A new culture's norms may seem odd, or even wrong, until you've lived there for a while and come to understand them . . . and I hope you already do, or you will. Ultimately, if I and other like-minded scientists are successful in substituting the new concepts for the old, well, that's a scientific revolution.

The theory of constructed emotion explains how you experience and perceive emotion in the absence of any consistent, biological fingerprints in the face, body, or brain. Your brain continually predicts and simulates all the sensory inputs from inside and outside your body, so it understands what they mean and what to do about them. These predictions travel through your cortex, cascading from the body-budgeting circuitry in your interoceptive network to your primary sensory cortices, to create distributed, brain-wide simulations, each of which is an instance of a concept. The simulation that's closest to your actual situation is the winner that becomes your experience, and if it's an instance of an emotion concept, then you experience emotion. This whole process occurs, with the help of your control network, in the service of regulating your body budget to keep you alive and healthy. In the process, you impact the body budgets of those around you, to help you survive to propagate your genes into the next generation. This is how brains and bodies create social reality. This is also how emotions become real.

Yes, that's a mouthful. And some details are still reasoned speculation, like the exact mechanisms of the concept cascade. But we can say confidently that the theory of constructed emotion is a viable way to think about how emotions are made. The theory accounts for all of the phenomena of the classical view, plus its anomalies such as the huge variability in emotional experiences, in emotion concepts, and in physical changes during emotion. It dissolves useless nature/nurture debates (e.g., what is hardwired versus what is learned) by using a single framework to understand both physical reality and social reality, moving us one step closer to a scientific bridge between the social and natural worlds. And this bridge, like all bridges, will lead us to a new place, as you'll see in the next chapter: a modern origin story of what it means to be human.

8

A New View of Human Nature

The theory of constructed emotion is not just a modern explanation of how emotions are made. It's also an ambassador for a radically different view of *what it means to be a human being*. This view is consistent with the latest research in neuroscience. It also gives you more control over your feelings and behavior than the classical view does, and it has deep implications for how to live your life. You are not a reactive animal, wired to respond to events in the world. When it comes to your experiences and perceptions, you are much more in the driver's seat than you might think. You predict, construct, and act. You are an architect of your experience.

Another compelling view of human nature comes from the classical view of emotion. It's been around for thousands of years and is still embedded in law, medicine, and other critical elements of society. The two views have in fact been at war with each other throughout recorded history. In previous battles, the classical view of human nature has consistently come out on top for reasons we'll see. But now, as we're in the midst of a revolution of mind and brain, modern neuroscience has given us the tools to settle the conflict, and based on overwhelming evidence, the classical view has lost.

In this chapter, I lay out the distinctive new view of human nature represented by the theory of constructed emotion and compare it to the traditional ideas espoused by the classical view. I also introduce you to a shadowy culprit that has kept the classical view so prominent for so long, entrenched in science and culture, despite a steady stream of contrary evidence.

. . .

Most of us think of the outside world as physically separate from ourselves. Events happen "out there" in the world, and you react to them "in here" in your brain.

In the theory of constructed emotion, however, the dividing line between brain and world is permeable, perhaps nonexistent. Your brain's core systems combine in various ways to construct your perceptions, memories, thoughts, feelings, and other mental states. You experienced this with the blobby bee picture, when you saw shapes that didn't physically exist, demonstrating that your brain models your world through simulation. Your brain issues a storm of predictions, simulates their consequences as if they were present, and checks and corrects those predictions against actual sensory input. Along the way, your interoceptive predictions produce your feelings of affect, influence every action that you perform, and determine which parts of the world you care about in the moment (your affective niche). Without interoception, you wouldn't notice or care about your physical surroundings or anything else, and you'd be unlikely to survive for long. Interoception enables your brain to construct the environment in which you live.

At the same time that your brain is modeling your world, the outside world helps to wire your brain. When you're an infant, awash in sensory input, the outside world seeds your earliest concepts, as your brain hardwires itself to the realities of the physical world around you. That's how babies' brains become wired to recognize human faces. As your brain develops and you begin learning words, your brain hardwires itself to the social world, and you begin creating purely mental concepts like "Things That Can Protect You from Stinging Insects" and "Sadness." These concepts from your culture appear to be in the outside world, but they are constructions of your conceptual system.

In this view, culture is not some gauzy, amorphous vapor that surrounds you. It helped to wire your brain, and you behave in certain ways that wire the brains of the next generation. For example, if a culture dictates that people with certain skin colors are less worthwhile, this social reality has a physical effect on the group: they have lower salaries and their children have poorer nutrition and living conditions. These factors change the structure of their children's brains for the worse, making school harder and increasing the odds that the children will earn lower salaries in the future.[1]

Your constructions aren't arbitrary — your brain (and the mind it creates) must keep in touch with the bits of reality that count in order to keep your body alive and healthy. Construction cannot make a solid wall un-solid (unless you have mutant superpowers), but you can redraw countries, redefine marriage, and decide who's worthwhile and who isn't. Your genes gave you a brain that can wire itself to its physical and social environment, and other members of your culture construct that environment with you. It takes more than one brain to create a mind.

The theory of constructed emotion also leads to a whole new way of thinking about personal responsibility. Suppose you're angry with your boss and lash out impulsively, slamming your fist on his desk and calling him an idiot. Where the classical view might attribute some blame to a hy-pothetical anger circuit, partially absolving you of responsibility, construc-tion extends the notion of responsibility beyond the moment of harm. Your brain is predictive, not reactive. Its core systems are constantly trying to guess what's coming next so you can survive. Therefore your actions, and the predictions that launched those actions, are shaped by all your past ex-periences (as concepts) that led up to that moment. You slam that desk be-cause your brain predicted an instance of anger, using your concept of "An-ger," and your past experience (whether direct, or from movies or books, etc.) includes an action of slamming the desk in a similar situation.

Your control network, you may recall, constantly shapes the course of your predictions and prediction error to help select among multiple actions, whether you experience yourself as in control or not. This network can only work with the concepts that you've got. So the question of responsibility be-comes, Are you responsible for your concepts? Not all of them, certainly. When you're a baby, you can't choose the concepts that other people put into your head. But as an adult, you absolutely do have choices about what you expose yourself to and therefore what you learn, which creates the con-cepts that ultimately drive your actions, whether they feel willful or not. So "responsibility" means making deliberate choices to change your concepts.[2]

As a real-world example, pick any extended conflict in the world: Israe-lis versus Palestinians, Hutus versus Tutsis, Bosnians versus Serbs, Sunni versus Shia. Climbing out on a limb here, I'd like to suggest that no living member of these groups is at fault for the anger that they feel toward each other, since the conflicts in question began many generations ago. But each individual today does bear some responsibility for continuing the conflict, because it's *possible* for each person to change their concepts and therefore their behavior. No particular conflict is predetermined by evolution. Conflicts persist due to social circumstances that wire the brains of the individuals who participate. *Someone* must take responsibility to change these circumstances and concepts. Who's going to do it, if not the people themselves?

To make this point, a scientific study provides some preliminary hope. Researchers trained a group of Israelis to think about various negative events, such as Palestinians' launching rockets and the kidnapping of an Israeli soldier, and recategorize them as less negative. The trainees were not only less angry afterward but they showed greater support for policies leading to more peaceful and conciliatory resolutions, such as providing aid to Palestinians, as well as less support for aggressive tactics toward Palestinians living in the Gaza Strip. Surrounding the recent Palestinian bid for membership in the United Nations, this training in recategorization led people to support giving up security control over neighborhoods in East Jerusalem in exchange for full peace, and to show less support for restrictive policies like prohibiting Palestinians from using the Israeli medical system. These latter changes persisted for five months after training.[3]

If you grow up in a society full of anger or hate, you can't be blamed for having the associated concepts, but as an adult, you can choose to educate yourself and learn additional concepts. It's certainly not an easy task, but it is doable. This is another basis for my frequent claim, "You are an architect of your experience." You are indeed partly responsible for your actions, even so-called emotional reactions that you experience as out of your control. It is your responsibility to learn concepts that, through prediction, steer you away from harmful actions. You also bear some responsibility for others, because your actions shape other people's concepts and behaviors, creating the environment that turns genes on and off to wire their brains, including the brains of the next generation. Social reality implies that we are all partly responsible for one another's behavior, not in a fluffy, let's-all-blame-society sort of way, but a very real brain-wiring way.

When I was a therapist, I worked with college-aged women who, as little girls, had suffered abuse at the hands of parents. I used to help my clients understand that they've been victimized twice: once in the moment and again because they've been left with emotional suffering that only they can resolve. Due to their trauma, their brains continue to model a hostile world, even after they've escaped to a better one. It was not their fault that their brains are wired for a specific, toxic environment. But each woman is the only one who can transform her conceptual system to make things better. That's the form of responsibility that I mean. Sometimes, responsibility means that you're the only one who can change things.

And now, we come to the question of human origin. We are accustomed to thinking about ourselves as the final destination of a long evolutionary journey. The theory of constructed emotion takes a more balanced perspective. Natural selection did not aim itself toward us. We are just another species with particular adaptations that help pass our genes to the next generation. Other animals have evolved plenty of powers that we don't have, like leaping great distances and scaling walls, which is why we're so fascinated by superheroes like Spider-Man. Humans are clearly the most talented at building rockets that reach other planets, and inventing and enforcing laws that exist in our minds and dictate how we treat each other. Something in the human brain gives us our unique abilities, but that "something" needn't be separate, dedicated brain circuitry for rocketry and law enforcement — or, for that matter, emotions — passed down from our non-human ancestors.

One of your most notable adaptations is that you needn't carry all the genetic material to create all the wiring in your brain. That would be tremendously expensive, biologically speaking. Instead, you have genes that let your brain develop in the context of the other brains around you, through culture. Just as an individual brain takes advantage of redundancy, compressing information into similarities and differences, multiple brains take advantage of one another's redundancies (that we're in the same culture and learned the same concepts) and wire each other. In effect, evolution improves its efficiency via human culture, and we pass culture to our offspring by wiring their brains.

The human brain, from the macro level to the micro level, is organized for variation and degeneracy. In its interacting networks, clusters of neurons are partly independent and share a lot of information efficiently. This arrangement allows ever-changing populations of neurons to form and dissolve in milliseconds, so that single neurons participate in different constructions in different situations, modeling a variable and only partly predictable world. Neural fingerprints have no place in such a dynamic environment. It would be highly inefficient for all of humanity to have one inherited set of mental modules, given that we live in such diverse geographic and social environments around the world. The human brain evolved to create different kinds of human minds, adapted to different environments.

We don't need one universal brain creating one universal mind to claim that we are all one species.[4]

On the whole, the theory of constructed emotion is a biologically informed, psychological explanation of who you are as a human being. It takes into account both evolution and culture. You are born with some brain wiring as determined by your genes, but the environment can turn some genes on and off, allowing your brain to wire itself to your experiences. Your brain is shaped by the realities of the world that you find yourself in, including the social world made by agreement among people. Your mind is a grand collaboration that you have no awareness of. Through construction, you perceive the world not in any objectively accurate sense but through the lens of your own needs, goals, and prior experience (as you did with the blobby bee). And you are not the pinnacle of evolution, just a very interesting sort of animal with some unique abilities.

· · ·

The theory of constructed emotion provides a very different outlook on human nature than the classical view does. Classical ideas about our evolutionary origins, our personal responsibility, and our relationship with the outside world have been dominant in Western culture for thousands of years. To understand this older view of human nature and why it's been so entrenched for so long, it's convenient to begin — as so many scientific stories do — with Charles Darwin.

In 1872, Darwin published *The Expression of the Emotions in Man and Animals,* where he wrote that emotions were passed down to us, unchanging through the ages, from an early animal ancestor. Emotions in modern humans are therefore caused by ancient parts of our nervous system, according to Darwin, and each emotion has a specific, consistent fingerprint.[5]

To borrow a term from philosophy, Darwin was saying that each emotion has an *essence.* If instances of sadness occur with a pout and a slowed heart rate, then a fingerprint of "pout and slowed heart rate" may be the essence of sadness. Alternatively, the essence might be an underlying cause that makes all the instances of sadness the emotion they are, such as a set of neurons. (I'll use the word "essence" to refer to both possibilities.)[6]

The belief in essences is called *essentialism.* It presupposes that certain categories — sadness and fear, dogs and cats, African and European Americans, men and women, good and evil — each have a true reality or nature. Within each category, the members are thought to share a deep, underlying property (an essence) that causes them to be similar, even if they have some superficial differences. There are many varieties of dog with differences in size, shape, color, gait, temperament, and so on, but these differences are considered superficial with regard to some essence that all dogs share. A dog is never a cat.

Likewise, all varieties of the classical view consider emotions like sadness and fear to have distinct essences. The neuroscientist Jaak Panksepp, for example, writes that an emotion's essence is a circuit in the subcortical regions of your brain. The evolutionary psychologist Steven Pinker writes that emotions are like mental organs, analogous to body organs for specialized functions, and that an emotion's essence is a set of genes. The evolutionary psychologist Leda Cosmides and the psychologist Paul Ekman assume that each emotion has an innate, unobservable essence, which they refer to as a metaphorical "program." Ekman's version of the classical view, called basic emotion theory, assumes that essences for happiness, sadness, fear, surprise, anger, and disgust are triggered automatically by objects and events in the world. Another version, called classical appraisal theory, inserts an additional step in between you and the world, saying that your brain first judges ("appraises") the situation and decides whether to trigger an emotion. All versions of the classical view agree that each emotion category has a distinct fingerprint; they just disagree on the nature of the essences.[7]

Essentialism is the culprit that has made the classical view supremely difficult to set aside. It encourages people to believe that their senses reveal objective boundaries in nature. Happiness and sadness look and feel different, the argument goes, so they must have different essences in the brain. People are almost always unaware that they essentialize; they fail to see their own hands in motion as they carve dividing lines in the natural world.

Darwin's belief in emotion essences, as revealed in *Expression,* helped to launch the modern classical view of emotion to prominence. That same belief also made Darwin unwittingly look like a hypocrite. It is no small task to criticize — let alone contradict — the ideas of one of the greatest scientists in history. But let's have a go, shall we?

Darwin's most famous book, *On the Origin of Species,* triggered a paradigm shift that transformed biology into a modern science. His greatest scientific achievement, so nicely summed up by the evolutionary biologist Ernst Mayr, was freeing biology from "the paralyzing grip of essentialism." Regarding emotion, however, Darwin made an inexplicable about-face thirteen years later by writing *Expression,* a book riddled with essentialism.

In doing so, he abandoned his remarkable innovations and returned to essentialism's paralyzing grip, at least where emotions are concerned.[8]

You see, before Darwin's theory from *Origin* became popular in the nineteenth century, essentialism ruled the animal kingdom. Each species was assumed to have an ideal form, created by God, with defining properties (essences) that distinguished it from all other species (each with their own essences). Deviations from the ideal were said to be due to error or accident. Think of this as the "dog show" version of biology. A dog show, in case you've never seen one, is a contest to identify the "best" dog in a field of competitors. The dogs do not directly compete with one another but are compared by judges to a hypothetical ideal dog to see who's the closest. When rating Golden Retrievers, for example, the judges compare each competitor to the ideal image of a Golden Retriever. Is the dog the right height? Are its limbs symmetrical? Is the muzzle straight, blending smoothly with the skull? Is the coat a rich, dense, lustrous gold? Any differences from the ideal dog are regarded as error, and the dog with the smallest amount of error wins. In the same manner, influential thinkers of the early nineteenth century saw the world of living creatures as one big dog show. If you looked at a Golden Retriever and observed that its stride was longer than average, then its stride was *too* long compared to the ideal, or even wrong.[9]

Then along came Darwin, who argued that variations within a species, such as length of stride, are not errors. Instead, variations are expected and are meaningfully related to the species' environment. Any population of Golden Retrievers has a variety of stride lengths, some of which provide a functional advantage for running, climbing, or hunting. The individuals with strides that best fit their environment will live longer and produce more offspring. This is Darwin's theory of evolution from *Origin* in action, known as natural selection and sometimes called "survival of the fittest." To Darwin, each species was a conceptual category — a population of unique individuals who vary from one another, with no essence at their core. The ideal dog doesn't exist: it's a statistical summary of many diverse dogs. No features are necessary, sufficient, or even typical of every individual in the population. This observation, known as population thinking, is central to Darwin's theory of evolution.[10]

Population thinking is based on variation, whereas essentialism is based on sameness. The two ideas are fundamentally incompatible. *Origin* is therefore a profoundly anti-essentialist book. So it is baffling that where emotion is concerned, Darwin reversed his greatest achievement by writing *Expression*.[11]

It is equally baffling, not to mention ironic, that the classical view of emotion is based on the very essentialism that Darwin is famous for vanquishing in biology. The classical view explicitly labels itself as "evolutionary" and assumes that emotions and their expressions are products of natural selection, yet natural selection is completely absent from Darwin's thinking on emotion. Any essentialist view that wraps itself in the cloak of Darwin is demonstrating a profound misunderstanding of Darwin's central ideas about evolution.

The compelling power of essentialism led Darwin to some beautifully ridiculous ideas about emotion. "Even insects," he wrote in *Expression*, "express anger, terror, jealousy, and love" when they rub their body parts together to make sounds. Think about that the next time you're chasing a fly in your kitchen. Darwin also wrote that emotional imbalance could cause frizzy hair.[12]

Essentialism is not only powerful but also infectious. Darwin's perplexing belief in unvarying emotion essences lived on after his death and distorted the legacy of other famous scientists. In the process, the classical view of emotion gained momentum. The most important example is that of William James, considered by many to be the father of American psychology. James might not be the household name that Darwin is, but he was, quite simply, an intellectual giant. His 1,200-page tome *Principles of Psychology* contains most of Western psychology's most important ideas and remains, after more than a century, the foundation of the field. His name graces the highest honor that can be bestowed on a scientist from the Association of Psychological Science, the William James Prize, and Harvard's psychology building is named William James Hall.

James is widely cited for saying that each type of emotion — happiness, fear, and so on — has a distinct fingerprint in the body. This essentialist idea is a key fact of the classical view, and generations of James-influenced researchers have searched for those fingerprints in heartbeats, respiration, blood pressure, and other bodily markers (and have written some bestselling books on emotion). James's statement has a catch, however: he never said it. The widely believed claim that he did comes from a hundred-year-old misinterpretation of his words through the lens of essentialism.

James actually wrote that each *instance* of emotion, not each category of emotion, comes from a unique bodily state. This is a wildly different statement. It means you can tremble in fear, jump in fear, freeze in fear, scream in fear, gasp in fear, hide in fear, attack in fear, and even laugh in the face of fear. Each occurrence of fear is associated with a different set of internal changes and sensations. The classical misinterpretation of James represents a 180-degree inversion of his meaning, as if he were claiming the existence of emotion essences, when ironically he was arguing against them. In James's words, "'Fear' of getting wet is not the same fear as fear of a bear."[13]

How did this widespread misunderstanding of James arise? I discovered that one of James's contemporaries sowed the confusion, a philosopher named John Dewey. He came up with his own theory of emotion by grafting Darwin's essentialist views from *Expression* onto James's anti-essentialist ideas, even though they are fundamentally incompatible. The result was a Frankenstein's monster of a theory that inverted James's meaning by assigning an essence to each emotion category. For the finishing touch, Dewey named his concoction after James, calling it "the James-Lange theory of emotion."* Today, Dewey's role in this jumble is forgotten, and countless publications attribute his theory to James. A prominent example is the writings of neurologist Antonio Damasio, author of *Descartes' Error* and other popular books on emotion. To Damasio, an emotion's unique physical fingerprint, which he calls a somatic marker, is a source of information used by the brain to make good decisions. These markers are like little bits of wisdom. Emotional experience, according to Damasio, occurs when somatic markers are transformed into conscious feelings. Damasio's hypothesis is actually a child of the James-Lange merger, not of James's actual views on emotion.[14]

Dewey's misinterpretation of James is one of the great mistakes in modern psychology, forged by essentialism in the name of Darwin. It is ironic, not to mention absurdly tragic, when Darwin's name is invoked to lend authority to essentialist scientific views, when his greatest scientific achievement was to vanquish essentialism in biology.

So why is essentialism so powerful that it can twist the words of great scientists and misdirect the path of scientific discovery?

The simplest reason is that essentialism is intuitive. We experience our

---

* "Lange" refers to physiologist Carl Lange, another contemporary of James and Dewey. His ideas on emotion were superficially similar to James's but retained the essentialist belief that each category of emotion had a distinct fingerprint. Lange was in the right place at the right time to have his name emblazoned on Dewey's theory.

own emotions as automatic reactions, so it's easy to believe that they spring forth from ancient, dedicated parts of the brain. We also see emotions in blinks, furrowed brows, and other muscle twitches, and we hear emotions in the pitch and lilt of voices, without any sense of effort or agency. Therefore, it's also easy to believe that we've been engineered by nature to recognize emotional displays and programmed to act on them. That's a dubious conclusion, however. Millions of people around the world can instantly, effortlessly recognize Kermit the Frog, but that doesn't mean the human brain is wired for Muppet recognition. Essentialism promises simple, single-cause explanations that reflect common sense, when in fact we live in a complex world.

Essentialism is also remarkably difficult to disprove. Since an essence can be an unobservable property, people are free to believe in essences even when they cannot be found. It's easy to come up with reasons why an experiment did not detect an essence: "we haven't looked everywhere yet," or "it's inside this complicated biological structure we can't see into yet," or "our tools today aren't sufficiently powerful to find the essence, but one day they will be." These hopeful thoughts are heartfelt but logically impossible to prove false. *Essentialism inoculates itself against counterevidence.* It also changes the way science is practiced. If scientists believe in a world of essences that are waiting to be discovered, then they devote themselves to finding those essences, a potentially endless quest.[15]

Essentialism also appears to be an inherent part of our psychological makeup. Humans create categories by inventing purely mental similarities, as you learned in chapter 5, and we name those categories with words. That's why a word like "pet" or "sadness" applies to a multitude of diverse instances. Words are an incredible achievement, but they are also a Faustian bargain for the human brain. On one hand, a word like "sadness," when applied to a collection of varied perceptions, invites you to search for (or invent) some underlying sameness that transcends their noticeable differences. That is, the word "sadness" guides you to create an emotion concept, which is a good thing. But the word also invites you to believe in a *reason* for that sameness: some deep, unobservable, or even unknowable quality that is responsible for their equivalence, giving them their true identity. That is, words invite you to believe in an essence, and that process is conceivably the psychological origin of essentialism. William James made a similar observation over a century ago when he wrote, "Whenever we have made a word

. . . to denote a certain group of phenomena, we are prone to suppose a sub-stantive entity existing beyond the phenomena, of which the word shall be the name." The very words that help us to learn concepts can also trick us into believing that their categories reflect firm boundaries in nature.[16]

Research with children illustrates how the human brain constructs a be-lief in essences. A scientist shows a child a red cylinder, calling it a nonsense name like "blicket," and demonstrates that it has a special function of light-ing up a machine. Next, the child is shown two more objects, a blue square that the scientist also calls a "blicket," and a second red cylinder that is not called a "blicket." The child will expect only the blue square to light up the machine, despite its visual differences from the original red "blicket." Chil-dren infer that each "blicket" contains an unseen causal force that lights the machine. This phenomenon, which scientists call induction, is an extremely efficient way for the brain to extend concepts by ignoring variation. How-ever, induction also encourages essentialism. As a child, when you saw a friend slumped on the ground, crying at the loss of a toy, and were told that the kid felt sad, your brain inferred that there was an unseen causal force inside the child causing the feeling of sadness, the slumped body posture, and the crying. You extended your belief in this essence to other instances of children who were pouting, throwing tantrums, gritting their teeth, and engaging in other behaviors, because adults labeled them for you as sad. Emotion words reinforce the fiction that the equivalences we create are ob-jectively real in the world, waiting to be discovered.[17]

Essentialism may also be a natural consequence of how your brain is wired. The same circuitry that allows you to form concepts and predict with them also makes essentializing easy. Your cortex learns concepts by sepa-rating similarities from differences, as you saw in chapter 6. It integrates in-formation across vision, hearing, interoception, and the other sensory do-mains, compressing them into efficient summaries. Each summary is like a little imaginary essence, invented by your brain to represent that a bunch of instances from your past are similar.[18]

So, essentialism is intuitive, logically impossible to disprove, part of our psychological and neural makeup, and a self-perpetuating scourge in sci-ence. It is also the basis for the classical view's most fundamental idea, that emotions have universal fingerprints. No wonder the classical view has such stamina — it's powered by a virtually unkillable belief.

When you embed essentialism in a theory of emotion, you get some-

thing more than just a doctrine on how and why you have feelings. You get
— yes — a compelling story of what it means to be a human being. A classical theory of human nature.

The classical story begins with your evolutionary origins. You are said
to be an animal at the core. You allegedly inherited various mental essences
from your non-human ancestors, including emotion essences buried deep
within your subcortex. To quote Darwin, "Man, with all his noble qualities
. . . with his god-like intellect . . . with all these exalted powers . . . still bears
in his bodily frame the indelible stamp of his lowly origin." Nevertheless,
the classical view considers you special because your animalistic essences
come gift-wrapped in rational thought. A uniquely human essence of reason supposedly lets you regulate your emotions by rational means, placing
you at the pinnacle of the animal kingdom.[19]

The classical view of human nature also speaks to personal responsibility. It says that your behavior is governed by internal forces beyond your
control: you are buffeted by the world and respond emotionally on impulse,
like an erupting volcano or a boiling pot. According to this view, sometimes
your emotion essences and cognitive essences vie for control of your behavior, and other times the two sets of essences work together to make you
wise. Either way, if you're at the mercy of strong emotions that can hijack
you, the argument goes, then you might be less culpable for your actions.
This assumption now sits at the foundation of Western legal systems, where
so-called crimes of passion are given special treatment. Additionally, if you
are completely devoid of emotion, then you are seen as more capable of in-
human acts. A serial killer who feels no remorse, some believe, is somehow
less human than a murderer who deeply regrets his actions. If this is the
case, then morality would be rooted in your ability to feel certain emotions.

The classical view also draws hard boundaries between you and the outside world. As you look around, you see objects like trees, rocks, houses,
snakes, and other people. These objects exist outside your anatomical body.
In this view, falling trees make a sound whether you're present or not. Your
emotions, thoughts, and perceptions, on the other hand, are said to exist
inside your anatomical body, each with its own essence. So, by implication,
your mind would be completely inside you and the world completely outside you.[20]

In a sense, the classical view wrenched human nature away from religion
and placed it into the hands of evolution. You are no longer an immortal
soul but a collection of specialized, distinct, inner forces. You come into the

world preformed, not in God's image but by your genes. You perceive the world accurately, not because God designed you this way but because the survival of your genes to the next generation depends on it. And your mind is a battleground, not of good and evil, righteousness and sin, but of rationality and emotionality, cortex over subcortex, inner versus outer forces, the thoughts in your brain versus the emotions in your body. You, with your animal brain wrapped in rational cortex, are distinct from other animals in nature, not because you have a soul but because you are the pinnacle of evolution, endowed with insight and reason.

Darwin embodied this essentialist view of human nature. Even though he vanquished essentialism from our understanding of the natural world, when it came to humans' place in that world, essentialism got the better of him. *Expression* covered all three parts of the classical view of human nature: that animals and humans share universal essences of emotion, that emotions seek expression in the face and body outside of our control, and that they are triggered by the outside world.

In the years that followed, however, Darwin's own essentialism came back to bite him in the behind. As Darwin's intellectual descendants adopted his views, shaping the classical view, they ironically misinterpreted (or twisted?) his own words to conform more fully to essentialism.

Darwin indeed stated in *Expression* that humans display universal facial expressions that evolved from a common ancestor:

> With mankind some expressions, such as the bristling of the hair under the influence of extreme terror, or the uncovering of the teeth under that of furious rage, can hardly be understood, except on the belief that man once existed in a much lower and animal-like condition. The community of certain expressions in distinct though allied species, as in the movements of the same facial muscles during laughter by man and by various monkeys, is rendered somewhat more intelligible, if we believe in their descent from a common progenitor.[21]

On first glance, you might think Darwin is saying that facial expressions are a useful and functional product of evolution, and, in fact, the classical view was founded on this idea. However, Darwin actually said the opposite. He wrote that smiles, frowns, eye-widening, and other expressions were useless, vestigial movements — products of evolution that no longer serve a function, like the human tailbone and appendix and the wings of the ostrich. He made this statement over a dozen times in *Expression*. Emotional expressions were primarily a compelling example for his broader arguments about evolution. If these expressions are useless in humans but shared with other animals, according to Darwin, they must exist because they were functional in a long-gone, common ancestor. Vestigial expressions would provide strong evidence that humans were animals, justifying his earlier views about natural selection from *On the Origin of Species* in 1859, which he then applied to human evolution in his next book, *The Descent of Man, and Selection in Relation to Sex,* in 1871.[22]

If Darwin didn't claim that emotional expressions evolved to serve a survival function, then why do so many scientists fervently believe that he claimed this? I discovered the answer in the manuscripts of an early-twentieth-century American psychologist, Floyd Allport, who wrote extensively on Darwin's ideas. In 1924, Allport made a sweeping inference from Darwin's writing that significantly changed the original meaning. Allport wrote that expressions begin as vestigial in newborns but quickly assume function: "Instead of the biologically useful reaction being present in the ancestor and the expressive vestige in the descendant, we regard both these functions as present in the descendant, the former serving as a basis from which the latter develops."[23]

Allport's modification obtained a certain authenticity and validity, despite being inaccurate, because it supported the classical view of human nature. It was eagerly adopted by like-minded scientists who could now claim to be the heirs of the unassailable Charles Darwin. In reality, they are merely the heirs of Darwin-hacking Floyd Allport.

As you can see, Darwin's name sometimes functions like a magical cloak that wards off the evil spirits of scientific criticism. It allowed Floyd Allport and John Dewey to transmute the words of William James and Darwin himself into their diametric opposites and shore up the classical view of emotion. The cloak is protective, for if you disagree with a Darwinian idea, you must be denying evolution. (Heck, you're probably a closet creationist.)

Darwin's magical cloak also helped to propagate the mistaken idea that the brain evolved as a bunch of blobs with distinct, dedicated functions. This key belief of the classical view led many scientists down the fruitless path of searching for emotion blobs in the brain. The path was paved by a Darwin-swaddled physician from the mid-nineteenth century, Paul Broca, who claimed to have discovered the brain blob for human language. He observed that patients with damage to a region of the left frontal lobe were rendered unable to speak fluently, a condition called nonfluent or expressive aphasia. When a person with Broca's aphasia tries to say something meaningful, the words come out jumbled: "Thursday, er, er, er, no, er, Friday . . . Bar-ba-ra . . . wife . . . and, oh, car . . . drive . . . purnpike [sic] . . . you know . . . rest and . . . TV." Broca inferred that he'd found the essence of language in the brain, much like classical view scientists point to amygdala lesions as proof of fear circuitry. The region has been known as Broca's area ever since.[24]

The thing is, Broca had scant evidence for his claims, and other scientists had plenty of evidence that he was wrong. They pointed out, for example, that other patients with nonfluent aphasia had a perfectly healthy Broca's area. But Broca's idea prevailed anyway because it was protected by Darwin's magical cloak reinforced by a healthy dose of essentialism. Thanks to Broca, scientists now had an evolutionary story for the origin of language — that it's located in "rational" cortex — countering the prevailing belief that language was given by God. Today's textbooks in psychology and neurology still hold up Broca's area as the clearest example of localized brain function, even as neuroscience has shown that the region is neither necessary nor sufficient for language.* Broca's area is actually a *failure* to localize a psychological function to a brain blob. Nevertheless, history was rewritten in Broca's favor, lending strength to essentialist views of the mind.[25]

Broca and his Darwinian cloak went on to reinforce the classical fiction that emotion and reason evolved as layers in the brain, which you encountered in chapter 4 as the "triune brain." Broca was inspired by Darwin's claims in *The Descent of Man* that the human mind, like the human body, was sculpted by evolution. Darwin wrote that "animals are excited by the same emotions as ourselves," surmising that human brains, like the rest of the human body, reflect our "lowly origin." So Broca and other neurologists

---

* A significant number of patients who suffer from Broca's aphasia have no damage in Broca's area, and conversely, about half the people with lesions in Broca's area do not have Broca's aphasia. Scientists continue to debate the function of Broca's area, which is better referred to as lateral prefrontal cortex, but few believe that it is specific to language production, grammatical abilities, or even general language processing. The current consensus is that it's part of several intrinsic networks, including the interoceptive and control networks. Where language is concerned, the control network helps your brain choose between conflicting options, such as the words "your" and "you're," but as we saw in chapter 6, this network participates in other non-language tasks.

and physiologists launched a grand search for animalistic emotion circuits — our inner beast. They focused on what they believed to be ancient parts of the brain, whose circuits were allegedly regulated by the more evolutionarily advanced cortex.[26]

Broca localized the "inner beast" in what he believed to be an ancient "lobe" deep within the human brain. He named it *le grand lobe limbique,* or "the limbic lobe." Broca did not brand his supposed lobe as the seat of emotion (actually, he thought it housed the sense of smell and other primitive survival circuitry), but he did treat limbic tissue as a single, unified entity, laying the first stone on a path toward essentializing it as the home of emotion. Over the next century, Broca's limbic lobe morphed into a unified "limbic system" for emotion, guided by other believers in the classical view. This so-called system was said to be evolutionarily old; to be virtually unchanged from its origin in non-human mammals; and to control the heart, lungs, and other internal organs of the body. It allegedly lay between ancient "reptilian" circuits in the brainstem for hunger, thirst, and so on, and the newer, uniquely human layers of cortex that regulate mankind's animalistic emotions. This illusory hierarchy embodied Darwin's ideas about human evolution — base appetites having evolved first, followed by wild emotional passions, with rationality as our crowning glory.[27]

Scientists inspired by the classical view have claimed to localize many different emotions to limbic brain regions, such as the amygdala, that are (allegedly) under the control of the cortex and cognition. Modern neuroscience, however, has shown that the so-called limbic system is a fiction, and experts in brain evolution no longer take it seriously, let alone consider it a system. Accordingly, it's not the home of emotion in the brain, which is unsurprising because no single brain area is dedicated to emotion. The word "limbic" still has meaning (when referring to brain anatomy), but the limbic system concept was just another example of applying an essentialist, Darwin-flavored ideology to the structure of the human body and brain.[28]

Long before Broca fashioned his first brain blob, the classical and construction views of human nature were at war. In Ancient Greece, Plato divided the human mind into three types of essences: rational thoughts, passions (which today we would call emotions), and appetites like hunger and sex drive. Rational thought was in charge, controlling the passions and appetites, an arrangement that Plato described as a charioteer wrangling two winged horses. A hundred years earlier, however, his countryman Heraclitus (chapter 2) was arguing that the human mind constructs perception in the moment, like constructing a river from countless drops of water. In Ancient Eastern philosophy, traditional Buddhism enumerated more than fifty discrete mental essences, called *dharmas,* some of which bear a striking resemblance to the so-called basic emotions of the classical view. Centuries later, a radical revision of Buddhism recast the *dharmas* as human constructions dependent on concepts.[29]

From those initial skirmishes, the war has continued throughout recorded history. The eleventh-century scientist Ibn al-Haytham, who made seminal contributions to developing the scientific method, held the constructionist view that we perceive the world through judgment and inference. Medieval Christian theologians were essentialists, associating different cavities in the brain with distinct essences of memory, imagination, and intelligence. Philosophers in the seventeenth century, such as René Descartes and Baruch Spinoza, believed in emotion essences and catalogued them, while eighteenth-century philosophers like David Hume and Immanuel Kant argued more for construction and perception-based explanations for human experience. The neuroanatomist Franz Joseph Gall in the nineteenth century founded phrenology, perhaps the ultimate essentialist view of the brain, to detect and measure mental essences as bumps on the skull (!!). Shortly thereafter, William James and Wilhelm Wundt espoused constructionist theories of the mind; as James wrote, "A science of the relations of mind and brain must show how the elementary ingredients of the former correspond to the elementary functions of the latter." James and Darwin were also casualties within this war over human nature, as their views of emotion were, shall we say, "adjusted," and the spoils went to scientists such as Broca who claimed a victory for evolution . . . or at least an essentialist sort of evolution.[30]

Plato's essences of the mind are still around today, though their names have changed (and we've dispensed with the horses). Nowadays we call them perception, emotion, and cognition. Freud called them the id, the ego, and the superego. The psychologist and Nobel laureate Daniel Kahneman metaphorically calls them System 1 and System 2. (Kahneman is *very careful* to say it's a metaphor, but many people seem to be ignoring him and essentializing Systems 1 and 2 as blobs in the brain.) The "triune brain" names them the reptilian brain, the limbic system, and the neocortex. Most recently, the neuroscientist Joshua Greene has used the intuitive analogy of a camera, which can operate quickly and effortlessly using its automatic settings, or more flexibly and deliberately in manual mode.[31]

On the other side of the fence, construction views of the mind are plentiful today. Psychologist and bestselling author Daniel L. Schacter has a construction theory of memory. And you can easily find construction theories for perception, the self, concept development, brain development (neuroconstruction), and of course the theory of constructed emotion.[32]

The battles today are all the more intense because it's easy for each side to view the other in caricature. The classical view often dismisses construction as saying everything is relative, as if the mind were merely a blank slate and biology can be disregarded. Construction blasts the classical view for ignoring the powerful effects of culture and justifying the status quo. In caricature, the classical view says "nature" and construction says "nurture," and the result has been a wrestling match between straw men.

Modern neuroscience, however, has burned down both caricatures. We are not blank slates, and our children are not "Silly Putty" to be shaped this way and that, but neither is biology destiny. When we peer into the workings of a functioning brain, we don't see mental modules. We see core systems that interact continuously in complex ways to produce many sorts of minds, depending on culture. The human brain is itself a cultural artifact because it is wired by experience. We have genes that are turned on and off by the environment, and other genes that regulate how sensitive to the environment we are. I'm not the first person to make these points. But I am perhaps the first one to point out how brain evolution, brain development, and its resulting anatomy point in a clear direction for the science of emotion and our view of human nature.[33]

Ironically, the millennia-long war over human nature has itself been tainted by essentialism. Both sides have assumed that a single, superior force must be shaping the brain and designing the mind. In the classical view, this force has been nature, God, and then evolution. In construction, it has been the environment and then culture. But neither biology nor culture is responsible alone. Others have made this point before me, but it's time to take it seriously. We don't know every detail about how the mind and brain work, but we know enough to say definitively that neither biological determinism nor cultural determinism is correct. The boundary of the skin is artificial and porous. As Steven Pinker so nicely writes, "It is now simply misguided to ask whether humans are flexible or programmed, whether behavior is universal or varies across cultures, whether acts are learned or innate." The devil is in the details, and the details give us the theory of constructed emotion.[34]

.   .   .

Now that the final nails are being driven into the classical view's coffin in this era of neuroscience, I would like to believe that this time, we'll actually push aside essentialism and begin to understand the mind and brain without ideology. That's a nice thought, but history is against it. The last time that construction had the upper hand, it lost the battle anyway and its practitioners vanished into obscurity. To paraphrase a favorite sci-fi TV show, *Battlestar Galactica,* "All this has happened before and could happen again." And since the last occurrence, the cost to society has been billions of dollars, countless person-hours of wasted effort, and real lives lost.

My cautionary tale begins in the early twentieth century, when scientists inspired by Darwin and the mutant James-Lange theory were searching in vain for the essences of anger, sadness, fear, and so on. Their repeated failures eventually led them to a creative solution. If we cannot measure emotions in the body and brain, they said, we'll measure only what happens before and after: the events that bring on an emotion and the physical reactions that result. Never mind what's happening inside that skull thing in the middle. Thus began the most notorious historical period in psychology, called *behaviorism.* Emotions were redefined as mere behaviors for survival: fighting, fleeing, feeding, and mating, collectively known as the "four F's." To a behaviorist, "happiness" equaled smiling, "sadness" was crying, and "fear" was the act of freezing in place. And so, the nagging problem of finding the fingerprints of emotional feelings was, with the flick of a pen, defined out of existence.[35]

Psychologists often recount stories of behaviorism in the same chilling tones as a ghost story around a campfire. It declared that thoughts, feelings, and the rest of the mind were unimportant to behavior or might not even exist. During this "dark ages" of emotion research, which lasted for several decades, nothing worthwhile was discovered on human emotion (supposedly). Ultimately, most scientists rejected behaviorism because it ignores a basic fact: that each of us has a mind, and in every waking moment of life, we have thoughts and feelings and perceptions. These experiences, and their relation to behavior, must be explained in scientific terms. Psychology emerged from the darkness in the 1960s, according to the official history, as a cognitive revolution reinstated the mind as a topic of scientific inquiry, likening emotion essences to modules or organs in a mind that was thought to function like a computer. With this transformation, the final pieces of the modern classical view fell into place, and the two main flavors of the classical view — basic emotion theory and classical appraisal theories — were officially anointed.[36]

That's what the history books say . . . but history books are written by the victors. The official history of emotion research, from Darwin to James to behaviorism to salvation, is a byproduct of the classical view. In reality, the alleged dark ages included an outpouring of research demonstrating that emotion essences don't exist. Yes, the same kind of counterevidence that we saw in chapter 1 was discovered seventy years earlier . . . and then forgotten. As a result, massive amounts of time and money are being wasted today in a redundant search for fingerprints of emotion.

I discovered this quite by chance in 2006 while cleaning my office, when I stumbled across a couple of old papers from the 1930s when emotion research was allegedly dead. These papers did not embrace behaviorism. They said that emotions do not have biological essences. Following a trail of references, I discovered a treasure trove of over a hundred publications, written across a span of fifty years, that most of my scientific colleagues had never heard of. The writers were nascent constructionists, though they did not use that term. They were running experiments to find physical fingerprints for distinct emotions, failing to do so, concluding that the classical view was unjustified, and speculating about constructionist ideas. I call this band of scientists the Lost Chorus because their work, published in prestigious journals, has been largely overlooked, ignored, or misunderstood since the supposed dark ages ended.[37]

Why did the Lost Chorus flourish for half a century and then vanish? My best guess is that these scientists did not offer a fully formed, alternative theory of emotion to compete with the compelling classical view. They presented solid counterevidence to be sure, but criticism alone was not enough to remain relevant. As philosopher Thomas Kuhn wrote about the structure of scientific revolutions: "To reject one paradigm without simultaneously substituting another is to reject science itself." So when the classical view reasserted itself in the 1960s, half a century of anti-essentialist research was swept into history's dustbin. And we are all the poorer for it, considering how much time and money are being wasted today in pursuit of illusory emotion essences. At press time, Microsoft is analyzing facial photographs in an attempt to recognize emotion. Apple has recently purchased

Emotient, a startup company using artificial intelligence techniques in an effort to detect emotion in facial expressions. Companies are programming Google Glass ostensibly to detect emotion in facial expressions in an effort to help autistic children. Politicians in Spain and Mexico are engaging in so-called neuropolitics to discern voter preferences from their facial expressions. Some of the most pressing questions about emotion remain unanswered, and important questions remain obscured, because many businesses and scientists continue practicing essentialism while the rest of us are figuring out how emotions are made.[38]

It's hard to give up the classical view when it represents deeply held beliefs about what it means to be human. Nevertheless, the facts remain that no one has found even a single reliable, broadly replicable, objectively measurable essence of emotion. When mountains of contrary data don't force people to give up their ideas, then they are no longer following the scientific method. They are following an ideology. And as an ideology, the classical view has wasted billions of research dollars and misdirected the course of scientific inquiry for over a hundred years. If people had followed evidence instead of ideology seventy years ago, when the Lost Chorus pretty solidly did away with emotion essences, who knows where we'd be today regarding treatments for mental illness or best practices for rearing our children.[39]

• • •

Every scientific journey is a story. Sometimes it's a story of gradual discovery: "Once upon a time, people didn't know very much, but we learned more and more over the years, and today we know lots of stuff." Other times, it's a tale of radical change: "Everyone used to believe something that seemed correct, but boy were we wrong! Now the fascinating truth is here."

Our journey is more of a story within a story. The inner story is how emotions are made, wrapped in an outer story of what it means to be human. "For two thousand years, people believed something about emotions, despite abundant counterevidence all around us. The human brain, you see, is wired to mistake its perceptions for reality. Today, powerful tools have yielded a more evidence-based explanation that's almost impossible to ignore . . . yet some people still manage."

The good news is that we're in a golden age of mind and brain research. Many scientists are now on a path forged by the data, rather than ideology, to understand emotion and ourselves. This new, data-driven understanding leads to innovative ideas about how to live a fulfilling and healthful life. If your brain operates by prediction and construction and rewires itself through experience, then it's no overstatement to say that if you change your current experiences today, you can change who you become tomorrow. The next few chapters delve into these implications in the areas of emotional intelligence, health, law, and our relationships with other animals.[40]

9

Mastering Your Emotions

Every time you bite into a juicy peach or munch a bag of crunchy potato chips, you're not simply replenishing your energy. You're having an experience that is pleasant, unpleasant, or something in between. You bathe not only to stave off disease but also to enjoy warm water against your skin. You seek out other people not to stand in a herd for protection from predators but to feel the glow of friendship or to unload when you're feeling burdened. And sex is clearly for more than propagating your genes.

These examples show that you have a special link between the physical and the mental. Each time you perform a physical act for your body budget, you're also doing something mental with concepts. Every mental activity has a physical effect as well. You can put this connection to work for you, to master your emotions, enhance your resilience, become a better friend or parent or lover, and even change your conception of who you are.

Change is not easy. Ask any therapist or Buddhist monk; they've trained for years to become aware of their experiences and control them. Even so, you can take small steps right now based on the theory of constructed emotion and the new view of human nature it implies.

Some of the suggestions I propose in this chapter will sound familiar, like getting enough sleep, but with new scientific justification to motivate you. Other advice will probably be entirely new, like learning words from a foreign language, which you've probably never associated with emotional health. Not every suggestion will be right for you; some will fit your lifestyle better than others. But the effort can lead to greater well-being and success. Students with a richer emotion vocabulary do better in school. People with a balanced body budget are less likely to develop serious illnesses like diabetes and heart disease, and as they age, their mental abilities will stay sharper for longer. And life may become more meaningful and fulfilling.

Can you snap your fingers and change your feelings at will, like changing your clothes? Not really. Even though you construct your emotional experiences, they can still bowl you over in the moment. However, you can take steps now to influence your *future* emotional experiences, to sculpt who you will be tomorrow. I don't mean that in some vague, pseudo-spiritual, let's-illuminate-your-cosmic-soul kind of way, but in a very real, predicting-brain way.

Everything you've read so far about interoception, affect, body budgets, prediction, prediction error, concepts, and social reality has broad and deep practical implications for who you are and how you live your life. That's our theme as we enter the final part of this book, which begins here with emotional well-being and then continues to health (chapter 10), the law (chapter 11), and non-human animals (chapter 12).

For the remainder of the book, we'll apply our new view of human nature, especially the porous boundary between the physical and the social, to architect a recipe for living. The major ingredients in that recipe are your body budget and your concepts. If you maintain a balanced body budget, you'll feel better in general, so that's where we'll start. And if you develop a rich set of concepts, you'll have a toolbox for a meaningful life.

• • •

Typical self-help books focus on your mind. If you think differently, they say, you will feel differently. You can regulate your emotions if you try hard enough. These books, however, don't give much consideration to your body. If there's one thing that (I hope) you've learned from the past five chapters, it's that your body and your mind are deeply interconnected. Interoception drives your actions. Your culture wires your brain.[1]

The most basic thing you can do to master your emotions, in fact, is to keep your body budget in good shape. Remember, your interoceptive network labors day and night, issuing predictions to maintain a healthy budget, and this process is the origin of your affective feelings (pleasantness, unpleasantness, arousal, and calmness). If you want to feel good, then your brain's predictions about your heart rate, breathing, blood pressure, temperature, hormones, metabolism, and so on, must be calibrated to your body's actual needs. If they aren't, and your body budget gets out of whack, then you're going to feel crappy no matter what self-help tips you follow. It's just a matter of which flavor of crap.

Modern culture, unfortunately, is engineered to screw up your body budget. Many of the products sold in supermarkets and chain restaurants are pseudo-food loaded with budget-warping refined sugar and bad fats. Schools and jobs require you to wake early and go to sleep late, leaving over 40 percent of Americans between the ages of thirteen and sixty-four regularly sleep-deprived, a condition that can lead to chronic misbudgeting and possibly depression and other mental illnesses. Advertisers play on your insecurities, suggesting you'll be judged badly by your friends unless you buy the right clothing or car, and social rejection is toxic for your body budget. Social media offers new opportunities for social rejection and adds ambiguity, which is even worse for your body budget. Friends and employers expect you to be surgically attached to your cell phone at all hours, which means you never truly relax, and late-night screen time disrupts your sleeping patterns. Your culture's expectations for work, rest, and socializing determine how easily you can manage that internal budget. Social reality transmutes into physical reality.[2]

Your body budget, you may remember, is regulated by predictive circuitry in your interoceptive network. If those predictions become chronically out of sync with your body's actual needs, it's hard to bring them back into balance. Your body-budgeting circuitry, the loudmouth of your brain, doesn't respond quickly to counterevidence (prediction error) from your body. Once the predictions have been off-base for long enough, you will feel chronically miserable.

When people feel crappy on a regular basis, quite a few of them self-medicate. Thirty percent of all medications consumed in the United States are taken to manage some form of distress. For these sufferers, their predictions are regularly not calibrated to their bodies' actual expenditures, likely because their brain is misestimating the cost. So they feel miserable and take medication, or they turn to alcohol or certain street drugs like opiates.[3]

That's the bad news. What can you do, practically speaking, to keep your predictions calibrated and body budget balanced? I apologize if I suddenly sound like your mother, but the road begins with eating healthfully, exercising, and getting enough sleep. I know, I know, it sounds mundane or even trite, but sadly there is no substitute, biologically speaking. A body budget, like a financial budget, is easier to maintain when you have a solid foundation. When you were a baby, your caretakers entirely managed your body budget. As you grew, they gradually transferred more and more responsibility for maintaining your budget to you. Today your friends and family might pitch in a little, but its nourishment is pretty much up to you. So to whatever extent you can, eat your greens, go easy on the refined sugars and bad fats and caffeine, work out vigorously and regularly, and get plenty of sleep.[4]

This advice might seem impossible without significant changes in the structure and habits of your life. For some people, the difficulty comes in resisting junk food and excessive TV time and other temptations of mainstream culture. Other people who struggle to make ends meet, who have to choose between eating and paying the bills, might not have the luxury of making lifestyle changes. But please do what you can. The science is crystal clear on healthful food, regular exercise, and sleep as prerequisites for a balanced body budget and a healthy emotional life. A chronically taxed body budget increases your chances of developing a host of different illnesses, as we'll see in the next chapter.

A next line of attack is to modify your physical comfort if you can. Try a massage from a lover, a close friend, or a paid massage therapist (if you can afford it). Human touch is good for your health — it improves your body budget by way of your interoceptive network. Massage is especially helpful after vigorous exercise. It limits inflammation and promotes faster healing of the tiny tears in muscle tissue that result from exercise, which you might otherwise experience as unpleasant.[5]

Yet another budget-balancing activity is yoga. People who practice yoga long-term are able to calm down more quickly and effectively, probably due to some combination of physical activity and the slow-paced breathing. Yoga also reduces levels of certain proteins, called proinflammatory cytokines, that over the long term promote harmful inflammation in your body. (We'll learn more about these proteins in the next chapter.) Regular exercise also increases the levels of other proteins, called anti-inflammatory cytokines, that reduce your chances of developing heart disease, depression, and other illnesses.[6]

Your physical surroundings also affect your body budget, so if possible, try to spend time in spaces with less noise and crowding, and more greenery and natural light. Not many of us can afford to sculpt our environment by moving into a new house or redecorating, but it is amazing what a simple houseplant will do. Environmental factors like these are so important to your body budget that they even appear to help psychiatric patients recover more quickly.[7]

Diving into a compelling novel is also healthful for your body budget. This is more than mere escapism; when you get involved in someone else's story, you aren't as involved in your own. Such mental excursions engage part of your interoceptive network, known as the default mode network, and keep you from ruminating (which would be bad for the budget). If you are not a reader, see a compelling film. If the story is sad, have a good cry, which is also beneficial to the budget.[8]

Here's another simple budget-booster: set up regular lunch dates with a friend and take turns treating each other. Research shows that giving and gratitude have mutual benefits for the body budgets involved, so when you take turns, you reap the benefits. (And over the long run, it costs the same as splitting the checks.)[9]

There are many more things you can try that I haven't mentioned yet. Adopt a pet, which gives you touch and unconditional adoration at the same time. Take walks in a public garden or park. Look online for research on your favorite hobbies, to see if they're beneficial for stress, or just try things out and see what works. Knitting works, apparently; for me, it's counted cross-stitch.[10]

Changing your habits to suit your body budget is never easy, and sometimes it's impossible, but try these techniques wherever you can. They will lift your mood and you'll feel less stressed more of the time.

• • •

After attending to your body budget, the next best thing you can do for emotional health is to beef up your concepts, otherwise known as "becoming more emotionally intelligent." People with a classical view mindset think about emotional intelligence as "detecting" other people's emotions "accurately," or experiencing happiness and avoiding sadness "at the right time." With our new understanding of emotions, however, we can think about emotional intelligence in a new way. "Happiness" and "Sadness" are each populations of diverse instances. Therefore, emotional intelligence (EI) is about getting your brain to construct the most useful instance of the most useful emotion concept in a given situation. (And also when *not* to construct emotions but instances of some other concept.)

Daniel Goleman, bestselling author of *Emotional Intelligence,* argues that higher EI leads to greater success in academics, business, and social relationships. "For star performance in all jobs, in every field," he writes, "emotional competence is twice as important as purely cognitive abilities." So you might be surprised to hear that science still has no generally accepted definition or measure of EI. Goleman's books offer a lot of reasonable, practical advice, but they don't properly explain *why* his advice works. Their scientific justification is heavily influenced by the outdated "triune brain" model — if you regulate your alleged emotional inner beast effectively, then you're emotionally intelligent.[11]

Emotional intelligence is better characterized in terms of concepts. Suppose you knew only two emotion concepts, "Feeling Awesome" and "Feeling Crappy." Whenever you experienced emotion or perceived someone else as emotional, you could categorize only with this broad brush. Such a person cannot be very emotionally intelligent. In contrast, if you could distinguish finer meanings within "Awesome" (happy, content, thrilled, relaxed, joyful, hopeful, inspired, prideful, adoring, grateful, blissful . . .), and fifty shades of "Crappy" (angry, aggravated, alarmed, spiteful, grumpy, remorseful, gloomy, mortified, uneasy, dread-ridden, resentful, afraid, envious, woeful, melancholy . . .), your brain would have many more options for predicting, categorizing, and perceiving emotion, providing you with the tools for more flexible and functional responses. You could predict and categorize your sensations more efficiently, and better tailor your actions to your environment.

What I'm describing is emotional granularity, the phenomenon (described in chapter 1) that some people construct finer-grained emotional experiences than others do. People who make highly granular experiences are emotion experts: they issue predictions and construct instances of emotion that are finely tailored to fit each specific situation. At the other end of the spectrum, there are young children who haven't yet developed adultlike emotion concepts, and who use "sad" and "mad" interchangeably to mean feeling unpleasant (as we discussed in chapter 5). My lab has shown that adults run the whole range from low to high emotional granularity. So, a key to EI is to gain new emotion concepts and hone your existing ones.[12]

There are many ways to gain new concepts: taking trips (even just a walk in the woods), reading books, watching movies, trying unfamiliar foods. Be a collector of experiences. Try on new perspectives the way you try on new clothing. These kinds of activities will provoke your brain to combine concepts to form new ones, changing your conceptual system proactively so you'll predict and behave differently later.

For example, in our household, my husband, Dan, is in charge of recycling because I am forever placing inappropriate items into the bin, like cellophane or wood, because by God, they *should* be recyclable. Instead of getting frustrated by the extra work I make for him, Dan applied a concept from his childhood, when he collected superhero comic books. My fruitless attempts at bucking reality became a "Superpower" that he calls wishful recycling. An irritating habit was thus transformed into an amusing foible.

Perhaps the easiest way to gain concepts is to learn new words. You've probably never thought about learning words as a path to greater emotional health, but it follows directly from the neuroscience of construction. Words seed your concepts, concepts drive your predictions, predictions regulate your body budget, and your body budget determines how you feel. Therefore, the more finely grained your vocabulary, the more precisely your predicting brain can calibrate your budget to your body's needs. In fact, people who exhibit higher emotional granularity go to the doctor less frequently, use medication less frequently, and spend fewer days hospitalized for illness. This is not magic; it's what happens when you leverage the porous boundary between the social and the physical.[13]

So, learn as many new words as possible. Read books that are outside of your comfort zone, or listen to thought-provoking audio content like National Public Radio. Don't be satisfied with "happy": seek out and use more specific words like "ecstatic," "blissful," and "inspired." Learn the difference between "discouraged" or "dejected" versus generically "sad." As you build up the associated concepts, you'll become able to construct your experiences more finely. And don't limit yourself to words in your native language. Pick another language and seek out its concepts for which your language has no words, like the Dutch emotion of togetherness, *gezellig,* and the Greek feeling of major guilt, *enohi.* Each word is another invitation to construct your experiences in new ways.[14]

Try also to invent your own emotion concepts, using your powers of social reality and conceptual combination. The author Jeffrey Eugenides presents a collection of amusing ones in his novel *Middlesex,* including "the hatred of mirrors that begins in middle age," "the disappointment of sleeping with one's fantasy," and "the excitement of getting a room with a minibar," though he does not assign them words. You can do the same thing yourself. Close your eyes and imagine yourself in a car, driving away from your hometown, knowing that you will never, ever return. Can you characterize that feeling by combining emotion concepts? If you can employ this technique day to day, you'll be better calibrated to cope with varied circumstances, and potentially more empathic to others, with improved skill to negotiate conflict and get along. You can even name your creations, like my word "chiplessness" in chapter 7, and teach them to your family and friends. Once you've shared your creations, they are just as real as any other emotion concept and bring the same benefits to your body budget.

An emotionally intelligent person not only has lots of concepts but also knows which ones to use and when. Just like painters learn to see fine distinctions in colors, and wine lovers develop their palettes to experience tastes that non-experts cannot, you can practice categorizing like any other skill. Suppose you see your teenage son heading out to school looking like he just rolled out of bed: hair unkempt, clothing wrinkled, and remnants of last night's dinner dotting his shirt. You could berate him and send him back to his room to change, but instead, ask yourself *what* you are feeling. Are you concerned that his teachers won't take him seriously? Disgusted by his greasy hair? Nervous that his attire will reflect badly on you as a parent? Irritated that you spend money on clothing he never wears? Or perhaps you're sad that your little boy has grown up and you miss the exuberance of his childhood. If all this introspection sounds implausible, realize that people pay good money to therapists and life coaches for exactly this purpose: to help them reframe situations, that is, find the most useful categorization in the service of action. You can do it yourself and become an expert categorizer of emotion with enough practice, and it gets easier with repetition.

Fine-grained categorizations have been shown to beat two other popular approaches for "regulating" emotions, in a study about fear of spiders. The first approach, called cognitive reappraisal, taught subjects to describe the spider in a nonthreatening way: "Sitting in front of me is a little spider, and it's safe." The second approach was distraction, having the subjects pay attention to something unrelated instead of the spider. The third was to categorize sensations with greater granularity, such as: "In front of me is an ugly spider and it is disgusting, nerve-wracking, and yet, intriguing." The third approach was the most effective in helping people with arachnophobia to be less anxious when observing a spider and to actually approach spiders. The effects lasted a week beyond the experiment, too.[15]

Higher emotional granularity has other benefits for a satisfying life. In a collection of scientific studies, people who could distinguish finely among their unpleasant feelings — those "fifty shades of feeling crappy" — were 30 percent more flexible when regulating their emotions, less likely to drink excessively when stressed, and less likely to retaliate aggressively against someone who has hurt them. For people who suffer from schizophrenia, those who exhibit higher emotional granularity report better relationships with family and friends, compared to those who exhibit lower granularity, and are better able to choose the correct action in social situations.[16]

In contrast, lower emotional granularity is associated with all sorts of afflictions. People who have major depressive disorder, social anxiety disorder, eating disorders, autism spectrum disorders, borderline personality disorder, or who just experience more anxiety and depressed feelings all tend to exhibit lower granularity for negative emotion. People who are diagnosed with schizophrenia exhibit low granularity for distinguishing positive from negative emotions. To be clear, nobody is claiming that low granularity causes these disorders, but it conceivably plays some role.[17]

After improving your emotional granularity, another way to hone your concepts, which is popular with therapists and self-help books, is to keep track of your positive experiences each day. Can you find anything that can make you smile, even briefly? Each time you attend to positive things, you tweak your conceptual system, reinforcing concepts about those positive events and making them salient in your mental model of the world. It's even better if you write about your experiences because, again, words lead to concept development, which will help you predict new moments to cultivate positivity.[18]

In contrast, when you ruminate about something unpleasant, you cause fluctuations in your body budget. Rumination is a vicious cycle: each time you dwell on (say) a recent breakup of a relationship, you add another instance to predict with, which expands your opportunity to ruminate. Certain concepts about your breakup, such as your final shouting match, or the look on your lover's face as he or she walked away for the last time, become entrenched in your model of the world. These concepts, as patterns of neural activity, get easier and easier for your brain to re-create, like well-trodden walking paths that grow deeper with each passerby's footsteps. You don't want them to become paved roads. Every experience you construct is an investment, so invest wisely. Cultivate the experiences you want to construct again in the future.

Sometimes it's helpful to construct instances of unpleasant emotion on purpose. Think about football players who cultivate anger before a big game. They shout and jump and pump their fists in the air to get themselves in the right frame of mind for crushing the competition. By elevating their heart rates, breathing more deeply, and generally influencing their body budgets, they create a familiar physical state and categorize it in the context of the sports stadium, based on their knowledge of past situations where a particular emotion helped with performance. Their aggression also strengthens bonds with their teammates and tells their opponents to beware. This is EI at work in a somewhat unlikely place.[19]

If you are a parent, you can help your children develop the skills to become emotionally intelligent. Speak to them about emotions and other mental states as early as you can, even if you think they are too young to understand. Remember that infants develop concepts well before you realize it is happening. So look children straight in the eye, widen your eyes to grab their attention, and speak about bodily sensations and movements in terms of emotions and other mental states. "See that little boy? He is crying. He is feeling pain from falling down and scraping his knee. He is sad and probably wants a hug from his parents." Elaborate on the feelings of storybook characters, on your children's own emotions, and on your emotions. Use a wide variety of emotion words. Talk about what causes emotions and what are their consequences to others. In general, think of yourself as your children's tour guide through the mysterious world of humans and their movements and sounds. Your detailed explanations help your children build a well-developed conceptual system for emotion.[20]

When you teach emotion concepts to children, you are doing more than communicating. You are *creating reality* for these kids — social reality. You're handing them tools to regulate their body budget, to make meaning of their sensations and act on them, to communicate how they feel, and to influence others more effectively. They will use these skills their whole lives.

As you teach your children about emotion, try not to limit yourself to essentialist stereotypes: smiling when happy, scowling when angry, and so on. (This may be difficult, as you're competing with TV cartoons that stick to Western stereotypes of emotion.*) Help them understand the variety of the real world, that a smile may mean happiness, embarrassment, anger, or even sadness depending on context. Try also to admit when you aren't sure

---

* Pixar's movies are impressive in how well they do not stick to the stereotypes. Even the characters in *Inside Out,* which is a thoroughly essentialist fantasy about emotions, show a broad range of subtle and fascinating facial and bodily configurations during emotional episodes.

how you feel, when you're guessing how someone else feels, or when you guess badly.

Carry on full conversations with your young child, taking turns, even when she is a baby who cannot respond verbally yet. By the time a child is a toddler, the conversational pattern matters as much as the words themselves for building emotion concepts. My husband and I never used "baby talk" with our daughter but spoke to her in fully formed, adult sentences from the time she was born, pausing afterward to let her "respond" in whatever way she could. People around us in the supermarket thought we were crazy, but we did wind up with an emotionally intelligent teenager who actually talks to adults. (And she can torture me with three-decimal precision. I'm so proud.)[21]

Do your children have screaming fits or throw tantrums? You can help them master their emotions and calm down by using social reality to your advantage. When my daughter, Sophia, was two and in her tantrum phase, telling her to calm down had no effect, of course. So we invented a concept called the "Cranky Fairy." Whenever Sophia launched into a tantrum (or if we were lucky, slightly beforehand), we'd explain to her, "Oh no, the Cranky Fairy is visiting. She's making you feel cranky. Let's try to make the Cranky Fairy go away." Then we directed her to a particular chair — a fuzzy red one with a picture of Elmo from Sesame Street — as her special place for calming down. (No, it didn't have little fuzzy red manacles.) At first we carried her to the chair, and sometimes she'd pitch a fit and kick the chair over, but eventually she would walk to it unasked and sit until her unpleasant feelings subsided. Sometimes she'd even announce that the Cranky Fairy was on her way. These practices might sound silly, but they have tangible effects. By inventing and sharing the concepts "Cranky Fairy" and "Elmo Chair" with Sophia, we created tools to help her calm herself. To her, these concepts were as real as money, art, power, and other constructions of social reality are to us.

In general, children with richer conceptual systems for emotion are poised for greater academic success. In one study conducted by the Yale Center for Emotional Intelligence, schoolchildren were taught to broaden their knowledge and use of emotion words for twenty to thirty minutes per week. The results were improved social behavior and academic performance. Classrooms that employed this educational model were also better organized and were rated by blind observers as having better instructional support for students.[22]

In contrast, if you don't talk to a child about his sensations in emotional terms, you can actually hamper his developing conceptual system. After four years of life, children in higher-income homes have seen or heard *four million* more words than their low-income counterparts, and they have better vocabulary and reading comprehension. Children with the fewest material advantages therefore lag in the social world. A simple intervention, like advising lower-income parents to communicate with their children more, improves the children's school performance. In the same manner, using more emotion words should improve children's EI.[23]

The same principles apply when you give your children feedback about their behavior. Studies show that children in low-income homes hear 125,000 more words of discouragement than praise, while their higher-income counterparts hear 560,000 more words of praise than discouragement, all by age four. That means children from lower-income homes have a more taxed body budget but fewer resources to deal with it.[24]

We all criticize our kids now and then, but try to make your feedback specific. If your daughter is whining incessantly, instead of yelling "Knock it off," try something like, "Your whining is irritating me, so stop it. If you are having a problem, use your words." When your son suddenly smacks your daughter in the head, don't call him "a bad boy." (That's not a concept you want him to develop.) Be specific: "Stop hitting your sister; it hurts her and makes her feel sad. Tell her you are sorry." The same rule holds for praise: don't call your daughter "a good girl." Praise her actions: "You made a good choice not hitting your brother back." This wording helps children to build more useful concepts. Your tone of voice matters too, since it easily communicates your affect and directly impacts the child's nervous system.[25]

By regulating your children's body budgets effectively, you guide them not only to a richer conceptual system for emotion but also better overall language development, which prepares them for better academic performance in school.

• • •

Okay, now you've done your best to revamp your lifestyle for a balanced budget, and you've beefed up your conceptual system to transform yourself into an emotion expert. You're still going to have ups and downs. You'll still have to deal with the compromises demanded by love, the ambiguities of your social life, the insincerity of the workplace, the fickleness of friendships, and your body slowly failing you as you age. What can you do to master your feelings in the moment?

The simplest approach, believe it or not, is to move your body. All animals use motion to regulate their body budgets; if their brain serves up more glucose than their body needs, a quick scamper up a tree will bring their energy level back into balance. Humans are unique in that we can regulate the budget without moving, using purely mental concepts. But when this skill fails you, remember that you too are an animal. Get up and move around, even if you don't feel like it. Turn on some music and dance around your home. Take a walk in a park. Why does this work? Moving your body can change your predictions and therefore your experience. Your movements may also help your control network to bring other, less bothersome concepts into the foreground.[26]

Another approach to mastering your emotions in the moment is to change your location or situation, which in turn can change your predictions. During the Vietnam War, for example, 15 percent of U.S. soldiers were addicted to heroin. When they came home as veterans, 95 percent of them stayed off the drug in their first year back — an astounding figure compared to the general population, where only 10 percent of users avoid relapse. The shift in location changed their predictions, which lessened their craving for the drug. (I sometimes wonder if midlife crisis is a drastic attempt to change one's predictions by changing the context.*)[27]

When changes in movement and context fail to help you master your emotions, the next big thing to try is recategorizing how you feel. This will require some explanation. Anytime you feel miserable, it's because you are experiencing unpleasant affect due to interoceptive sensations. Your brain will dutifully predict causes for those sensations. Perhaps they are a message from your body, like "I have a stomachache." Or perhaps they're saying, "Something is seriously wrong with my life." This is the distinction between *discomfort* and *suffering*. Discomfort is purely physical. Suffering is personal.

Imagine what your body looks like to an invading virus. You are just a big bag of DNA, proteins, water, and whatever other biological stuff it must steal to replicate itself. An influenza virus doesn't care about your beliefs, qualities, or values when it infects your cells. It does not make moral judgments on your character, like "Oooh, she's a snob with a bad haircut . . . let's

---

* My friend Kevin, who cultivated the pink Queen Anne's lace in chapter 7, has a saying: "Honey, when all else fails, put on a beautiful flowing scarf and a chic pair of sunglasses, buy a convertible, and drive across the country."

infect her!" No, a virus is egalitarian toward its victims. It brings discomfort, but it's nothing personal. All humans who haven't slept enough, with a nice wet set of lungs, can apply for the job of host.

Affect, on the other hand, transforms interoceptive sensation into something about *you*, with your particular strengths and faults. Now the sensations are personal — they reside inside your affective niche. When you feel wretched, the world seems like an awful place. People are judging you. Wars are raging. The polar ice caps are melting. You are suffering. Most of us devote a lot of time to relieving suffering. We often eat for pleasure or to soothe ourselves, rather than for the nutrients. I think drug addiction is often a misguided attempt to relieve the suffering from a body budget that's chronically out of whack.[28]

It's tricky to distinguish discomfort and suffering in the moment. Are you feeling irritated or just having caffeine withdrawal? If you are a woman, you probably have ambiguous physical symptoms related to your menstrual cycle or during menopause, and you may categorize the sensations as having emotional meaning when they do not. I remember in 2010 when my whole lab was moving from one university to another, including twenty researchers and hundreds of thousands of dollars of equipment. Everything seemed to be going wrong, plus I was about to leave for a two-week trip. Somehow I was holding myself together, extinguishing each fire as it ignited ... and then my laptop died. I sank to the floor in the middle of my kitchen and started sobbing. At just that moment, my husband walked in, noticed my state, and asked innocently, "Are you premenstrual?" Oh. My. God. I lashed out at him, the goddamn sexist pig and how *dare* he be so smug when I'm barely holding my life together?? My fury shocked us both. And three days later, I discovered that he was right.

With practice, you can learn to deconstruct an affective feeling into its mere physical sensations, rather than letting those sensations be a filter through which you view the world. You can dissolve anxiety into a fast-beating heart. Once you can deconstruct into physical sensations, then you can recategorize them in some other way, using your rich set of concepts. Perhaps that pounding in your chest is not anxiety but anticipation, or even excitement.

Look around right now and find an object to focus on. Try recategorizing it not as a three-dimensional visual object but as the individual pieces of differently colored light that your perception is constructed from. Tough, isn't it? Nevertheless, you can train yourself to do it. Pick the shiniest part of the object and try tracing its outlines with your eye. With a lot of practice, you can learn how to deconstruct objects like this. Great artists like Rembrandt could do it and realistically render objects in paint on a canvas. In a similar manner, you can deconstruct your emotions.

Recategorization is a tool of the emotion expert. The more concepts that you know and the more instances that you can construct, the more effectively you can recategorize in this manner to master your emotions and regulate your behavior. For instance, if you're about to take a test and feel affectively worked up, you might categorize your feeling as harmful anxiety ("Oh no, I'm doomed!") or as helpful anticipation ("I'm energized and ready to go!"). The head of my daughter's karate school, Grandmaster Joe Esposito, advises his nervous students before their black belt test: "Make your butterflies fly in formation." He is saying yes, you feel worked up right now, but don't perceive it as nervousness: construct an instance of "Determination."

Recategorization of this kind can bring tangible benefits to your life. Numerous studies have looked at performance on math tests such as the GRE and found that students achieve higher scores when they recategorize anxiety as merely a sign that the body is coping. People who recategorize anxiety as excitement show similar effects, with better performance and fewer classic symptoms of anxiety when speaking in public and even when singing karaoke. Their sympathetic nervous system still creates those jittery butterflies, but with fewer of the proinflammatory cytokines that lower performance and generally make people feel crappy, so they perform better. Studies have shown that remedial math students at community colleges can improve their exam grades and their final course grade through effective recategorization. This significant development can change the trajectory of a person's life, given that a college degree can be the difference between financial success and a lifelong struggle to make ends meet.[29]

If you can categorize your discomfort as helpful, say, when you're exercising hard, you can cultivate greater stamina. The U.S. Marine Corps has a motto that embodies this principle: "Pain is weakness leaving the body." Whenever you exercise just until you feel unpleasant and then stop, you're categorizing your physical sensations as exhaustion. You'll always exercise below your threshold, despite the health benefits of continuing. Through recategorization, however, you can continue exercising and feel even better later, as you reap the benefits of a stronger, healthier body. The more you do it, the more you tune your conceptual system toward longer exercise in the future.[30]

Lower back pain, sports injuries, soreness from arduous medical treatments, and other ailments offer similar opportunities to distinguish between physical discomfort and affective distress. People who live with chronic pain, for example, commonly have catastrophic thoughts that appear to impact their lives even more than the intensity of the pain does. When they learn to separate their physical sensations from their unpleasant affect, they may use fewer opiate drugs and crave them less. This is a significant finding considering that nearly 6 percent of Americans use prescription medication for chronic pain each year, mostly addictive opiates that are now known to enhance pain symptoms with long-term use. According to Deborah Barrett, author of *Paintracking* (and my sister-in-law), when you can categorize pain as physical, the pain need not be a personal catastrophe.[31]

The notion of recategorizing suffering as discomfort, or deconstructing the mental into the physical, has ancient origins. In Buddhism, some forms of meditation help to recategorize sensations as physical symptoms to reduce suffering, a practice Buddhists call *deconstructing the self*. Your "self" is your identity — a collection of characteristics that somehow define you, like your assorted memories, beliefs, likes, dislikes, hopes, life choices, morals, and values. You can also define yourself by your genes, your physical characteristics (weight, eye color), your ethnicity, your personality (funny, trustworthy), the relationships you have with other people (friend, parent, child, lover), the roles you hold (student, scientist, salesperson, factory worker, physician), your geographic or ideological community (American, New Yorker, Christian, Democrat), even the car that you drive. A common core runs through all these views: the self is your sense of who you are, and it's continuous through time, as if it were the essence of you.[32]

Buddhism considers the self to be a fiction and the primary cause of human suffering. Whenever you crave material things like expensive cars and clothes, or desire compliments to enhance your reputation, or seek positions of status and power to benefit your life, Buddhism says you are treating your fictional self as real (*reifying* the self). These material concerns may bring immediate gratification and pleasure but they also entrap you, like golden handcuffs, and cause persistent suffering, which we would call prolonged unpleasant affect. To a Buddhist, a self is worse than a passing physical illness. It is an enduring affliction.[33]

My scientific definition of the self is inspired by the workings of the brain yet is sympathetic to the Buddhist view. The self is part of social reality. It's not exactly a fiction, but neither is it objectively real in nature like a neu-tron. It depends on other people. In scientific terms, your predictions in the moment, and your actions that derive from them, depend to some extent on the way that others treat you. You can't be a self by yourself. We can un-derstand why Tom Hanks's character in the movie *Cast Away,* who was ma-rooned alone on a desert island for four years, needed to create a compan-ion named Wilson out of a volleyball.[34]

Certain behaviors and preferences are consistent with your self and some are not. There are foods you enjoy and others you'd prefer not to eat. You might call yourself a "dog person" or a "cat person." These behaviors and preferences vary quite a bit: your favorite food might be French fries, but not at every meal. The most enthusiastic dog lovers know a couple of dogs that they can't stand and are secretly fond of a few cats. Overall, your self is like a collection of dos and don'ts that summarizes your likes, dislikes, and habits in the moment.

We've seen something like this before. These dos and don'ts are like the features of a concept. So in my view, the self is a plain, ordinary concept just like "Tree," "Things That Protect You from Stinging Insects," and "Fear." I am quite sure you don't go around thinking of yourself as a concept, but just go with me for a bit on this.[35]

If the self is a concept, then you construct instances of your self by sim-ulation. Each instance fits your goals in the moment. Sometimes you cat-egorize yourself by your career. Sometimes you're a parent, or a child, or a lover. Sometimes you're just a body. Social psychologists say that we have multiple selves, but you can think of this repertoire as instances of a sin-gle, goal-based concept called "The Self" in which the goal shifts based on context.[36]

How does your brain keep track of all the varied instances of your "Self" as an infant, a young child, an adolescent, a middle-aged adult, and an older adult? Because one part of you has remained constant: you've always had a body. Every concept you have ever learned includes the state of your body (as interoceptive predictions) at the time of learning. Some concepts involve a lot of interoception, such as "Sadness," and others have less, such as "Plas-tic Wrap," but they're always in relation to the same body. So every catego-rization you construct — about objects in the world, other people, purely mental concepts like "Justice," and so on — contains a little bit of you. This is the rudimentary mental basis of your sense of self.[37]

The fiction of the self, paralleling the Buddhist idea, is that you have some enduring essence that makes you who you are. You do not. I speculate that your self is constructed anew in every moment by the same predictive, core systems that construct emotions, including our familiar pair of networks (interoceptive and control), among others, as they categorize the continuous stream of sensation from your body and the world. As a matter of fact, a portion of the interoceptive network, called the default mode network, has been called the "self system." It consistently increases in activity during self-reflection. If you have atrophy in your default mode network, as happens in Alzheimer's disease, you eventually lose your sense of self.[38]

Deconstructing the self offers a new inspiration for how to become the master of your emotions. By tweaking your conceptual system and changing your predictions, you not only change your future experiences; you can actually change your "Self."

Suppose you are feeling bad—worried because you are struggling with your finances, angry that you did not receive the promotion you deserved, dejected because your teacher believes you are not as intelligent as other students, or heartbroken because your lover abandoned you. A Buddhist mindset would describe these feelings as the suffering that results from clinging to material wealth, reputation, power, and security in an effort to reify the self. In the language of the theory of constructed emotion, wealth, reputation, and the rest are firmly within your affective niche, impacting your body budget, which ultimately leads you to construct instances of unpleasant emotions. Deconstructing the self for a moment allows you to reduce the size of your affective niche so concepts like "Reputation," "Power," and "Wealth" become unnecessary.[39]

Western culture has some common wisdom associated with these ideas. Don't be materialistic. What doesn't kill us makes us stronger. Sticks and stones. But I am asking you to take this one step further. When you are suffering from some ill or insult that has befallen you, ask yourself: Are you really in jeopardy here? Or is this so-called injury merely threatening the social reality of your self? The answer will help you recategorize your pounding heartbeat, the knot in the pit of your stomach, and your sweaty brow as purely physical sensations, leaving your worry, anger, and dejection to dissolve like an antacid tablet in water.[40]

I'm not saying this kind of recategorization is easy, but with practice it's possible, and it's also healthful. When you categorize something as "Not About Me," it exits your affective niche and has less impact on your body budget. Similarly, when you are successful and feel proud, honored, or gratified, take a step back and remember that these pleasant emotions are entirely the result of social reality, reinforcing your fictional self. Celebrate your achievements but don't let them become golden handcuffs. A little composure goes a long way.

If you are interested in taking this strategy further, try meditation. Mindfulness meditation, just one type of many, teaches you to stay alert and present in the moment but to observe sensations as they come and go, nonjudgmentally.* This state (which requires tremendous practice) reminds me of the quiet, alert state of newborn babies when they observe the world, their brains comfortably awash in prediction error, with no anxiety in sight. They experience sensations and release them. Meditation achieves something similar. This state may take years of practice to achieve, so the next best thing is to recategorize your thoughts, feelings, and perceptions as physical sensations, which are easier to let go of. You can use meditation, at least at first, to prioritize categorizations that focus on the physical, and deprioritize those that add more psychological meaning about you or your place in the world.

Meditation has a potent effect on brain structure and function, though scientists have not sorted out the exact details yet. Key regions in the interoceptive and control networks are larger for meditators, and connections between these regions are stronger. This matches what we might expect, since the interoceptive network is critical to constructing mental concepts and representing physical sensations from the body, and the control network is critical to regulating categorization. In some studies, we see stronger connections even after only a few hours of training. Other studies find that meditation reduces stress, improves the detection and processing of prediction error, facilitates recategorization (termed "emotion regulation"), and reduces unpleasant affect, although the findings are often inconsistent from one study to the next because not all the experiments have been well-controlled.[41]

Sometimes deconstructing the self is too challenging. You can achieve some of the same benefits more simply by cultivating and experiencing awe, the feeling of being in the presence of something vastly greater than yourself. It helps you get some distance from your self.[42]

---

* From a Buddhist perspective, we might say that deconstructing the self helps to "suspend categorization." From a neuroscience perspective, however, the brain never stops predicting so you can't turn off concepts.

I experienced these benefits firsthand when my family spent a few summer weeks at a beach house in Rhode Island. A symphony of crickets surrounded us each evening, resonating with an intensity I'd never heard before. I hadn't paid much attention to crickets before that, but now they entered my affective niche. I began to look forward to them every evening and to find their song comforting while falling asleep. When we returned from our vacation, I discovered that I could hear crickets through the thick walls of my home if I lay quietly enough. Now, whenever I wake in the middle of a summer night, feeling anxious after a stressful day in the lab, the crickets help me drift back to sleep. I developed an awe-inspired concept of being enveloped within nature and feeling like a tiny speck. This concept helps me change my body budget whenever I want. I can notice a tiny weed forcing its way through a crack in the sidewalk, proving yet again that nature cannot be tamed by civilization, and employ the same concept to take comfort in my insignificance.[43]

You can experience similar awe when hearing ocean waves crash against rocks on a beach, gazing at the stars, walking under storm clouds in the middle of the day, hiking deep into uncharted territory, or taking part in spiritual ceremonies. People who report feeling awe more frequently also have the lowest levels of those nasty cytokines that cause inflammation (though nobody has proved cause and effect).[44]

Whether you cultivate awe, meditate, or find other ways to deconstruct your experience into physical sensations, recategorization is a critical tool for mastering your emotions in the moment. When you feel bad, treat yourself like you have a virus, rather than assuming that your unpleasant feelings mean something personal. Your feelings might just be noise. You might just need some sleep.

· · ·

At this point you've seen how to work on becoming more emotionally intelligent about your experiences. Now let's turn to perceiving emotion intelligently in other people around you, and the subsequent benefits for your well-being.

My husband, Dan, went through a brief, difficult time a few decades ago, before we knew each other, and was referred to a psychiatrist. About thirty seconds into the first session, Dan knitted his brow and scowled, as he often does when he is concentrating, and the psychiatrist, trusting his perceptions as accurate, pronounced that Dan was "filled with pent-up anger." The thing is, Dan is one of the calmest people I know. When Dan assured the psychiatrist that he wasn't angry, the psychiatrist, confident in his ability to read his patients, insisted, "Yes, you are." Well, Dan was out the door before the second hand had completed its first revolution. He may well hold the world record for the shortest therapy session.

My point here isn't to knock the mental-health profession but to illustrate the false confidence that one's perceptions of other people's mental states are — or ever can be — "right." It comes from the classical view, which proposes that Dan broadcasts anger with a distinct fingerprint and the therapist detects it, even if Dan is unaware. If you want to gain mastery at perceiving other people's emotional experiences, you must let go of this essentialist assumption.

What happened during Dan's minute in therapy? He constructed an experience of concentration, and the therapist constructed a perception of anger. Both constructions were real, not in the objective sense but in the social sense. Perceptions of emotion are guesses, and they're "correct" only when they match the other person's experience; that is, both people agree on which concept to apply. Anytime you think you know how someone else feels, your confidence has nothing to do with actual knowledge. You're just having a moment of affective realism.[45]

To improve at emotion perception, we must all give up the fiction that we *know* how other people feel. When you and a friend disagree about feelings, don't assume that your friend is wrong like Dan's ex-therapist did. Instead think, "We have a disagreement," and engage your curiosity to learn your friend's perspective. Being curious about your friend's experience is more important than being right.

So, if our perceptions are just guesses, how do we ever communicate with each other? If you tell me that you're proud of your child's accomplishments in school, and "Pride" is a population of diverse instances with no consistent fingerprint, how can I know which "Pride" you mean? (This question doesn't arise in the classical view, where pride has a distinct essence; you simply broadcast pride and I recognize it.) You and I communicate emotion, in the face of huge variability, by way of the brain's predictive machinery. Your emotions are guided by your predictions. And as I observe you, the emotions I perceive are guided by my predictions. Emotional communication happens, therefore, when you and I predict and categorize *in synchrony.*[46]

Scientists and bartenders know that people synchronize in various ways when they communicate, especially if they like or trust each other. I nod, then you nod. You touch my arm and a moment later I touch yours. Our nonverbal behaviors coordinate. There's also biological synchrony; a mother's and child's heart rates will synchronize if they are securely bonded, and the same can happen to anyone during an engaging conversation. The mechanism is still a mystery. I suspect it's because their breathing synchronizes as they unconsciously observe each other's chests rising and falling. When I was a training therapist, I learned to intentionally synchronize my breathing with my clients' to prepare them for hypnosis.[47]

We likewise synchronize our concepts for emotion. My emotions are guided by my predictions. And as you observe me, the emotions you perceive are guided by your predictions. The sound of my voice and the motions of my body, as they are perceived by your brain, either confirm your predictions or become prediction error for you.

Suppose you tell me, "My son got the lead in the school play. I'm so proud." Your words and actions launch a population of predictions in my brain, helping to coordinate a shared concept of "Pride" between us in the moment. My brain computes probabilities based on past experience and winnows down its predictions to a winning instance, perhaps leading me to say, "Congratulations." Then the process repeats in the other direction as you perceive me. We'll be more in sync if we share a cultural background or other past experiences, and if we agree that certain facial configurations, body movements, vocal acoustics, and other cues have certain meanings in certain contexts. Little by little, we co-construct an emotional experience that we both identify with the word "proud."

In this scenario, our concepts don't need to match exactly for me to understand how you feel; they just must have reasonably compatible goals. On the other hand, if I construct an instance of the unpleasant kind of pride, in which you're arrogant and dismissive, I might obtusely fail to comprehend what you are saying, because you've used a concept that does not match mine in that instance. Note that our mutual construction is a continuous process with both brains in constant activity, even though I'm portraying it here as a simple back-and-forth sequence of events.

The co-construction of experience also allows us to regulate each other's body budgets; this is one of the great benefits that we get from living in groups. All members of a social species regulate each other's body budgets — even bees, ants, and cockroaches. But we are the only species who can do so by teaching each other purely mental concepts, and then using them in synchrony. Our words allow us to enter each other's affective niches, even at extremely long distances. You can regulate your friend's body budget (and he yours) even if you are an ocean apart — by phone or email or even just by thinking about one another.[48]

Your choice of words has a huge impact on this process, as those words shape other people's predictions. Parents who ask a child, "Are you upset?" instead of the more general question, "How are you feeling?" are influencing the answer, co-constructing emotion and honing the child's concepts toward being upset. Doctors who ask a patient, "Are you feeling depressed?" likewise make a positive response more likely than if they'd said, "Tell me how you've been." These are leading questions, the same sort that attorneys utilize (and object to) with witnesses on the stand. In everyday life, as in the courtroom, you need to be mindful of influencing people's predictions by your words.

Likewise, if you want someone else to know what you're feeling, you need to transmit clear cues for the other person to predict effectively and for synchrony to occur. In the classical view of emotion, the responsibility is all on the perceiver's end because emotions are supposedly displayed universally. In a construction mindset, you also bear the responsibility to be a good sender.[49]

• • •

Suppose you hadn't read this book, and someone said to you, "Pssst! Wanna be the master of your emotions? Then eat less junk food and learn lots of new words." I admit, it sounds unintuitive. But healthful eating leads to a body budget that is easier to balance and to more calibrated interoceptive predictions, and new words seed new concepts that are a basis for constructing emotional experiences and perceptions. Many things that seem unrelated to emotion actually have a profound impact on how you feel, because of the porous boundary between the social and the physical.

You are a remarkable animal who can create purely mental concepts that influence the state of your body. The social and the physical are intimately linked via your body and your brain, and your ability to move effectively between social and physical depends on a set of skills that you can learn. So grow your emotion concepts. Cultivate opportunities for your brain to wire itself to the realities of your social world. If you feel unpleasant in the moment, then deconstruct or recategorize your experiences. And realize that your perceptions of others are just guesses and not facts.

Some of these new skills are supremely difficult to cultivate. It's one thing for a scientist like me to tell you, "That's how the brain works." It's another thing entirely to up-end your whole lifestyle to take advantage of the science. Who has time to revamp their eating and sleeping habits and get more exercise, let alone learn new concepts, practice categorizing, and occasionally step back from the fiction of the self? We all have jobs and schoolwork and time constraints and all sorts of personal and home situations. Also, some of these suggestions require an investment of time or money, which might be in short supply for the people who could benefit most. But . . . everyone can find *something* they can try in this chapter, even if it's just taking walks or combining some emotion concepts before you go to sleep. Or giving up potato chips. (Okay, maybe not completely.)

Emotion concepts and body budgeting can improve your health and well-being, as you've just seen, but they can also be a catalyst for illness. Emotions are said to influence a variety of debilitating medical disorders like depression, anxiety, and unexplained chronic pain, as well as metabolic dysfunctions that lead to type-2 diabetes, heart disease, and even cancer. At the same time, new discoveries about the nervous system are dissolving the sacred boundary between what we think of as physical and mental illness, in the same way that the theory of constructed emotion blurs the boundary between the physical and the social. That is the next topic we'll visit.

10

Emotion and Illness

Think about the last time you had a cold. You probably had a runny nose, cough, fever, and other diverse symptoms. Most people attribute colds to a single cause, namely, a cold virus. And yet, when scientists place a cold virus into the noses of one hundred people, only 25–40 percent get sick. So a cold virus cannot be the essence of a cold — something more complex must be going on. The virus is necessary but not sufficient.[1]

The diverse set of symptoms that you collectively call "a cold" involves not just your body but also your mind. For example, if you are an introverted or negative-minded person, you're more likely to develop a cold from a noseful of germs.[2]

Our new view of human nature, inspired by the theory of constructed emotion, dissolves the boundaries between the mental and physical, including where illness is concerned. Old, essentialist thinking, in contrast, keeps those dividing lines sharp. Having a problem with your brain? Then see a neurologist. If the problem is with your mind, well, you need a psychiatrist. A more modern view integrates mind and brain and offers guidance on how better to understand human illness.

For example, if you look at the diverse symptoms found in illnesses like anxiety, depression, chronic pain, and chronic stress, they don't fit into a handful of neat compartments, like a silverware drawer. Each illness has tremendous variability, and all of their sets of symptoms have tremendous overlap. This situation should sound familiar. You've already learned that emotion categories like happiness and sadness have no essences; they're made by core systems in your body and brain, in the context of other bodies and brains. Now I'll suggest that some illnesses that seem distinct are like-wise constructions: human-made ways of carving up the same highly vari-able biological pie.

A construction approach to understanding illness can answer some per-plexing questions that have never been resolved. Why do so many disorders share the same symptoms? Why are so many people both anxious and de-pressed? Is chronic fatigue syndrome a distinct illness, or merely depression in disguise? Are people who suffer from chronic pain with no identifiable tissue damage mentally ill? And why do so many people with heart disease develop depression? If differently named illnesses are related to the same set of core causes, muddying the dividing lines between those illnesses, then such questions cease to be mysteries.

This is the most speculative chapter in the book, but it's informed by data, and I hope you'll find the ideas intriguing and provocative. In the pages that follow, I demonstrate that phenomena like pain and stress, and illnesses such as chronic pain, chronic stress, anxiety, and depression, are more in-tertwined than you might think, and they're constructed in the same man-ner as emotion. A key component of this viewpoint is a better understand-ing of the predictive brain and your body budget.

• • •

Your body budget fluctuates normally throughout the day, as your brain anticipates your body's needs and shifts around your budgetary resources like oxygen, glucose, salt, and water. When you digest food, your stomach and intestines "borrow" resources from your muscles. When you run, your muscles borrow from your liver and kidneys. During these transfers, your budget remains solvent.

Your body budget tilts out of balance when your brain estimates badly. This is a fairly normal occurrence. When something psychologically mean-ingful happens, like seeing your boss or coach or teacher walking toward you, your brain may predict unnecessarily that you need fuel, activating survival circuits that impact your budget. In general, these short-term im-balances are nothing to worry about, as long as you pay back your with-drawals by eating and sleeping.

When a budget imbalance becomes prolonged, however, your inter-nal dynamics change for the worse. Your brain mispredicts that your body needs energy over and over and over, driving your budget into the red. The effects of chronic misbudgeting can be devastating to your health and sum-mon your body's "debt collectors," which are part of your immune system.

Usually, your immune system is one of the good guys in your body, since it protects you from invaders and injury. It helps you by causing inflammation, like the swelling you get from banging your finger by accident with a hammer, or from a bee sting or an infection. The inflammation comes from little proteins called proinflammatory cytokines, which I mentioned briefly in the previous chapter. When you have an injury or illness, your cells secrete cytokines that draw blood to the affected region, raising its temperature and causing swelling.* These cytokines can make you feel fatigued and generally sick while they go about their job of helping you heal.

Proinflammatory cytokines can also become bad guys, however, given the right conditions for debt collection. This is particularly true when your body budget is chronically unbalanced, say, if you live in a dangerous neighborhood and hear gunfire every night. In such a harsh environment, your brain might regularly predict that you need more energy than your body requires. These predictions cause your body to release cortisol more often and in greater amounts than you need. Cortisol normally suppresses inflammation (that's why hydrocortisone cream relieves itching, and cortisone shots reduce swelling). When you have too much cortisol in your blood for a long time, inflammation flares up. You feel devoid of energy. You might run a fever. If someone placed a cold virus into your nose, you'd be one of the people who gets sick.[3]

Now a vicious cycle can ensue. When you feel fatigued due to inflammation, you don't move as much, in order to conserve (what your brain mistakenly believes to be) your limited energy resources. You start eating and sleeping poorly and neglect exercise, which throws your budget out of balance even more, and you start to feel seriously like crap. You might gain weight, which enhances your problems because certain fat cells actually produce the proinflammatory cytokines that make inflammation worse. You might also start avoiding other people, who then cannot help balance your body budget, and people with fewer social connections also have more proinflammatory cytokines and might even get sick more often.[4]

About ten years ago, scientists discovered — to their astonishment — that proinflammatory cytokines can cross from the body into the brain. We also now know that the brain has its own inflammatory system with cells that

---

* Not all types of inflammation involve cytokines, and not all cytokines cause inflammation. We're concerned only with chronic inflammation, which is caused by proinflammatory cytokines. For simplicity I just say "cytokines."

secrete these cytokines. These little proteins, with their capacity to induce feelings of such misery, reshape the brain. Inflammation in the brain causes changes in brain structure, particularly within your interoceptive network; it interferes with neural connections, and even kills neurons. Chronic inflammation can also make it harder for you to pay attention and remember things, lowering performance on IQ tests.[5]

So consider what happens if you're in a stressful social situation, like when a clique of coworkers suddenly stops inviting you to join them at lunch, or when friends read your text messages but don't answer. As per normal, your brain predicts you need fuel that your body doesn't require, temporarily impacting your budget. But what if the social situation doesn't resolve quickly? What if this social rejection is your life every day? Your body stays on alert, flush with cortisol and cytokines. Now your brain starts treating your body as if it were sick or damaged, and chronic inflammation sets in.[6]

Inflammation in your brain is very bad. It affects your predictions, in particular those that manage your body budget, sending your budget into overdraft. Remember that your body-budgeting circuitry is hard of hearing — it can be mostly deaf to corrections from your body. Inflammation moves the needle toward "completely deaf." Your body-budgeting regions become insensitive to your situation, making it more likely that your budget will remain overdrawn. You can become consumed with fatigue and unpleasant feelings. The chronic misbudgeting depletes your resources, causes wear and tear on your body, and eventually builds up more proinflammatory cytokines. When that happens, you are really, truly in trouble.[7]

A chronically imbalanced body budget acts like fertilizer for disease. In the last twenty years, it has become clear that the immune system is an ingredient in far more illnesses than you might expect, including diabetes, obesity, heart disease, depression, insomnia, reduced memory, and other "cognitive" functions related to premature aging and dementia. For example, if you already have cancer, inflammation makes tumors grow faster. The cancer cells also become more likely to survive the perilous journey through the bloodstream to infect other sites in the body, a process called metastasis. Death from cancer comes sooner.[8]

Inflammation has been a game-changer for our understanding of mental illness. For many years, scientists and clinicians held a classical view of mental illnesses like chronic stress, chronic pain, anxiety, and depression. Each ailment was believed to have a biological fingerprint that distinguished it from all others. Researchers would ask essentialist questions that assume each disorder is distinct: "How does depression impact your body? How does emotion influence pain? Why do anxiety and depression frequently co-occur?"[9]

More recently, the dividing lines between these illnesses have been evaporating. People who are diagnosed with the same-named disorder may have greatly diverse symptoms — variation is the norm. At the same time, different disorders overlap: they share symptoms, they cause atrophy in the same brain regions, their sufferers exhibit low emotional granularity, and some of the same medications are prescribed as effective.

As a result of these findings, researchers are moving away from a classical view of different illnesses with distinct essences. They instead focus on a set of common ingredients that leave people vulnerable to these various disorders, such as genetic factors, insomnia, and damage to the interoceptive network or key hubs in the brain (chapter 6). If these areas become damaged, the brain is in big trouble: depression, panic disorder, schizophrenia, autism, dyslexia, chronic pain, dementia, Parkinson's disease, and attention deficit hyperactivity disorder are all associated with hub damage.[10]

My view is that some major illnesses considered distinct and "mental" are all rooted in a chronically unbalanced body budget and unbridled inflammation. We categorize and name them as different disorders, based on context, much like we categorize and name the same bodily changes as different emotions. If I'm correct, then questions like, "Why do anxiety and depression frequently co-occur?" are no longer mysteries because, like emotions, these illnesses do not have firm boundaries in nature. I present more justification for this view as we discuss the details of stress, pain, depression, and anxiety.

· · ·

Let's begin with stress. You might think that stress is something that happens to you, like when you try to juggle five tasks at once, or your boss tells you that tomorrow's work was due yesterday, or you lose a loved one. But stress doesn't come from the outside world. You construct it.

Some stress is positive, like the challenge of learning a new subject in school. Some is negative but tolerable, like having a fight with your best friend. And some is toxic, like the chronic stress of prolonged poverty, abuse, or loneliness. In other words, stress is a population of diverse instances. It is a concept, just like "Happiness" or "Fear," that you apply to construct experiences from an imbalanced body budget.[11]

You construct instances of "Stress" via the same brain mechanisms that construct emotion. In each case, your brain issues predictions about your body budget in relation to the outside world and makes meaning. These predictions issue from your interoceptive network and descend along the same pathways from the brain to the body. In the opposite direction, the ascending pathways that carry sensory inputs from the body to the brain are also the same for stress and emotion. And the same pair of networks, interoceptive and control, play their same roles. (Emotion and stress researchers rarely recognize these similarities, and tend to ask how stress influences emotion and vice versa, as if stress and emotion are independent.) From the viewpoint of construction, what differs is the end result, whether your brain categorizes your sensations as stressful or emotional.[12]

Why does the predicting brain construct instances of stress or emotion in a given situation? No one knows. Maybe the longer your body budget is out of whack, the more likely you are to categorize with the concept "Stress," but this is pure speculation.

If your body budget is unbalanced for a long time, you may experience chronic stress. (Chronic misbudgeting is often diagnosed as stress, which is why people think stress causes illness.) Chronic stress is dangerous to your physical health. It literally eats away at your interoceptive and control networks, causing them to atrophy, as your chronically imbalanced body budget remodels the very brain circuitry that regulates the budget. So much for the classical division between mental and physical illness.[13]

Scientists are still figuring out the puzzle of immune system, stress, and emotion, but we do know a few things right now. Cumulative imbalance in the body budget — say, from growing up in adversity, where you don't feel safe or are deprived of basic necessities like nutritious food, quiet time to sleep, and so on — also changes the structure of your interoceptive network, rewiring your brain and reducing its ability to accurately regulate your body budget. All it takes are a couple of highly negative experiences for children to feel like they are living in a combat zone, reducing the size of their body-budgeting regions by the time they reach adulthood. Growing up in a family that is harsh or chaotic, with a lot of conflict or verbal criticism, increases inflammation in adolescent girls and places kids on a trajectory toward chronic disease; it's almost as bad for the development of these networks as childhood abuse or neglect. Ditto for suffering as the target of a bully. Kids who were bullied as children show low-grade inflammation that persists into adulthood, which predisposes them to a host of psychiatric and physical diseases. These are the myriad ways that an imbalanced body budget sculpts your brain, translating into a higher lifetime risk of heart disease, arthritis, diabetes, cancer, and other diseases.[14]

On the positive side, the link between emotion and stress suggests that you can reduce inflammation by applying techniques from the previous chapter. More emotionally intelligent people with cancer, for example, appear to have lower levels of proinflammatory cytokines. In studies, when patients said that they frequently categorize, label, and understand their emotions, they were less likely to have increased cytokines during recovery from prostate cancer, or after a stressful event, and the highest levels of circulating cytokines were found in men who expressed a lot of affect that they didn't label. Female breast cancer survivors who explicitly label and understand their emotions also have better health and fewer medical visits for cancer-related symptoms. This means that over time, people who effectively categorize their interoceptive sensations as emotion might be better protected against chronic inflammatory processes that lead to poor health.[15]

• • •

Pain, like stress and emotion, is a word that describes a population of diverse experiences — the ache of a twisted ankle, the steady pounding of a headache, the irritation of a mosquito bite, and, of course, the agony of pushing a thirty-five-centimeter head through a ten-centimeter cervix.

You might think that when your body is harmed, information simply radiates from the afflicted area to your brain, leading you to swear loudly and reach for the ibuprofen and bandages. It's true that your nervous system sends sensory input to your brain when your muscles or joints are injured, or your body tissues are damaged by excessive heat or cold, or in response to a chemical irritation like a pinch of pepper in your eye. This process is called *nociception*. And in the past, scientists believed that your brain simply received and represented nociceptive sensations and, *voilà*, you experience pain.

But the inner workings of pain are more complex in a predictive brain. Pain is an *experience* that occurs not only from physical damage but also when your brain predicts damage is imminent. If nociception works by prediction, as does every other sensory system in the brain, then you construct instances of pain out of more basic parts using your concept of "Pain."[16]

The way I see it, pain is constructed in the same way that emotions are made. Suppose you're at your doctor's office receiving a tetanus shot. Your brain constructs an instance of "Pain" by issuing predictions about the needle piercing your skin, since you have prior experience with shots. You might feel the pain even before the needle touches your arm. Your predictions are then corrected by actual nociceptive input from the body — the injection occurs — and once any prediction errors are dealt with, you have categorized the nociception sensations and made them meaningful. The pain you experience as coming from the shot is really in your brain.[17]

My prediction-based explanation of pain is backed up by a couple of observations. When you are expecting pain, like the moment just before an injection, your brain regions that process nociception change their activity. That is, you simulate pain and therefore feel it. This phenomenon is called the nocebo effect. You're probably more familiar with its counterpart, the placebo effect, which relieves pain using a medically ineffective treatment like a sugar pill. If you believe you'll feel less pain, your beliefs influence your predictions and tune down your nociceptive input so you do feel less pain. Both placebos and nocebos involve chemical changes in the brain regions that process nociception. These chemicals include opioids that relieve pain and work similarly to morphine, codeine, heroin, and other opiate drugs. Opioids increase during placebo and turn down nociception, and likewise decrease during nocebo effects, earning them the moniker of "your internal medicine cabinet."[18]

I watched my daughter experience the nocebo effect when she was a baby and had thirteen ear infections in nine months. The first time we visited the pediatrician's office for treatment, she wailed in discomfort as he peered into her ears (though he is a caring and careful physician). The second time, she cried in the waiting room. The third time, she began sobbing in the building lobby, and the fourth time, as we entered the parking garage. After that, she would whimper anytime we passed the street where the doctor's office was located. This is the predicting brain in action; little Sophia was likely simulating ear pain. It took many months, after Sophia was past the infections and well into toddlerhood, for her to stop asking, "Go to dottor? Kekk Sophie's ears?" whenever we were in the vicinity.

Pain, like emotion and stress, appears to be a whole-brain construction. It involves our familiar pair of networks, the interoceptive and control networks. And the similarities don't stop here. The pathways sending nociceptive predictions down to the body, and those bringing nociceptive input up to the brain, are closely related to interoception. (It's even possible that nociception *is* a form of interoception.) Overall, the body sensations that are categorized as pain, stress, and emotions are fundamentally the same, even at the level of neurons in the brain and spinal cord.* Distinguishing be-tween pain, stress, and emotion is a form of emotional granularity.[19]

It's easy to show that interoception and nociception are in bed with each other. If I made you feel unpleasant affect in my lab while applying pain-ful heat to your arm, you'd report feeling more pain. This happens because your body-budgeting regions issue predictions that can dial pain up and down like a volume control. Those predictions can influence your brain's simulation of pain, and they also reach down to your body and can amplify or dampen its status reports to your brain. Your body-budgeting regions can therefore trick your brain into believing that there is tissue damage, regardless of what is happening in your body. So, when you're feeling un-pleasant, your joints and muscles might hurt more, or you could develop a stomachache. When your body budget's not in shape, meaning your intero-ceptive predictions are miscalibrated, your back might hurt more, or your headache might pound harder — not because you have tissue damage but because your nerves are talking back and forth. This is not imaginary pain. It is real.[20]

When people experience ongoing pain without any damage to their body tissue, it's called *chronic pain*. A few well-known examples are fibromyalgia, migraine headaches, and chronic back pain. Over 1.5 billion people suffer from chronic pain, including 100 million in the United States who collec-tively pay $500 billion per year for treatment. When you include lost pro-ductivity in the price tag, pain costs the United States $635 billion each year. It is also frustratingly hard to treat, as the currently prescribed pain medi-cations, analgesics, are ineffective more than half the time. This worldwide epidemic of chronic pain is one of today's great medical mysteries.[21]

How and why do so many people experience ongoing pain when their bodies appear to have no physical damage? To answer that question, think about what would happen if your brain issued unnecessary predictions of pain and then ignored prediction error to the contrary. You would genu-inely experience pain for no discernable reason. This is much like your ex-perience when the blobby picture in chapter 2 became a bee, as you gen-uinely perceived lines that didn't exist. Your brain ignored sensory input, maintaining that its predictions are reality. Apply this example to pain and

---

* For the sake of this discussion, I will continue referring to interoception and nocicep-tion separately.

the result is a plausible model of chronic pain: errant predictions without correction.

Scientists now consider chronic pain to be a brain disease with its roots in inflammation. It's possible that the brain of a chronic pain sufferer received intense nociceptive input sometime in the past, and as the injury healed, the brain didn't get the memo. It keeps predicting and categorizing anyway, generating chronic pain. It's also possible that predictions about inner-body movements are turning up the volume for nociceptive input as it heads from the body to the brain.[22]

If you're unlucky enough to suffer from chronic pain, then you've probably faced skeptics who don't understand what you're going through. They try to explain away your pain by saying, "It's in your head," by which they mean, "You have no tissue damage, so go see a psychiatrist." I'm saying that you're not crazy. There *is* something wrong with you. Your predictive brain, which is indeed located "in your head," is *generating authentic pain* that continues past the point when your body has already healed. It's similar to phantom limb syndrome, when an amputee can still feel his missing arm or leg because his brain keeps issuing predictions about it.[23]

We already have intriguing evidence that some types of chronic pain work by prediction. Animals who have stress or injury early in life become more likely to develop persistent pain. Human infants who have surgery are more likely to have heightened pain in later childhood. (Incredibly, infants prior to the 1980s were routinely *not anesthetized* during major surgery, on the belief that they couldn't feel pain!) There's also a medical condition called complex regional pain syndrome, in which pain from an injury spreads inexplicably to other areas of the body, which appears to be linked to bad nociceptive predictions.[24]

So "Pain," like "Stress," is another concept with which you make meaning of physical sensations. You could characterize pain and stress as emotions, or even emotion and stress as types of pain. I'm not saying that instances of emotion and pain are indistinguishable in the brain, but neither has a fingerprint. If I scan your brain while you're having a toothache and when you're angry, the scans will look somewhat different. But then, if I scan your brain during different instances of anger, they look somewhat different too. Different instances of dental pain likely vary as well. This is degeneracy; variation is the norm.[25]

Emotion, acute pain, chronic pain, and stress are constructed in the same networks, the same neural pathways to and from the body, and most likely the same primary sensory region of cortex, so it is completely plausible that we distinguish emotion and pain by concept — that is, via the concepts the brain applies to make sense of bodily sensations. Chronic pain is likely a misapplication of the concept "Pain" by your brain, as it constructs the experience of pain without injury or threat to your tissue. Chronic pain seems to be a tragic case of predicting poorly *and* receiving misleading data from your body.[26]

• • •

Keeping in mind what you've just learned about chronic stress and chronic pain, let's turn our attention to depression, which is another debilitating condition that can overwhelm a life. Also known as major depressive disorder, depression is far beyond the everyday distress that people feel when they groan, "I'm like sooo depressed." Marvin the Paranoid Android, in Douglas Adams's *The Hitchhiker's Guide to the Galaxy*, was truly depressed. Sometimes he was so despondent about life that he shut himself down. A major depressive episode is similarly incapacitating. "The pain of severe depression is quite unimaginable to those who have not suffered it," recalled the novelist William Styron in his memoir, "and it kills in many instances because its anguish can no longer be borne."[27]

To many scientists and physicians, depression remains a disease of the mind. It's classified as a disorder of affect and often blamed on negative thinking: You're too hard on yourself, or have too many self-defeating, catastrophic thoughts. Or perhaps traumatic events trigger depression, particularly if your genes make you vulnerable. Or maybe you don't regulate your emotions well, making you too responsive to negative events and too unresponsive to positive ones. All of these explanations assume that thinking controls feeling — the old "triune brain" idea. Change your thoughts or regulate your emotions better, the logic goes, and depression will lift. The mantra seems to be: "Don't worry, be happy; and if that doesn't work, try antidepressants."[28]

Twenty-seven million Americans take daily antidepressants, yet more than 70 percent continue to experience symptoms anyway, and psychotherapy is not effective for everyone either. Often the symptoms begin in adolescence to early adulthood and then recur throughout life. The World Health Organization projects that by 2030, depression will cause more premature deaths and years of disability than cancer, stroke, heart disease, war, or accidents. Those are pretty dreadful outcomes for a "mental" illness.[29]

A lot of research seeks to find the universal genetic or neural essence of depression. But most likely, depression is not just one thing. Depression is — you guessed it — a concept. It is a population of diverse instances, so there are many degenerate paths to depression, many of which begin with an imbalanced body budget. If depression is a disorder of affect, and affect is an integrated summary of how your body budget is doing (answer: pretty poorly), then depression may actually be a disorder of misbudgeting and prediction.[30]

We know that your brain continually predicts your body's energy needs based on past experience. Under normal circumstances, your brain also corrects its predictions based on actual sensory information from your body. But what if this correction wasn't working properly? Your momentary experience would be constructed from the past but *not corrected by the present.* In general terms, that's what I think is happening in depression. Your brain is continually mispredicting your metabolic needs. Your body and brain therefore act as if you were fighting off an infection or healing from a wound when none exists, as in chronic stress or pain. As a result, your affect is out of whack: you experience debilitating misery, fatigue, or other symptoms of depression. Simultaneously, your body is quickly metabolizing unnecessary glucose to meet those high yet nonexistent energy needs, leading to weight problems and leaving you at risk for other metabolic-related illnesses that co-occur with depression, including diabetes, heart disease, and cancer.[31]

The traditional view of depression is that negative thoughts cause negative feelings. I'm suggesting it's the other way around. Your feelings right now drive your next thought, as well as your perceptions, as predictions. So a depressed brain relentlessly keeps making withdrawals from the budget, basing its predictions on similar withdrawals from the past. This means constantly reliving difficult, unpleasant events. You wind up in a cycle of budgeting imbalances, unbroken by prediction error because it is ignored, gets tuned down, or doesn't make it to the brain. In effect, you're locked into a cycle of uncorrected predictions, trapped in an adverse past when your metabolic needs were high.

A depressed brain is effectively locked into misery. It's like a brain in chronic pain, ignoring prediction error, but on a much larger scale that shuts you down. It puts your budget chronically in debt, so your brain tries to cut spending. What's the most efficient way to do that? Stop moving and don't pay attention to the world (prediction error). That is the unrelenting fatigue of depression.[32]

If depression is a disorder caused by chronic misbudgeting, then it's not, strictly speaking, exclusively a psychiatric disease. It's also a neurological, metabolic, and immunologic disease. Depression is an imbalance of many entwined parts of the nervous system that we can understand only by treating the whole person, not by treating one system in isolation like the parts of a machine. The tipping point into a major depressive episode can come from many different sources. You could suffer prolonged stress or abuse, particularly in childhood, leaving you carrying around a model of the world built from toxic past experiences. You could have physical conditions like chronic heart disease or insomnia that lead to bad interoceptive predictions. Your genes could leave you sensitive to your environment and every little problem. Also, if you're a woman of reproductive age, the connectivity within your interoceptive network changes throughout the month, leaving you more vulnerable, at certain points in your cycle, to unpleasant affect, rumination, and perhaps even increased risk of mood disorders such as depression and post-traumatic stress disorder. "Thinking positive thoughts" or taking antidepressants might not be enough to bring your body budget back into balance: other lifestyle changes or system adjustments might be necessary.[33]

The theory of constructed emotion suggests that we can treat depression by breaking the cycle of misbudgeting, that is, by changing interoceptive predictions to be more in line with what's going on around you. Scientists have found evidence that this is the case. As treatments like antidepressants and cognitive behavioral therapy start to work and you feel less depressed, your activity in a key body-budgeting region returns to normal levels, and connectivity in your interoceptive network is restored. These changes are consistent with the idea of reducing the excessive predictions. We might also treat depression by letting in more prediction error, say, by asking people to keep a diary of their positive experiences, which can ease the drain on the body budget. The problem, of course, is that no treatment works for everyone, and there are some people for whom no treatments work.[34]

One of the most promising avenues for treatment I've seen is the groundbreaking work of neurologist Helen S. Mayberg (chapter 4), who electrically stimulates the brains of unrelentingly depressed patients. Her technique instantly relieves the agony of depression, if only while the current is on, as the patient's brain shifts from all-consuming internal focus to the external world, so it can predict and process prediction error normally. Let's hope that these preliminary yet encouraging results will ultimately lead scientists to a more lasting treatment for depression. At the very least, these results should help spread the word that depression is a brain disease and not just a shortage of happy thoughts.

• • •

Anxiety is a condition that seems very different from chronic pain and depression. When you're anxious, you feel worried or worked up, like you don't know what to do with yourself, and generally miserable. This is a stark contrast with depression, in which you feel sluggish, like you can't go on with life, and also generally miserable, and with chronic pain, which is, well, painful.

So far, we've learned that emotion, chronic pain, chronic stress, and depression all involve the interoceptive and control networks. Those same networks are critical to anxiety as well. Anxiety is still a puzzle being unraveled,* but one thing seems certain: it is yet another disorder of prediction and prediction error across these two networks. The neural pathways studied in anxiety for prediction and prediction error are also the same ones as for emotion, pain, stress, and depression.[35]

Traditional research on anxiety disorders is founded on the old "triune brain" model, that cognition controls emotion. Your allegedly emotional amygdala is overactive, they say, and your so-called rational prefrontal cortex is failing to regulate it. This approach is still influential, even though the amygdala is not the home of any emotion, the prefrontal cortex does not house cognition, and emotion and cognition are whole-brain constructions that cannot regulate each other. So, how is anxiety made? We don't know all the details yet, but we have some tantalizing clues.[36]

I speculate that an anxious brain, in a sense, is the opposite of a depressed brain. In depression, prediction is dialed way up and prediction error way down, so you're locked into the past. In anxiety, the metaphorical dial is stuck on allowing too much prediction error from the world, and too many predictions are unsuccessful. With insufficient prediction, you don't know what's coming around the next corner, and life contains a lot of corners. That's classic anxiety.[37]

---

* In this chapter, I discuss all anxiety disorders as a group (unless otherwise indicated), because it's well known that these disorders have common causes. For many years, a variety of anxiety disorders were presumed to be biologically distinct, but (as you should not be surprised to learn by now) there is a lot of overlap in their symptom profiles, making it challenging to study one disorder in the absence of the others.

Anxiety sufferers, for whatever reason, have weakened connections between several key hubs in the interoceptive network, including the amygdala. Some of these hubs also happen to sit in the control network. These weakened connections likely translate into an anxious brain that is clumsy at crafting predictions to match the immediate circumstances, and that fails to learn effectively from experience. You might predict threats needlessly, or create uncertainty by predicting imprecisely or not at all. In addition, your interoceptive inputs become even more noisy than usual when your body budget has been in the red for a while; as a consequence, your brain ignores them. These situations leave you open to a lot of uncertainty and a lot of prediction error that you can't resolve. And uncertainty is more unpleasant and arousing than assured harm, because if the future is a mystery, you can't prepare for it. For example, when people are seriously ill but have an excellent chance of recovery, they are less satisfied with life than people who know their disease is permanent.[38]

Based on the evidence, it appears that anxiety, like depression, is a constructed category in the same fashion as emotion, pain, and stress. The misery you feel in anxiety and depression tells you that something is seriously wrong with your body budget. Either your brain is trying to secure a deposit, ramping up unpleasant affect, or it's attempting to reduce your need for the deposit by remaining still, resulting in fatigue. Your brain may categorize these sensations as anxiety, depression, or, for that matter, pain or stress or emotion.

To be clear, I am not saying that major depressive disorder and anxiety disorders are interchangeable. I'm suggesting that every category of mental illness is a diverse population of instances, and certain collections of symptoms could reasonably be categorized equally well as an anxiety disorder or as depression. There's also the issue of severity — some of Helen Mayberg's severely depressed patients, such as those who are near-catatonic, would clearly not be diagnosed with an anxiety disorder. However, some of her other patients who are in agony might reasonably be diagnosed with anxiety, chronic stress, or even chronic pain. In general, moderately severe depression and anxiety can have overlapping symptom profiles with one another, and with chronic stress and chronic pain, and also with chronic fatigue syndrome.[39]

These observations provide a solution to the mystery that opened chapter 1: why did test subjects in my graduate school experiments seem unable to distinguish between anxious and depressed feelings? One reason we've covered already is emotional granularity: some of my subjects could prob-ably construct more finely tailored emotions than others could. But now a second reason comes to light: that "Anxiety" and "Depression" are concepts for categorizing similar sensations.

When my subjects were feeling unpleasant, I handed them rating scales to report their feeling, but only in terms of anxiety and depression. Peo-ple will use whatever measure you give them to describe how they feel. If someone feels crappy and you give her only an anxiety scale, she'll report her feelings using words for anxiety. She might even come to feel anxious as the words prime her to simulate an instance of "Anxiety." Alternatively if you hand her a depression scale, she'll report her feeling using words for depression and might likewise end up feeling depressed. This would ex-plain my mysterious results. Concepts like "Anxiety" and "Depression" are highly variable and malleable. Words on questionnaires can influence peo-ple's categorizations, just like the basic emotion method influences percep-tions with its list of emotion words.[40]

I encountered something similar in a physician's office not long ago. I'd been feeling fatigued for some time and had gained some weight, and the doctor asked, "Are you depressed?" I responded, "Well, I don't have sad feel-ings, but I do feel dead tired much of the time." He countered with, "Maybe you're depressed and you don't know it." My doctor did not realize that un-pleasant affect can have a physical cause, which in my case was probably lack of sleep from running a lab of a hundred people, staying up late work-ing on this book, and being a mother to my teenage daughter, plus a little thing called menopause. (I wound up explaining interoception and body budgets to him.) But here's the thing: If he had simply diagnosed me with depression, he could have actually cultivated a feeling of depression in me in that instant. Sure, I was fatigued, and I probably had some inflamma-tion going on due to a bit of chronic stress. If I hadn't resisted, I could have come away with a prescription for antidepressants and a belief that some-thing was seriously wrong with my life or myself for being unable to cope. This belief might have worsened my miscalibrated body budget, if I started to search for problems in my life . . . and you can always find something if you look. Instead, my doctor and I uncovered a body-budgeting issue and looked for ways to repair it. My doctor didn't realize it, but he was co-con-structing my experience. He wanted to construct one social reality, and I had another.

• • •

When prediction error from the world dominates prediction, you can have anxiety. Suppose you couldn't predict *at all, ever*. What would happen?

For starters, your body budget would be screwed up because you couldn't predict your metabolic needs. You'd have difficulty integrating sensory input from vision, hearing, smell, interoception, nociception, and your other sensory systems into a cohesive whole. You'd therefore have impaired statistical learning, making it difficult for you to learn basic concepts, even to recognize the same person from different angles. Many things would be outside your affective niche. If you were an infant in that situation, you'd most likely be disinterested in other humans; you'd stop looking at the faces of your caregivers, making it harder for them to regulate your highly disrupted body budget, breaking a crucial bond. You would also have trouble learning purely mental concepts of social reality because they're learned with words, but you're disinterested in humans so you probably have difficulty learning language. You'd never grow a proper conceptual system.

In the end, you'd exist in a constant stream of ambiguous sensory input with few concepts to help you make sense of it. You'd be anxious all the time because sensations are unpredictable. In effect, you'd have a total breakdown of interoception, concepts, and social reality. In order to learn at all, you'd need your sensory input to be very consistent, even stereotyped, with as little variation as possible. I don't know about you, but to me, this collection of symptoms sounds just like autism.[41]

Clearly, autism is an incredibly complex condition and a gigantic area of research, and it can't be summed up in a handful of paragraphs. Autism is also hugely variable, a term applied to a wide spectrum of symptoms that probably have multiple, complex causes. All I'm saying is: the possibility is intriguing that autism is a disorder of prediction.[42]

People with autism who can describe their experiences say things consistent with the idea. Temple Grandin, one of the most famous and outspoken individuals with autism, writes clearly about her lack of prediction and her overwhelming prediction error. "Sudden loud noises hurt my ears like a dentist's drill hitting a nerve," she writes in "An Inside View of Autism." Grandin eloquently describes how she struggled to form concepts: "When I was a child, I categorized dogs from cats by sorting the animals by size. All the dogs in our neighborhood were large until our neighbors got a Dachshund. I remember looking at the small dog and trying to figure out why she was not a cat." Naoki Higashida, a thirteen-year-old boy with autism who wrote *The Reason I Jump*, notes his efforts to categorize: "First, I scan my memory to find an experience closest to what's happening now. When I've found a good close match, my next step is to try to recall what I said the last time. If I'm lucky, I hit upon a usable experience and all is well." In other words, lacking a properly functioning conceptual system, Higashida has to work hard to do what other brains do automatically.[43]

Other researchers too are now speculating that autism is a failure of prediction. Some believe that autism is primarily caused by a dysfunction of the control network, producing a model of the world that is too specific to each situation. Others see the problem as a deficit in the neurochemical called oxytocin, leading to problems in the interoceptive network. I suspect that there isn't just one network problem in autism but a menu of different possibilities, owing to degeneracy. In fact, autism is characterized as a neurodevelopmental disorder that is extremely variable in its genetics, neurobiology, and symptoms. I speculate that the problems begin with body-budgeting circuitry because it's present at birth, and all statistical learning is grounded in body-budget regulation (chapters 4 and 5). Alterations in the circuitry will change the trajectory of brain development. Without a fully loaded predictive brain, you'd be at the mercy of your environment. You'd have a brain driven by stimulus and response, when the nervous system is optimized for a more metabolically efficient brain organization. That might explain the experiences of people with autism.[44]

• • •

You've now seen that several notable and serious disorders may all be related to your immune system, which links your mental and physical health within your predicting brain. When bad predictions go unchecked, they may lead to a chronically unbalanced body budget, which contributes to inflammation in the brain and corrupts your interoceptive predictions even further in a vicious cycle. In this manner, the same systems that construct emotion also can contribute to illness.

I'm not saying that body-budget debt is the single cause of all mental illness. Nor am I suggesting that rebalancing the budget is the golden cure. I'm just saying that, thanks to our new view of human nature, we can understand that a body budget is a common factor in diseases that are traditionally considered separable.

When you have too much prediction and not enough correction, you feel bad, and the flavor of badness depends on the concepts you use. In small amounts, you might feel angry or shameful. In extreme amounts, you get chronic pain or depression. In contrast, too much sensory input and ineffective prediction yields anxiety, and in extreme amounts, you might develop an anxiety disorder. With no prediction at all, you'd have a condition comparable to autism.

All of these disorders appear to be rooted in misbudgeting. Now imagine with me, for a moment, the myriad ways that a young person can develop a budget that's chronically in overdraft. There's overt abuse and neglect, of course, but also an avalanche of smaller events. The steady stream of violence they witness on TV and in movies, videos, and computer games. The degrading language they hear in popular music and casually mimic as they greet peers with "Hey, bitch." (Is it a friendly hello, an insult, or a threat?) The rise of bullying as a form of joking because on television, people say horrible things to each other to the sound of a laugh track. Add to this the almost limitless opportunities for social rejection that texting and some forms of social media provide, combined with not enough sleep and exercise, plus too much pseudo-food of dubious nutritional quality, and you have a cultural recipe for a generation of adults with chronic body misbudgeting.[45]

Could the misery of chronic misbudgeting be one reason why the United States is in the midst of an opiate crisis? Your brain's natural opioids reduce pain because they regulate affect (not nociception), and opiate drugs mimic these effects—which might explain their widespread abuse. From 1997 to 2011, the number of U.S. adults who are addicted to prescription drugs increased by 900 percent. Many others have resorted to heroin, methamphetamines, and other street drugs that reduce distress. We also know that a significant portion of the population isn't sleeping enough, eating well, or exercising regularly. With opiate drugs, people are probably self-medicating the discomfort that stems from a chronically imbalanced body budget. They begin taking opiates for a variety of reasons, but they keep using and even abusing, I suspect, because they are regulating their out-of-whack affect to feel better. Their body budgets are too messed up for their brain's natural opioids to do their job.[46]

The wretchedness of chronic misbudgeting can also be temporarily reduced with food, which stimulates some of the same brain receptors that respond to opiate drugs. In experiments on rats, this stimulation leads the rats to binge on high-carbohydrate foods, even when they are not hungry. In people, eating sugar triggers the brain's opioids to increase production. So eating junk food or white bread actually *feels* good. No wonder I love a crusty French loaf. And sugar may actually act as a mild analgesic. So, when people talk about our society being addicted to sugar, they might not be far off. I wouldn't be surprised if people are employing high-carbohydrate food as a drug to manage their affect and feel better. Hello, obesity epidemic.[47]

A population of citizens with imbalanced body budgets doesn't just cost billions of dollars in health care. It costs people their well-being, their relationships, and even their lives. People who study these illnesses are beginning to set aside the essentialism that creates categories like "Anxiety" and "Depression" and "Chronic Pain," and looking to common underlying factors instead. If we could add interoception, body-budget balancing, and emotion concepts to the list of those common factors, I suspect we'd make more progress against these debilitating disorders. In the meantime, your own knowledge of these common factors may help you avoid illness and communicate more effectively with your doctors.[48]

We all walk a tightrope between the world and the mind, and between the natural and the social. Many phenomena that were once considered purely mental — depression, anxiety, stress, and chronic pain — can, in fact, be explained in biological terms. Other phenomena that were believed to be purely physical, like pain, are also mental concepts. To be an effective architect of your experience, you need to distinguish physical reality from social reality, and never mistake one for the other, while still understanding that the two are irrevocably entwined.

11

Emotion and the Law

Every society has rules for which emotions are acceptable, when they are acceptable, and how to express them. In my American culture, it's appropriate to feel grief when someone dies, and inappropriate to chuckle as the casket is lowered into the ground. A surprise party is a time to feel surprised and then joyful, and if you know about your own party in advance, it's appropriate to feign surprise when you arrive. Members of the Ilongot tribe in the Philippines may feel the emotion *liget* when acting as a team to behead an enemy, in celebration of a job well done.[1]

If you violate your culture's rules of social reality, punishment may follow. Laughter at a funeral may get you shunned. Failure to be surprised at your own party may yield disappointed guests. And most cultures no longer prize decapitation.

The ultimate rules for emotion in any society are set by its legal system.* That might seem like a surprising claim, but consider this. In the United States, if your accountant steals your life savings, or a banker sells you a bad mortgage, it's considered unacceptable to kill them; but if you murder your spouse in a fit of rage for cheating on you with a secret lover, the law might cut you some slack, especially if you're a man. It's unacceptable to make your neighbor feel fear that you will harm him bodily — that is considered a form of assault — but in some states it's okay for you to "stand your ground"

---

* My comments in this chapter are limited to the legal system in the United States, though they may be true for legal systems in other countries. All phrases like "the law" and "the legal system" refer to the United States.

and harm someone first, even if you kill the person. It's acceptable for you to profess romantic love, but not (at various times in U.S. history) toward people whose sex is the same as yours or whose skin color isn't. Violate these norms, and you might lose your money, your freedom, or your life.

For centuries, laws in the United States have been shaped by the classical view of emotion, steeped in the essentialist view of human nature. Judges, for example, attempt to set emotion aside to render a decision by pure reason, a belief that assumes emotion and reason are distinct entities. Violent defendants plead that they were hijacked by their anger, assuming that anger is one single, unitary cauldron that, when unconstrained by clear thought, bubbles over to unleash a torrent of aggression. Juries look for remorse in a defendant, as if remorse had a single, detectable expression in the face and body. Expert witnesses testify that a defendant's bad behavior was caused by one errant brain blob, an example of baseless blob-ology.

The law is a social contract that exists in a social world. Are you responsible for your actions? Yes, says the essentialist view of human nature, as long as you haven't been commandeered by your emotions. Are other people responsible for your actions? No, you are an individual with free will. How do you determine what a defendant is feeling? By detecting his or her emotions in expressions. How do you make a just, moral decision? By setting your emotions aside. What is the nature of harm? Physical harm, that is, tissue damage, is worse than emotional harm, which is considered to be separate from the body and less tangible. All of these assumptions — born of essentialism — are baked into the law at its deepest levels, driving verdicts of guilt and innocence and gauging punishments on a massive scale, even as neuroscience has been quietly debunking them as myths.[2]

Simply put, some people are punished undeservedly, and others escape punishment, based on an outdated theory of the mind that is rooted in belief rather than science. In this chapter, we'll explore some common myths about emotion in the legal system and ask whether a biologically richer theory of the mind, especially one that is grounded in realistic neuroscience, can improve society's pursuit of justice.

• • •

As every budding adolescent discovers, freedom is great. You can decide to stay out past midnight with your friends. You can decide not to do your homework. You can choose to eat cake for dinner. But as we all learn, choices come with consequences. The law is founded on the simple idea that you can choose to treat others well or badly. Choice bestows responsibility. If you treat others badly and consequently they suffer some harm, then you must be punished, particularly if you intended that harm. This is how society shows its respect for you as an individual. Your value as a human being, some legal scholars say, is rooted in the fact that you choose your actions and are responsible for them.[3]

If something interferes with your ability to choose your actions freely, the law says that you might be less responsible for the harm you caused. Take the case of Gordon Patterson, who caught his wife, Roberta, "in a state of semiundress" with her boyfriend, John Northrup. Patterson shot Northrup twice in the head, killing him. Patterson confessed to the shooting but argued that he was less culpable due to his "extreme emotional disturbance" at the time of the crime. According to U.S. law, Patterson's sudden burst of rage caused him not to be fully in control of his actions, and he was therefore found guilty of second-degree murder — rather than first-degree murder, which requires premeditation and carries a harsher punishment. In other words, rational killing is considered worse than emotional killing, all other circumstances being equal.[4]

The U.S. legal system assumes that emotions are part of our supposed animal nature and cause us to perform foolish and even violent acts, unless we control them with our rational thoughts. Centuries ago, legal minds decided that people, when provoked, sometimes kill because they haven't "cooled off" yet, and anger erupts unbidden. Anger steams, boils, explodes, and leaves a wake of destruction in its path. Anger makes people unable to conform their actions to the law, and so partially mitigates a person's responsibility for his actions. The argument is known as a *heat-of-passion* defense.[5]

The heat-of-passion defense depends on some familiar assumptions from the classical view of emotion. The first assumption is that there is one universal type of anger, with a specific fingerprint, that justifies such a defense to a charge of murder. It supposedly includes a flushed face, clenched jaw, flared nostrils, and increased heart rate, blood pressure, and perspiration. As you've already learned, this alleged fingerprint is merely a Western cultural stereotype that's not supported by data. On average, people's heart rates go up when angry, but there's tremendous variation, and similar increases are also part of the stereotypes for happiness, sadness, and fear. And yet, most killings are not committed in happiness or sadness; and if they were, the law does not consider these emotional episodes to be a mitigating factor.[6]

What's more, most instances of anger do not lead to killing. I can state quite definitively that in twenty years of creating anger in my lab, we've never seen a test subject kill anybody. We see a far greater repertoire of action: swearing, threatening, pounding the table, leaving the room, crying, trying to resolve whatever conflict they're having, or even smiling while wishing ill upon their oppressor. So the idea of anger as a trigger for uncontrolled murder is at best questionable.[7]

When I explain to people in the legal profession that anger has no biological fingerprint, they often assume I am claiming emotions don't exist. That's not at all the case. Of course anger exists. You just can't point to a spot in a defendant's brain, face, or EKG, and say, "Look, anger is right here," let alone draw legal conclusions.

The legal system's second assumption behind the heat-of-passion defense is that "cognitive control" in the brain is synonymous with rational thought, deliberate actions, and free will. For you to be considered culpable, it is not enough that you performed a harmful action (known by the legal term *actus reus*). You also had to mean it. You caused harm of your own free will with a guilty mind (*mens rea*). Emotions, on the other hand, are seen as rapid, automatically triggered reactions spewing from your ancient, inner beast. The human mind is considered a battleground for reason and emotion, so when you fail to exercise sufficient cognitive discipline, emotions are said to burst forth to hijack your behavior. They interfere with your choice of action, and therefore make you less culpable. This narrative of emotion as the primitive part of human nature, to be controlled by the more advanced and uniquely human rational parts, is the "triune brain" myth (chapter 4) whose roots go all the way back to Plato.

The distinction between emotion and cognition hinges on their alleged separation in the brain, with one regulating the other. Your emotional amygdala spies an open cash register, but then, as the story goes, you rationally consider your likelihood of jail time, which causes your prefrontal cortex to slam on the brakes and stop your arm from dipping into the drawer. But as you've learned by now, thinking and feeling are not distinct in the brain. Your desire for easy cash and your decision to pass it up are both constructed across your entire brain by interacting networks. Whenever you carry out an action — whether it feels automatic, like recognizing an object as a gun, or more deliberate, like aiming one—your brain is always a whirlwind of parallel predictions that compete with one another to determine your actions and your experience.

At different times, you have different experiences of agency. Emotion sometimes can feel uncontrollable, like a burst of anger that arrives without warning, but you can also act in anger with intent, methodically plotting someone's demise. In addition, non-emotions like memories or ideas can pop into your head unbidden. And yet we never hear of defendants who commit murder "in a fit of thinking."

You can even work yourself up deliberately into a frothing anger. Accused mass murderer Dylann Roof, who shot nine people in a Bible study meeting in South Carolina in June 2015, appeared to cultivate his anger toward African Americans deliberately for many months before the day he walked into that church. Roof said that he almost didn't go through with his plan because everyone was so nice to him, and he appeared to work himself up to the heinous deed in the meeting, uttering repeated phrases like "I have to do it" and "You have to go." So, overall, moments of emotion are not synonymous with moments that you're out of control.[8]

Anger is a population of diverse instances, not a single automatic reaction in the true sense of the phrase. The same holds for every other category of emotion, cognition, perception, and other type of mental event. It might seem like your brain has a quick, intuitive process and a slower, deliberative one, and that the former is more emotional and the latter more rational, but this idea is not defensible on neuroscience or behavioral grounds. Sometimes your control network plays a large role in the construction process, and other times its role is less, but it is always involved, and the latter times are not necessarily emotional.[9]

Why does the fiction of the two-system brain survive, beyond the usual reason of essentialism? Because most psychology experiments unwittingly perpetuate this fiction. In real life, your brain predicts nonstop, with each brain state dependent on those that came before. Laboratory experiments break this dependency. Test subjects view images or listen to sounds presented in random order, responding after each one, say, by pressing a button. Such experiments disrupt the brain's natural process of prediction. And the results come out looking like the subject's brain makes a rapid, automatic response, followed by a controlled choice about 150 milliseconds later, as if the two responses came from distinct systems in the brain.[10] The illusion of a two-system brain is a byproduct of a century-old, flawed experimental design, and our laws maintain the illusion.*

The legal system, with its essentialized view of the mind and brain, mixes up volition — whether your brain actually played a role in controlling your behavior — and *awareness* of volition — whether you experience having a choice. Neuroscience has quite a bit to say about this distinction. If you sit in a chair with your legs bent, toes not touching the floor, and tap your knee just below your kneecap, the bottom half of your leg gives a little kick. Hold your hand to a flame and your arm recoils. Present a puff of air to your cornea and you blink. Each of these examples is a reflex: sensation leading directly to motion. Reflexes in your peripheral nervous system have sensory neurons wired directly to motor neurons. We call the resulting actions "involuntary" because there is one, and only one, specific behavior for a specific sensory stimulation due to the direct wiring.[11]

Your brain, however, is not wired like a reflex. If it were, you'd be at the mercy of the world, like a sea anemone that reflexively stabs whatever fish happens to brush against its tentacles. The anemone's sensory neurons, which receive input from the world, are directly connected to its motor neurons for movement. It has no volition.

A human brain's sensory and motor neurons, however, communicate through intermediaries, called *association neurons,* and they endow your nervous system with a remarkable ability: decision-making. When an association neuron receives a signal from a sensory neuron, it has not one possible action but two. It can stimulate *or inhibit* a motor neuron. Therefore, the same sensory input can yield different outcomes on different occasions. This is the biological basis of choice, that most prized of human possessions. Thanks to association neurons, if a fish brushes against *your* skin, you can react with indifference, laughter, violence, or anything in between. You might feel like a sea anemone at times, but you have much more control over your harpoon than you might think.[12]

Your brain's control network, which helps select your actions, is composed of association neurons. This network is always engaged, actively selecting your actions; you just don't always feel in control. In other words, your experience of being in control is just that — an experience.[13]

---

* In my more cynical moments, I also think the "two-system brain" survives as a convenient scapegoat, an animalistic, emotional part of the brain on which we can blame our bad behavior.

Here's where the law is out of sync with science, thanks to the classical view of human nature. The law defines deliberate choice — free will — as whether you *feel* in control of your thoughts and actions. It fails to distinguish between your ability to choose — the workings of your control network — and your subjective experience of choice. The two are not the same in the brain.[14]

Scientists are still trying to figure out how the brain creates the experience of having control. But one thing is certain: there is no scientific justification for labeling a "moment without awareness of control" as emotion.[15]

What does all this mean for the law? Remember that the legal system decides guilt or innocence based on intent — whether someone meant to commit harm. The law should continue to punish based on how intentional harm is, not on whether emotion is involved or whether a person experiences himself as an agent with volition.

Emotions are not temporary deviations from rationality. They are not alien forces that invade you without your consent. They are not tsunamis that leave destruction in their wake. They are not even your reactions to the world. They are your constructions of the world. Instances of emotion are no more out of control than thoughts or perceptions or beliefs or memories. The fact is, you construct many perceptions and experiences and you perform many actions, some that you control a lot and some that you don't.

• • •

The legal system has a standard called the *reasonable person* who represents the norms of society, that is, the social reality within your culture. Defendants are measured against this standard. Consider the legal argument at the heart of the heat-of-passion defense: would a reasonable person have committed the same killing if he'd been similarly provoked without a chance to cool off?

The standard of the reasonable person, and the social norms behind it, is not merely reflected in the law — it is created by the law. It is a way of saying, "Here is what we expect a human person to act like, and we will punish you if you don't conform." It's a social contract, a guide to behavior for the average person in a population of diverse individuals. And like all averages, the reasonable person is a fiction that doesn't apply exactly to any single individual. It's a stereotype, and it encompasses stereotyped ideas about emotional "expression," feeling, and perception that are part of the classical view of emotion and the theory of human nature that supports it.

A legal standard based on emotion stereotypes is especially problematic for the equitable treatment of men and women. The prevailing belief in many cultures is that women are more emotional and empathic, whereas men are more stoic and analytical. Shelves full of popular books portray this stereotype as fact: *The Female Brain; The Male Brain; His Brain, Her Brain; The Essential Difference; Brain Sex; Unleash the Power of the Female Brain;* and on and on. This stereotype affects even powerful women who are widely respected. Madeleine Albright, the first female U.S. secretary of state, wrote in her memoir that "many of my colleagues made me feel that I was overly emotional, and I worked hard to get over that. In time, I learned to keep my voice flat and unemotional when I talked about issues that I considered important."[16]

Take a moment and reflect on your own emotions. Do you tend to feel things intensely or more moderately? When we ask these types of questions in my lab to male and female test subjects — to describe their feelings from memory — the women report feeling more emotion than the men do on average. That is, the women believe they are more emotional than men, and the men agree. The one exception is anger, as subjects believe that men are angrier. However, when the same people record their emotional experiences as they occur in everyday life, there are no sex differences. Some men and women are very emotional, and some are not. Likewise, the female brain is not hardwired for emotion or empathy, and the male brain is not hardwired for stoicism or rationality.[17]

Where do these gender stereotypes come from? In the United States at least, women routinely "express" more emotion when compared to men. For example, women move their facial muscles more when watching films than men do, but women don't report more intense experiences of emotion while watching. This finding, if nothing else, might explain why the stereotypes of the stoic man and the emotional woman leak into the courtroom and have a significant influence on judges and juries.[18]

Because of these stereotypes, heat-of-passion defenses — and legal proceedings in general — are often applied differently to male versus female defendants. Consider two murder cases that are pretty similar except for the sex of the defendant. In the first case, a man named Robert Elliott was convicted of killing his brother, allegedly because of "extreme emotional disturbance" that included "an overwhelming fear of his brother." The jury found him guilty of murder but the decision was overturned by the Supreme Court of Connecticut, citing that Elliott's "intense feelings" about his brother overwhelmed his "self-control" and "reason." In the second case, a woman named Judy Norman killed her husband after he had systematically beaten and abused her for years. The Supreme Court of North Carolina rejected the defense's claim that Norman was acting in self-defense out of "a reasonable fear of imminent death or great bodily harm," and she remained convicted of voluntary manslaughter.[19]

These two cases match several stereotypes about emotion in men versus women. Anger is stereotypically normal for men because they are supposed to be aggressors. Women are supposed to be victims, and good victims shouldn't become angry; they're supposed to be afraid. Women are punished for expressing anger — they lose respect, pay, and perhaps even their jobs. Whenever I see a savvy male politician play the "angry bitch card" against a female opponent, I take it as an ironic sign that she must be really competent and powerful. (I have yet to meet a successful woman who hasn't paid her dues as a "bitch" before she was accepted as a leader.)[20]

In courtrooms, angry women like Ms. Norman lose their liberty. In fact, in domestic violence cases, men who kill get shorter and lighter sentences, and are charged with less serious crimes, than are women who kill their intimate partners. A murderous husband is just acting like a stereotypical husband, but wives who kill are not acting like typical wives, and therefore they are rarely exonerated.[21]

Emotion stereotyping is even worse when the female victim of domestic violence is African American. The archetypal victim in American culture is fearful, passive, and helpless, but in African American communities, women sometimes violate this stereotype by defending themselves vigorously against their alleged batterers. By fighting back, they reinforce a different stereotype of female emotion, the "angry black woman," which is also pervasive in the U.S. legal system. These women are more likely to be charged with domestic violence themselves, even when their actions were in self-defense and were less severe than the original assault. (No "stand your ground" allowed here!) And if they injure or kill their alleged batterer, they usually fare worse than a European American woman in the same situation.[22]

For example, consider the case of Jean Banks, an African American woman who stabbed and killed her live-in partner, James "Brother" McDonald, after he had beaten her for years, sometimes so severely that she required medical attention. On this particular day, both had been drinking, and during an argument, McDonald pushed Banks to the ground and attempted to slice her with a glass cutter. Banks grabbed a knife to defend herself and stabbed him through the heart. She claimed self-defense but nonetheless was convicted of second-degree murder. (Compare this to light-skinned Judy Norman, who was convicted of voluntary manslaughter, a lesser charge.)[23]

Angry women do not fare well outside of domestic violence cases either. Judges infer all sorts of negative personality characteristics in angry female rape victims that they tend not to attribute to angry male crime victims. When a woman has been raped, for instance, judges (and juries and the police) expect to see her express grief on the witness stand, which tends to bring the rapist a heavier sentence. When a female victim expresses anger, judges evaluate *her* negatively. These judges are falling prey to another version of the "angry bitch" phenomenon. When people perceive emotion in a man, they usually attribute it to his situation, but when they perceive emotion in a woman, they connect it to her personality. She's a bitch, but he's just having a bad day.[24]

Outside the courtroom, we find laws where gender stereotypes prescribe the acceptable emotions we must feel and express. Abortion laws, as written, signal which emotions are appropriate for a woman to feel, namely, remorse and guilt, whereas relief and happiness go unmentioned. The debate over the legality of gay marriage was, in a way, whether the law should sanction the emotion of romantic love between two people of the same sex. Adoption laws governing gay men raise the question of whether a father's love is equal to that of a mother.[25]

Overall, there is no scientific justification for the law's view of men's and women's emotions. They are merely beliefs that come from an outdated view of human nature. The examples I've chosen represent only a small slice of the issue, both on the legal side and on the science side. I've barely scratched the surface of emotion stereotypes of ethnic groups, for example, who face similar struggles in and out of court. As long as the law codifies emotion stereotypes, people will continue to be the target of inconsistent rulings.[26]

• • •

When Stefania Albertani pled guilty to drugging and killing her own sister, not to mention setting the corpse on fire, her defense team took a bold step and blamed her brain.

Brain imaging revealed that two regions of Albertani's cortex contained fewer neurons than a control group of ten other healthy women. The regions were the insula, which the defense claimed was associated with aggression, and the anterior cingulate gyrus, which allegedly was associated with lowering one's inhibitions. Two expert witnesses concluded that a "causal relationship" between her brain structure and her crime was possible. After this testimony, Albertani's jail sentence was reduced from life imprisonment to twenty years.[27]

Legal decisions like this one, which was a media sensation in Italy in 2011, are becoming more common as lawyers employ neuroscience findings in their defense strategy. But are these decisions justified? Can brain structure explain why someone committed a crime? Can a region of a certain size or connectivity actually cause murderous behavior, and in the process, make a defendant less responsible for a crime?[28]

Legal arguments like those made by Albertani's defense team grossly misrepresent neuroscience findings and the conclusions that can be drawn from them. It is just not possible to localize a complex, psychological category like "Aggression" to one set of neurons, because of degeneracy; "Aggression," like any other concept, may be implemented differently in the brain each time it's constructed. Even simple actions like hitting or biting have not been localized to a single set of neurons in the human brain.[29]

The brain regions mentioned by Albertani's defense team are among the most highly connected hubs in the entire brain. They show increased activation for just about every mental event you can list, from language to pain to math skills. So, sure, they might play a role in aggression and impulsivity in some instances. But it's a stretch to claim any specific causal relationship between these regions and the extreme aggression of murder . . . if Albertani's motive was even aggression in the first place.[30]

It's also a stretch to claim that variation in brain size translates into variation in behavior. No two brains are exactly alike. They generally have the same parts, roughly in the same place, connected together in pretty much the same way, but at a fine-grained level, in their microcircuitry, they have vast differences. Some may translate into behavioral differences, but many do not. Your insula might be larger or more highly connected than mine without any discernable effect on your behavior when compared to my behavior. Even if we examine many brains and find a statistically significant difference in insula size between people who are more or less aggressive, that doesn't mean that a larger insula *causes* aggression, let alone murder. (Plus, even if a larger insula did cause aggression, how big does it need to be to produce a killer?) In rare cases, a tumor can press against the brain and cause severe personality changes, but in general, it is not scientifically justified to try a brain region for murder.[31]

Perhaps the most surprising thing about Albertani's case is that the expert witnesses and the judge thought that the brain was an "extenuating explanation" for Albertani's murderous behavior. *All* behavior stems from the brain. No human actions, thoughts, or feelings exist apart from firing neurons. The wrong way to use neuroscience in court is to argue that a biological explanation automatically releases someone from responsibility. You are your brain.[32]

The law often looks for simple, single causes, so it's tempting to blame a brain aberration for criminal behavior. But behavior in real life is anything but simple. It's a culmination of multiple factors, including predictions from your brain, prediction error from your five senses plus interoceptive sensation, and a complex cascade involving billions of prediction loops. And that's just the story inside a single person. Your brain is also surrounded by other brains in other bodies. Whenever you speak or act, you influence the predictions of others around you, who in turn influence your predictions right back. A whole culture collectively plays a role in the concepts you build and the predictions you make, and therefore in your behavior. People can argue over how large a role culture plays, but the fact of its role is not debatable.

Bottom line: Sometimes a biological problem can interfere with your brain's ability to choose your actions with intent. Maybe you grow a brain tumor, or some neurons begin to die in just the wrong places. But mere variability in the brain — in its structure, function, chemistry, or genetics — is not an extenuating circumstance for a crime. Variation is the norm.

• • •

Dzhokhar Tsarnaev, the Boston Marathon bomber, was convicted in 2015 and sentenced to death. Tsarnaev received a trial by jury, a right guaranteed to all Americans by the U.S. Constitution. According to the BBC, who reported on the sentencing, "Only two of the jurors believed Tsarnaev has felt remorse. The other 10, like many in Massachusetts, think he has no regrets." Jurors formed these opinions of Tsarnaev's remorse by observing him closely during the trial, where he reportedly sat "stone-faced" throughout most of the proceedings. Slate.com noted that Tsarnaev's defense attorney "did not — or could not — present evidence [that] Dzhokhar Tsarnaev has felt any of the remorse that the prosecution says he is devoid of."[33]

Trial by jury is considered the gold standard for fairness in a criminal case. Jurors are instructed to make decisions based only on the evidence presented. In a predicting brain, however, this is an impossible task. The jurors perceive every defendant, plaintiff, witness, judge, attorney, courtroom, and iota of evidence through the lens of their own conceptual system, which makes the idea of the impartial juror an implausible fiction. In effect, a jury is a dozen subjective perceptions that are supposed to yield one fair and objective truth.

The idea that jurors can somehow detect remorse in a defendant, from his facial configurations or bodily movements or words, is steeped in the classical view, which assumes that emotions are universally expressed and recognized. The legal system assumes that remorse, like anger and other emotions, has a single, universal essence with a detectable fingerprint. However, remorse is an emotion category composed of many diverse instances, each one made for a specific situation.

A defendant's construction of remorse depends on his concept for "Remorse," culled from his prior experiences within his culture, which exists as cascades of predictions that guide his expression and his experience. On the other side of the courtroom, a juror's perception of remorse is a mental inference — a guess based on cascades of predictions in her brain that make sense of the defendant's facial movements, body posture, and voice. For that juror's perceptions to be "accurate," she and the defendant must categorize with similar concepts. This kind of synchrony, with one person feeling remorse and the other perceiving it, even without words ever being spoken, is more likely to occur when two people have similar backgrounds, age, sex, or ethnicity.[34]

In the Boston Marathon Bombing case, if Tsarnaev felt remorse for his deeds, what would it have looked like? Would he have openly cried? Begged his victims for forgiveness? Expounded on the error of his ways? Perhaps, if he were following American stereotypes for expressing remorse, or if this were a trial in a Hollywood movie. But Tsarnaev is a young man of Muslim faith from Chechnya. He lived in the United States and had close American friends, but Tsarnaev had also (by his defense team's account) spent a lot of time with his older, Chechen brother. Chechen culture expects men to be stoic in the face of adversity. If they lose a battle, they should bravely accept defeat, a mindset known as the "Chechen wolf." So if Tsarnaev felt remorse, he might well have remained stony-faced.[35]

Tsarnaev did reportedly become tearful for a moment when his aunt took the stand to plead for his life. Chechnya has a culture of honor, where it is painful to shame your family. If Tsarnaev saw a loved one publicly shamed, say, an aunt begging on his behalf, a few tears would be consistent with Chechen cultural norms for honor.[36]

We — and jurors — can only guess when constructing a perception to explain Tsarnaev's impassive stance. Using our Western cultural concepts of remorse, we perceived him as coolly indifferent or full of bravado, rather than stoic. So it's possible that our guesswork, in this case, produced a cultural misunderstanding in the courtroom, ultimately leading to his death sentence. Or maybe he really is remorseless.[37]

As it turns out, Tsarnaev actually did convey remorse for his actions in a letter of apology he wrote in 2013, just a few months after the bombing, two years before he went to trial. Jurors never saw the letter, however. It was sealed as confidential under the U.S. Government's Special Administrative Measures, citing an "international security issue," and excluded as evidence from the trial.[38]

On June 25, 2015, Tsarnaev finally spoke at his sentencing hearing. He confessed to the bombing and stated that he understood the impact of his crime. "I am sorry for the lives that I've taken," he apologized quietly and calmly, "for the suffering that I've caused you, for the damage that I've done. Irreparable damage." The range of responses from victims and the press covering the trial was predictably variable. Some were stunned. Some were upset. Some were outraged. Some accepted his apology. And many just could not decide whether it was sincere.

We can never know whether Tsarnaev experienced remorse for his terrible actions, nor if his letter could have affected his sentence. But one thing is certain: At a death penalty proceeding, a defendant's remorse is a critical feature that jurors must rely on, according to the law, to make a decision between imprisonment and death. And those perceptions of remorse, like all perceptions of emotion, are not detected but constructed.[39]

At the other end of the spectrum, a show of remorse can mean absolutely nothing. Take the case of Dominic Cinelli, a violent criminal with a thirty-year history of armed robberies, assaults, and prison escapes. Cinelli was serving three consecutive life sentences when he appeared before the Massachusetts Parole Board in 2008. A parole board is made up of psychologists, corrections officers, and other knowledgeable professionals who decide whether an inmate will serve beyond his minimum sentence or be released. They witness a virtual parade of remorse, some genuinely experienced and some faked, and their profound responsibility to the public rests on their ability to tell the difference.

In November 2008, Cinelli convinced the parole board that he was no longer a criminal with darkness in his soul. The board unanimously voted to free him. It didn't take long for Cinelli to embark on a new series of robberies and fatally shoot a police officer. Cinelli was later killed during a shootout with the police. The governor of Massachusetts, Deval Patrick, saw five of the seven members of the parole board resign. He seemed to think that they lacked the ability to detect authentic remorse.[40]

It's possible that Cinelli was putting on an act. It's also possible that Cinelli authentically felt remorse in the moment while he was testifying, but once he was out of prison, his old model of the world resurfaced, with his old predictions, creating his old self, and his remorse evaporated. Since there is no objective criterion for feelings of remorse, we will never know for sure. There is likewise no objective criterion for anger, sadness, fear, or any other emotion relevant to a trial.

U.S. Supreme Court Justice Anthony Kennedy once said that juries must "know the heart and mind of the offender" in order for a defendant to have a fair trial. Emotions, however, have no consistent fingerprints in facial movements, body posture and gestures, or voice. Jurors and other perceivers make educated guesses about what those movements and sounds mean in emotional terms, but there is no objective accuracy. At best, we can measure whether jurors agree with one another in the emotions they perceive, but when the defendant and the jurors have different backgrounds, beliefs, or expectations, agreement is a poor substitute for accuracy. If a defendant's demeanor cannot reveal emotion, then the legal system is left to grapple with a difficult question: under what circumstances can a trial be completely fair?[41]

• • •

When jurors or judges see smugness in a defendant's smile, or when they hear a witness's quavering voice as fear, they are making a mental inference, employing their emotion concepts to guess that the action (smiling or quavering) was caused by a particular state of mind. Mental inference, you'll remember, is how your brain gives meaning to other people's actions through a cascade of predictions (chapter 6).[42]

Mental inference is so pervasive and automatic, at least in cultures of the West, that we're usually unaware of doing it. We believe that our senses provide an accurate and objective representation of the world, as if we had X-ray vision for deciphering another person's behavior to discover his intent ("I can see right through you"). In these moments, we experience our perceptions of other people as an obvious property of *them* — a phenomenon we've called affective realism — rather than a combination of their actions and the concepts in our own brain.

When someone is on trial for a crime, and liberty and life are at stake, there can be a gaping chasm between appearance and reality. Deep down we know this, but at the same time we are supremely confident that *we* can discern truth from fiction more accurately than the other schmucks in the room. And herein lies the problem in court.

Jurors and judges are charged with an almost impossible task: to be a mind reader, or if you'd rather, a lie detector. They must decide if a person *intended* to cause harm. According to the legal system, intent is a fact that is as plain as the nose on a defendant's face. But in a predicting brain, a judgment about someone else's intent is always a *guess* you construct based on the defendant's actions, not a fact you detect; and just as with emotions, there is no objective, perceiver-independent criterion of intent. Seventy years of psychological research confirms that judgments like these are mental inferences, that is, guesses. Even if DNA evidence connects a defendant to the scene of a crime, it does not determine whether he had criminal intent.[43]

Judges and jurors infer intent, usually in line with their own beliefs, stereotypes, and current body states. Here is just one example of how this works. Test subjects watched a video of protestors being dispersed by police. They were told the protestors were pro-life activists picketing an abortion clinic. Those who were liberal Democrats, who tend to be pro-choice, inferred that the activists had violent intentions, whereas socially conservative subjects inferred peaceful intentions. The researchers also showed the same video to a second set of subjects, describing the protestors this time as gay rights activists objecting to the military's Don't Ask, Don't Tell policy. This time, those who were liberal Democrats, who tend to support gay rights, inferred that the activists had peaceful intentions, whereas socially conservative subjects inferred violent intentions.[44]

Now imagine that this video were evidence at a trial. All jurors would watch the same scenes, with exactly the same behaviors onscreen, but through affective realism, they would come away with only perceptions, not facts, constructed in line with their own beliefs, entirely without their awareness. My point is that bias is not advertised by a glowing sign worn around jurors' necks; we are all guilty of it, because the brain is wired for us to see what we believe, and it usually happens outside of everyone's awareness.

Affective realism decimates the ideal of the impartial juror. Want to increase the likelihood of a conviction in a murder trial? Show the jury some gruesome photographic evidence. Tip their body budgets out of balance and chances are they'll attribute their unpleasant affect to the defendant: "I feel bad, therefore you must have done something bad. You are a bad person." Or permit family members of the deceased to describe how the crime has hurt them, a practice known as a victim impact statement, and the jury will tend to recommend more severe punishments. Crank up the emotional impact of a victim impact statement by recording it professionally on video and adding music and narration like a dramatic film, and you've got the makings of a jury-swaying masterpiece.[45]

Affective realism intertwines with the law outside the courtroom as well. Imagine that you are enjoying a quiet evening at home when suddenly you hear loud banging outside. You look out the window and see an African American man attempting to force open the door of a nearby house. Being a dutiful citizen, you call 911, and the police arrive and arrest the perpetrator. Congratulations, you have just brought about the arrest of Harvard professor Henry Louis Gates, Jr., as it happened on July 16, 2009. Gates was trying to force open the front door of his own home, which had become stuck while he was traveling. Affective realism strikes again. The real-life eyewitness in this incident had an affective feeling, presumably based on her concepts about crime and skin color, and made a mental inference that the man outside the window had intent to commit a crime.[46]

A similar bout of affective realism gave birth to Florida's controversial "Stand Your Ground" law. This law permits the use of deadly force in self-defense if you reasonably believe you're in imminent danger of death or great bodily harm. A real-life incident was the catalyst for the law, but not in the way that you might think.

Here's how the story is usually told: In 2004, an elderly couple was asleep in their trailer home in Florida. An intruder tried to break in, so the husband, James Workman, grabbed a gun and shot him. Now here's the true, tragic backstory: Workman's trailer was in a hurricane-damaged area, and the man he shot, Rodney Cox, was an employee of the Federal Emergency Management Agency (FEMA). Workman, mostly likely under the influence of affective realism, perceived that Cox meant him harm and opened fire on an innocent man. Nevertheless, the inaccurate *first* story became a primary justification for Florida's law.[47]

The very history of stand your ground laws is, ironically, potent evidence against their value. It's impossible to determine reasonable fear for one's life in a society where racist stereotypes abound and affective realism literally transforms how people see each other. The whole line of reasoning for stand your ground is gutted by affective realism.

If stand your ground doesn't scare the crap out of you, think about the impact of affective realism on people who legally carry concealed weapons. Affective realism indisputably influences people's perceptions of threat; therefore it virtually assures that innocent people will be shot by accident. It's simple: you predict a threat, sensory information from the world says otherwise, but then your control network downplays the prediction error to maintain the prediction of threat. Bam, you've shot a harmless fellow citizen. Human brains are built for this sort of delusion, through the same process that produces daydreams and imagination.

I will not wade any further into the national debate about firearms for now, but from a purely scientific perspective, consider this. The founding fathers of the United States had good reasons for protecting a "right of the people to keep and bear Arms" in the Second Amendment of the Constitution, but they were not neuroscientists. Nobody in 1789 knew that the human brain constructs every perception and is ruled by interoceptive predictions. Right now, over 60 percent of people in the United States believe that crime is on the rise (though it's historically low), and they also believe that owning a gun will make them safer. These beliefs are ripe to lead people, through affective realism, to genuinely see a deadly threat where there is none and to act accordingly. Now that we know definitively that our senses don't reveal objective reality, shouldn't this critical knowledge influence our laws?[48]

As a general rule, the legal system has had a lot of difficulty coming to grips with the mountains of scientific evidence that our senses don't provide a literal readout of the world. For hundreds of years, eyewitness reports used to be considered one of the most reliable forms of evidence. When a witness said, "I saw him do it" or "I heard her say it," these statements were considered to be facts. The law also treated memories as if they entered the brain pristinely, were stored whole, and were later retrieved and played back like a movie.[49]

Just as jurors cannot pull back the curtain of their own beliefs for direct access to some unblemished version of reality, witnesses and defendants do not report a collection of facts but a description of their own perceptions. One can glance at Serena Williams's triumphant face at the beginning of chapter 3 and later, on the witness stand, swear on a Bible that Williams was screaming in terror. Any words spoken by eyewitnesses are based on recollections that are constructed in the moment, using past experiences that were themselves constructed.

Psychologist Daniel L. Schacter, one of the world's experts on memory, tells the story of a brutal rape that took place in Australia in 1975. The victim told police that she'd seen her attacker's face clearly, identifying him as Donald Thomson, a scientist. Police picked up Thomson the next day based on this eyewitness evidence, but Thomson had an iron-clad alibi: he was being interviewed on television at the time of the rape. It turned out that the victim's TV was on when the intruder broke into her house, and it was tuned to Thomson's interview, which ironically was about Thomson's research on memory distortion. The poor woman had somehow, in her trauma, fused Thomson's face and identity onto her attacker.[50]

Most men falsely accused are not so lucky. Jurors place a lot of weight on eyewitness testimony, yet they accept mistaken identifications just as frequently as correct ones, as long as the witnesses sound confident. In one study of convictions that were later overturned by DNA evidence, 70 percent of the accused were convicted based on eyewitness testimony.[51]

Eyewitness reports are perhaps the least reliable evidence one can have. Memories are not like a photograph — they are simulations, created by the same core networks that construct experiences and perceptions of emotion. A memory is represented in your brain in bits and pieces as patterns of firing neurons, and "recall" is a cascade of predictions that reconstruct the event. Your memories are therefore highly vulnerable to reshaping by your current circumstances, like having your body all worked up in the witness stand, or if you're being badgered by a persistent defense attorney.

The law has been slow to accept that memories are constructed, but the situation is gradually changing. The Supreme Courts of New Jersey, Oregon, and Massachusetts are leading the way in this regard. Their jurors now receive instructions that provide step-by-step details — based on years of psychological research — explaining all the ways in which memory can go wrong in eyewitness testimony. They read how memories are constructed and infused with beliefs that can result in distortions and illusions, how the instructions given by lawyers and police can introduce biases, how confidence is unrelated to accuracy, how stress can impair memory, and how eyewitness testimony was a factor in falsely convicting more than three quarters of the people who were exonerated by DNA evidence for crimes that they did not commit.[52]

Unfortunately, no such guidelines exist to explain to jurors what an emotional expression is, what a mental inference is, or how they are constructed.

• • •

The figure of the dispassionate judge, who renders emotionless decisions in strict accordance with the law, is an archetype in many societies. The law expects judges to be neutral, as emotion would presumably get in the way of fair decisions. "Good judges pride themselves on the rationality of their rulings and the suppression of their personal proclivities," wrote the late U.S. Supreme Court Justice Antonin Scalia, "including most especially their emotions."[53]

In some ways, a purely rational approach to legal decision-making sounds compelling and even noble, but as we've seen so far, the brain's wiring doesn't divide passion from reason. We needn't work hard to poke holes in this argument; it comes with its own holes pre-drilled.

Let's start with the idea that a judge can be dispassionate, which should be interpreted as "having no affect" (rather than "having no emotion"). This idea is a biological impossibility unless that person has suffered brain damage. As we discussed in chapter 4, no decision can ever be free of affect as long as loudmouthed body-budgeting circuitry is driving predictions throughout the brain.

Affectless decision-making from the bench is a fairy tale. Robert Jackson, another former Supreme Court justice, described "dispassionate judges" as "mythical beings" like "Santa Claus or Uncle Sam or Easter bunnies." Direct scientific evidence shows him to be pretty much on target. Remember how judges' impartiality was easily swayed in parole cases held right before lunchtime, when they attributed their unpleasant affect to the prisoner instead of to hunger (chapter 4)? In another series of experiments, over 1,800 state and federal judges from the United States and Canada were handed scenarios of civil and criminal cases and asked what their rulings would be. Some scenarios were identical except the defendants were portrayed as more likeable or unlikeable. The experimenters found that judges tended to rule in favor of more likeable or sympathetic people.[54]

Even the U.S. Supreme Court is not immune to leaking passion from the bench. A team of political scientists examined 8 million words spoken

by the members of the Court during oral arguments, and their questioning, over thirty years. They found that when judges focus "more unpleasant
language" toward an attorney, that side is more likely to lose. You can predict the loser by simply counting the justices' negative words during questioning. Not only that, but by examining the affective connotations in the
judges' words during oral arguments, you can predict their votes.[55]

Common sense dictates that judges experience strong affect in the courtroom. How could they not? They hold people's futures in their hands. Their
working hours are filled with heinous crimes and grievously harmed victims. I know how draining this can be, having been a therapist for victims
of rape and childhood sexual abuse, and sometimes working with the perpetrators. Judges also encounter defendants who are more likable than the
people they have preyed on, a situation that surely is challenging to grapple
with, especially in a courtroom full of whispering spectators and bicker-
ing attorneys. And sometimes a judge must shoulder the affect of an entire
country. Former U.S. Supreme Court Justice David Souter suffered so much
while deciding *Bush v. Gore* that he wept because of its deliberations (along
with half of the United States). All this mental effort taxes a judge's body
budget. The judge's life is one of intense and continual emotional labor under the fiction of equanimity.[56]

Nevertheless, the law continues to hold dear the fiction of the dispassionate judge, even at the highest levels. When Supreme Court Justice Elena
Kagan, as a nominee in 2010, was asked whether it was ever appropriate for
feelings to help decide a case, she replied to the contrary, "It's the law all the
way down." Justice Sonia Sotomayor also ran into opposition during her
confirmation hearings, as some senators feared that her emotions and empathy were in direct opposition to her abilities to judge fairly. Her take on all
this, for the most part, was that judges do have feelings but should not make
decisions based on them.

Nonetheless, the evidence is clear that judges are not affectless in their
rulings. The next question is: *should* they be? Is pure reason really the best
way to render a wise decision? Imagine a person who is very calmly and
coolly weighing the pros and cons about whether or not another person
should die. There's not a trace of emotion in sight. Like Hannibal Lecter in
The Silence of the Lambs, or Anton Chigurh in *No Country for Old Men.* I
am being a bit facetious here, but this kind of dispassionate decision-making is essentially what the law instructs in the sentencing portion of criminal cases. Rather than pretend that affect is absent, it's better to use affect

wisely. As U.S. Supreme Court Justice William Brennan once expressed, "Sensitivity to one's intuitive and passionate responses, and awareness of the range of human experience, is therefore not only an inevitable but a desirable part of the judicial process, an aspect more to be nurtured than feared." The key is emotional granularity: having a wide and deep range of concepts (emotion, physical, or otherwise) to make sense of the onslaught of bodily sensations that are the hazards of the job.[57]

Consider, for example, a judge faced with a defendant like James Holmes, who murdered twelve moviegoers and injured seventy more during a midnight screening of a Batman movie in Aurora, Colorado, in 2012. Such a judge might reasonably construct an experience of anger, but that feeling alone could be problematic; anger could prompt the judge to punish the defendant too harshly for the sake of retribution, threatening the moral order that the trial is founded on. To balance his view, some legal scholars argue, the judge could try to cultivate empathy for the defendant, who perhaps is insane or a victim of some sort himself. Anger is a form of ignorance; in this case, ignorance of the defendant's perspective. Holmes clearly struggled with serious mental illness for years. He tried to kill himself for the first time when he was eleven years old, and has attempted suicide several times in jail. Empathy is extremely difficult to cultivate for someone who opens fire on innocents in a movie theater. Even remembering that the defendant is a human being, no matter how severe or gruesome the crime, might be a struggle at times, but this is when empathy might be most important. It may prevent a judge from going too far in punishing the offender during sentencing, and help to ensure the morality of penal decision-making and retributive justice. This is the type of emotional granularity that makes for wise use of emotion in the courtroom.[58]

When it comes right down to it, the most useful emotions for a judge to feel depend on the judge's goals during the trial. What, for example, is the goal of punishment? Is it retribution? Deterrence to avoid future harm? Rehabilitation? This depends on the law's theory of the human mind. Whatever the goal, punishment must be enacted so that the defendant's humanity is preserved, while the victim's humanity is honored, even if the defendant commits an unspeakable act. To do otherwise puts the legal system itself in jeopardy.

• • •

Why is it that you can sue someone for breaking your leg but not for breaking your heart? The law considers emotional damage to be less serious than physical damage and less deserving of punishment. Think about how ironic this is. The law protects the integrity of your anatomical body but not the integrity of your mind, even though your body is just a container for the organ that makes you who you are — your brain. Emotional harm is not considered real unless accompanied by physical harm. Mind and body are separate. (Let's all raise a glass to René Descartes here.)

If there is one thing you can take away from this book, it is that the boundaries between mental and physical are porous. Chapter 10 explained a bit about the ways in which emotional harm from chronic stress, parental emotional abuse and neglect, and other psychological ills can ultimately cause physical illness and injury. And we've seen how stress and proinflammatory cytokines lead to numerous health problems, including brain atrophy, and increase the likelihood of cancer, heart disease, diabetes, stroke, depression, and a host of other illnesses.[59]

But that's not the whole story. Emotional harm can shorten your life. Inside your body, you have little packets of genetic material that sit on the ends of your chromosomes like protective caps. They're called telomeres. All living things have telomeres — humans, fruit flies, amoebas, even the plants in your garden. Every time one of your cells divides, its telomeres get a little shorter (although they can be repaired by an enzyme called telomerase). So generally their size slowly decreases, and at some point, when they are too short, you die. This is normal aging. But guess what else causes your telomeres to get smaller? Stress does. Children who experience early adversity have shorter telomeres. In other words, emotional harm can do more serious damage, last longer, and cause more future harm than breaking a bone. This means the legal system might be misguided when it comes to understanding and gauging the degree of lasting injury that can come from emotional harm.[60]

As another example, consider chronic pain. The law treats chronic pain by and large as "emotional" because there's no observable tissue damage. In these cases, the law usually concludes that the suffering is not real enough to merit compensation. People who suffer from chronic pain are often diagnosed as mentally ill, and even more so if they opt for an invasive operation to try and reduce their "illusory" suffering. Medical insurance companies deny treatment since chronic pain is considered psychological, not physical. The sufferer cannot work, yet no compensation is provided. But as we saw in the preceding chapter, chronic pain is likely a brain disease of prediction gone wrong. The suffering is real. The law is missing the point that prediction and simulation are the normal way that the brain works, and chronic pain is a difference of degree, not kind.[61]

Interestingly, the law does accept that other types of harm can be absent now but show up in the future. A prominent example is chemical harm such as Gulf War Syndrome, a chronic, multi-symptom illness allegedly caused by unknown factors during the Gulf War, whose effects did not appear until later. Gulf War Syndrome is controversial; there is no consensus on whether it's actually a distinct medical condition. Regardless, thousands of veterans have taken their claims of Gulf War Syndrome to court. There is no analogous legal avenue for stress or other harm seen as emotional. (Awards for pain and suffering are relatively rare.)

Having made this observation, I must point out that the law is deeply inconsistent and even ironic in its view of emotional harm when you consider international norms for torture. The Geneva Conventions prohibit psychological harm to prisoners of war, and the U.S. Constitution likewise forbids "cruel and unusual punishment." So it's illegal for a government to torture a prisoner psychologically, but it's perfectly legal to place a prisoner in solitary confinement for long periods, even though the stress of confinement may shorten the prisoner's telomeres and therefore his life.[62]

It's also perfectly legal for a high school bully to insult, torment, and humiliate your children even though this will shorten their telomeres and potentially their lifespan. When a group of middle-school girls deliberately excludes another girl, they are acting with intent and motivation to cause suffering, yet legal action is rare. In one highly publicized case, fifteen-year-old Phoebe Prince hanged herself in 2010 after months of verbal aggression and physical threats. Six teenagers were criminally prosecuted for harassment, stalking, assault, and assorted civil rights violations after they bullied her and then posted crude comments on her Facebook memorial page. This case prompted Massachusetts to pass anti-bullying laws. These laws are a start, but they punish only the most extreme cases. How do you regulate the playground in a legal context?[63]

Bullies intend to cause suffering, but is the intent to cause harm? We cannot know for sure, but in most cases I doubt it. Most kids are unaware that the mental anguish they inflict can translate into physical illness, atrophied brain tissue, reduced IQ, and shortened telomeres. Kids will be kids, we say. But bullying is a national epidemic. In one study, over 50 percent of children nationwide reported being verbally or socially bullied at school, or having participated in bullying another child at school, at least once in two months. Over 20 percent reported being the victim or perpetrator of physical bullying, and over 13 percent reported involvement with electronic bullying. Bullying is considered a serious enough childhood risk, with potential lifelong health consequences, that at press time, the U.S. Institute of Medicine and the National Research Council's Committee on Law and Justice are producing a comprehensive report on its biological and psychological ramifications.[64]

If you suffer mental anguish in the moment, whether from bullying or another cause, should your suffering count as harm, and should the perpetrators be punished? A recent legal case implies the answer is sometimes yes. A company in Atlanta demanded DNA samples from its employees because someone was contaminating its warehouse with feces. It's illegal to take genetic information from someone without his consent (it violates the Genetic Information Nondiscrimination Act), but the case was won largely on emotional grounds. The two plaintiffs were awarded about $250,000 each to compensate them for feeling humiliated and bullied, plus a remarkable $1.75 million in punitive damages for "emotional distress and mental anguish." The large award was not for the plaintiffs' actual emotional suffering but their *potential* emotional suffering in the future. After all, their personal health information could be used against them at any time for the rest of their lives. This fear of the future was easy for jurors to simulate and therefore empathize with. In a chronic pain case, it's harder: how do you see the invisible? There are no injuries to look at, and nothing to help your brain create the simulation, so empathy suffers and consequently so does compensation.[65]

The legal system has difficulty dealing with mental anguish for purely practical reasons. How do you measure it objectively if emotions have no essences or fingerprints? Also, physical harm like a broken leg is usually more economically predictable than emotional harm, which is far more variable. And how do you distinguish everyday emotional pain from lasting harm?[66]

Perhaps the most important question here is: Whose suffering counts as harm? Who deserves our empathy and therefore the full protection of the law? If you negligently or intentionally break my arm, you owe me. But if you negligently or intentionally break my heart, you don't, even if we were close for a long time, regulating each other's body budgets, and the breakup will put me through a physical process that can be as excruciating as withdrawal from an addictive drug. You can't sue someone for heartbreak, no matter how much you might want to (or how much they deserve it). The law is about creating and enforcing social reality. Empathic claims about pain are fundamentally claims about whose rights matter . . . and whose humanity matters.[67]

. . .

As you've seen, the law embodies the classical view of emotion and the view of human nature from which it derives. This essentialist story is a folktale that is not respected by the brain and its connection to the body. Therefore, based on today's scientific view of the brain, I'm going to go out on a limb with some recommendations for jurors, judges, and the legal system in general. I am not a legal scholar, and I realize that the concerns of science are not the same as those of the law. I realize also that it's one thing to speculate about basic dilemmas of humanity in the pages of a book but quite another to establish legal precedent on them. But it's important to try to build bridges between disciplines. Neuroscience and the legal system are seriously out of sync on fundamental issues of human nature. These discrepancies must be addressed if the legal system is to remain one of our most impressive achievements of social reality and continue protecting people's inalienable rights to life, liberty, and the pursuit of happiness.

I'd begin by educating judges and jurors (and other legal actors like attorneys, police officers, and parole officers) about the basic science of emotion and the predictive brain. The New Jersey, Oregon, and Massachusetts Supreme Courts are taking steps in the right direction by formally instructing jurors that human memory is constructed and fallible. We need a similar approach for emotion. Toward that end, I propose a set of five teaching points. You might call it an affective science manifesto for the legal system.

The first teaching point in the manifesto concerns so-called expressions of emotion. Emotions are not expressed, displayed, or otherwise revealed in the face, body, and voice in any objective way, and anyone who determines innocence, guilt, or punishment needs to know this. You cannot recognize or detect anger, sadness, remorse, or any other emotion in another person — you can only guess, and some guesses are more informed than others. A fair trial depends on synchrony between experiencers (defendants and witnesses) and perceivers (jurors and judges), and this can be difficult to achieve in many circumstances. For example, some defendants are better at using their nonverbal movements to communicate information about their emotions, such as remorse. Some jurors will be better at synchronizing their concepts with a defendant than others will. That means jurors might need to work harder to perceive emotions in challenging situations, like when they disagree with a defendant or witness on a political issue, or when the other person is of a different ethnicity. Jurors should try to put themselves in the other person's shoes to facilitate this synchrony and cultivate empathy.[68]

The second point is about reality. Your sight, hearing, and other senses are always colored by your feelings. Even the most objective-sounding evidence is colored by affective realism. Jurors and judges must be educated about the predictive brain and affective realism, how their feelings literally alter what they see and hear in court. Perhaps the protestor video study I mentioned, where political beliefs caused people to perceive violent intent or not, could serve as an educational example. Jurors must also understand how affective realism influences eyewitnesses. Even a simple statement like "I saw him holding the knife" is a perception infused with affective realism. Eyewitness testimony does not relay cold, hard facts.

The third point is about self-control. Events that feel automatic are not necessarily completely outside your control and are not necessarily emotional. Your predicting brain provides the same range of control when you construct an emotion as when you construct a thought or a memory. The defendant in a murder trial is not a man-shaped sea anemone at the mercy of his environment, triggered by anger to pursue an inevitable, aggressive act. Most instances of anger, no matter how automatic they feel, don't lead to murder. Anger can also unfold very deliberately over a long time, so there is nothing inherently automatic about it. You have relatively more responsibility for your actions when you have relatively more control, regardless of whether the event is an emotion or a cognition.

Fourth, beware the "my brain made me do it" defense. Jurors and judges should be skeptical of claims that certain brain regions directly cause bad behavior. That is junk science. Every brain is unique; variation is normal (think degeneracy) and not necessarily meaningful. Unlawful behavior has never been definitively localized to any brain region. I am not referring here to foreign growths like tumors or obvious signs of neurodegeneration, which in some cases, such as certain types of frontotemporal dementia, can make it harder for people to conform their actions to the law. Even so, many tumors and neurodegenerative damage cause no run-ins with the legal system at all.

The final teaching point is to be mindful of essentialism. Jurors and judges need to know that every culture is full of social categories like sex, race, ethnicity, and religion. These must not be mistaken for physical, biological categories with deep dividing lines in nature. Also, emotion stereotypes don't belong in a courtroom. Women should not be punished for feeling anger rather than fear toward their aggressors, and men should not be punished for feeling helpless and vulnerable rather than brave and aggressive. The law's reasonable person standard is a fiction based on stereotypes, and it is inconsistently applied. Perhaps it's time to bury the reasonable person and conceive some other standard for comparison.[69]

Beyond the affective science manifesto, we also have the longstanding myth of the dispassionate judge, which is both propagated and questioned by members of the U.S. Supreme Court and other legal experts. Scholars may debate in legal journals about the value of emotion in judicial action, but the anatomy of the human brain makes it implausible for any human, including a judge, to escape the influence of interoception and affect when making decisions. Emotions are neither the enemy nor a luxury but a source of wisdom. Judges need not reveal their emotions (just as therapists learn not to), but they must be aware of them and explicitly use them to the best of their ability.

To employ emotions wisely, I suggest that judges learn to experience emotion with high granularity. If they feel unpleasant, they'll be helped if they can categorize finely to experience (say) anger distinctly from irritation or hunger. Anger can be a reminder to cultivate empathy toward an unsympathetic defendant, a gullible plaintiff, a belligerent witness, or a particularly intrusive attorney. Without empathy, anger can foster the type of retributive punishment that risks undermining the very notion of justice at the foundation of the legal system. Judges can cultivate higher granularity using the exercises I recommended in chapter 9: collecting experiences, learning more emotion words, using conceptual combination to invent and explore new emotion concepts, and deconstructing and recategorizing their emotional experiences in the moment. It sounds like a lot of work, but like any skill, it becomes habitual with practice. Also, it would not hurt for judges who face defendants from other cultures to be briefed on the different cultural norms for emotional experience and communication.

Judges might also be educated to reduce the influence of affective realism when selecting jurors (a process known as *voir dire*). Often, judges and attorneys weed out jurors by asking them direct, transparent questions such as "Can you be objective, fair, and impartial in this case?" or "Do you know the defendant?" They also try to assess superficial similarities between jurors and defendants. For example, if a financial advisor stands accused of embezzling millions of dollars of his clients' retirement investments, the judge might ask potential jurors whether they themselves have been victims of embezzlement, or whether a close relative works in the financial industry. But surface markers of similarities and differences are only the tip of the iceberg. It might be wise to examine a juror's affective niche to understand how the juror might predict during a trial, which could indicate biases that shape perception. For example, a judge could ask what magazines the jurors read, what movies they prefer to see, or whether they play first-person shooter games, using standard assessment techniques from psychology. Such information would allow a judge to consider the potential biases of jurors based on how they spend their time, rather than just asking jurors directly about their biases (since such self-reports are not necessarily valid).[70]

My suggestions so far address low-hanging fruit. Now we're ready for the really difficult stuff — scientific considerations that could change fundamental assumptions in the law.

We already know that our senses do not reveal reality, and judges and jurors necessarily suffer from affective realism. These factors, along with the rest of our knowledge of mind and brain, lead to a fairly radical idea (I'm almost afraid to say it): perhaps it is time to reevaluate trial by jury as the basis for determining guilt and innocence. Yes, it's enshrined in the U.S. Constitution, but the writers of that landmark document had no inkling of how the human brain works, nor that one day we could detect a defendant's DNA under a victim's fingernails. Before DNA evidence, the law could not say whether a judgment of guilt was true or false. The legal system could only decide whether or not the judgment was rendered fairly, meaning that the rules and procedures of law were followed consistently. The law was therefore not about truth but consistency. Due process was about avoiding procedural errors in rendering a decision of guilt or innocence, not about the validity of the decision itself. Today's legal system works only if we assume that consistency produces a just outcome. DNA testing is changing all that. It's not perfect, but it's immeasurably more objective than the affect-laden perceptions of human jurors.[71]

When DNA evidence is unavailable or irrelevant, perhaps trials might dispense with a jury and instead feature the collective wisdom of multiple judges working together, randomly drawn from a larger pool of judges. As I've said already, I'm not a legal scholar, just a scientist, so perhaps wiser legal minds can construct a balanced judicial panel system in better ways. A panel of skilled judges who are trained to be self-aware and emotionally granular might avoid affective realism more effectively than a jury would. It's not a perfect solution by any means: in the United States at least, judges tend to be on the older side, predominantly European American, and may overrepresent a particular set of beliefs while maintaining the illusion that they are free of them. Judges are also more likely to hand out maximum sentences. But one thing is certain: every day in America, thousands of people appear before a jury of their peers and hope they will be judged fairly, when in reality they are judged by human brains that always perceive the world from a self-interested point of view. To believe otherwise is a fiction that is not supported by the architecture of the brain.[72]

And now we get to the toughest issue of all: what it means to control your behavior and therefore be responsible for your actions. The law (like much of psychology) usually considers responsibility in two parts: actions caused by you, where you have more responsibility, and actions caused by the situation, where you have less. This simple dichotomy of internal versus external does not mesh with the reality of the predictive brain.

In a construction view of human nature, every human action involves three types of responsibility, not two. The first is traditional: your *behavior* in the moment. You pull the trigger. You grab the money and run. (The legal system names this behavior *actus reus*, the harmful action.)

The second type of responsibility involves your specific *predictions* that brought about the unlawful act (known as *mens rea*, the guilty mind). Your behavior is not caused in a single moment; it is always driven by prediction. When you steal money from an open cash register, you are an agent in the moment, but the ultimate cause of your behavior also includes concepts like "Cash Register," "Money," "Ownership," and "Stealing." Each of these concepts is associated with a large and diverse population of instances in your brain, and based on them, you issued predictions that led to your action. Now, if other people with similar concepts in the same situation (i.e., the reasonable person) would also steal the cash, well, you might be less culpable for your actions. However, they may well have left the cash untouched, in which case your responsibility is greater.

The third type of responsibility relates to the *content* within your conceptual system, separately from how your brain uses that system to predict when breaking the law. A brain does not compute a mind in a vacuum. Every human being is the sum of his or her concepts, which become the predictions that drive behavior. The concepts in your head are not purely a matter of personal choice. Your predictions come from the cultural influences you were pickled in. When a European American police officer shoots an unarmed African American civilian, and the officer honestly saw a gun in the civilian's hands due to affective realism, the event has roots in something outside the moment. Even if the officer were overtly racist, his actions were partly caused by his concepts, formed by a lifetime of experience, which includes American stereotypes about race. The victim's concepts and actions are likewise informed by a lifetime of experience, which includes American stereotypes of cops. All of your predictions are shaped not just by direct experience but also indirectly by television, movies, friends, and the symbols of your culture. While it's exciting to escape into a world of urban crime in a movie, or to retreat from the stress of the day by watching an hour or two of a police drama on TV, routine depictions of police conflicts have a cost. They fine-tune our predictions about the danger posed by people of certain ethnicities or socioeconomic status. Your mind is not only a function of your brain but also of the other brains in your culture.[73]

This third domain of responsibility cuts two ways. Sometimes it's trivialized as "society is to blame," a phrase lampooned as bleeding-heart liberal sentiment. I am saying something more nuanced. If you commit a crime, you are indeed to blame, but your actions are rooted in your conceptual system, and those concepts don't just appear in a puff of magic. They are forged by the social reality you live in, which gets under your skin to turn genes on and off and wire your neurons. You learn from your environment like any other animal. Nevertheless, all animals shape their own environment. So as a human being, you have the ability to shape your environment to modify your conceptual system, which means that you are ultimately responsible for the concepts that you accept and reject.

As we discussed in chapter 8, the predictive brain expands the horizon of self-control beyond the moment of action and therefore broadens your responsibility in a complicated way. Your culture might teach you that people of a certain skin color are more likely to be criminals, but you have the ability to mitigate the harm that such beliefs can cause, and hone your predictions in a different direction. You can befriend people of different skin tones and see for yourself that they're law-abiding citizens. You can choose not to watch TV shows that reinforce racist stereotypes. Or you can blindly follow the norms of your culture, accept the stereotyped concepts bestowed upon you, and increase the chances that you'll treat certain people badly.

Dylann Roof, the man who shot African American members of a Bible study group, chose to surround himself with symbols of white supremacy. Sure, he grew up in a society struggling with racism, but so did most adults in the United States, and most of us don't go around shooting people. So at the level of neurons, you and your society jointly cause certain predictions to become more likely in your brain. However, you still bear responsibility to overcome harmful ideology. The difficult truth is that each of us, ultimately, is responsible for our own predictions.

The law has precedent for this prediction-based view of responsibility. For example, if you drive drunk and hit someone with your car, you are responsible for the harm you caused, even though you could not control your limbs effectively in your inebriated state. You should have known better, because every adult in our society knows that drunkenness carries a risk of bad decision-making, so you are culpable for bad things that happen downstream.

The law calls this a foreseeability argument. It doesn't matter whether you intended to cause harm or not: you are liable. And we now have enough scientific evidence to extend the foreseeability argument from large-scale common sense to the millisecond predictions of the brain. You know full well that some of your concepts, such as racial stereotypes, can lead you into trouble. If your brain predicts that an African American youth in front of you is holding a weapon, and you perceive a gun where none is present, you have some degree of culpability even in the face of affective realism, because it is your responsibility to change your concepts. If you educate yourself and inoculate yourself against such stereotypes, expanding your conceptual system with the goal to change your predictions, you still might mistakenly see a gun where none is present, and a tragedy still might occur. But your culpability is diminished somewhat, because you've acted responsibly to change what you can.

Eventually, the legal system must come to grips with the tremendous influence of culture on people's concepts and predictions, which determine their experiences and actions. After all, the brain wires itself to the social reality it finds itself in. This ability is one of the most important evolutionary advantages we have as a species. So we bear some responsibility for the concepts we help wire into future generations of little human brains. But this is not an issue for criminal law. It is actually a policy issue relevant to the First Amendment, which guarantees the right to free speech. The First Amendment was founded on the notion that free speech produces a war of ideas, allowing truth to prevail. However, its authors did not know that culture wires the brain. Ideas get under your skin, simply by sticking around for long enough. Once an idea is hardwired, you might not be in a position to easily reject it.

• • •

The science of emotion is a convenient flashlight for illuminating some of the law's long-held assumptions about human nature—assumptions that we now know are not respected by the architecture of the human brain. People don't have a rational side and an emotional side, with the former regulating the latter. Judges can't set aside affect to issue rulings by pure reason. Jurors can't detect emotion in defendants. The most objective-looking evidence is tainted by affective realism. Criminal behavior can't be isolated to a blob in the brain. Emotional harm is not mere discomfort but can shorten a life. In short, every perception and experience within the courtroom — or anywhere else — is a culturally infused, highly personalized belief, corrected by sensory inputs from the world, rather than the result of an unbiased process.

We're at a turning point where the new science of mind and brain can begin to shape the law. By educating judges, jurors, attorneys, witnesses, police officers, and other participants in the legal process, we should be able to produce a legal system that is ultimately more fair. Perhaps we cannot move away from trial by jury anytime soon, but even simple steps, like educating jurors that emotions are constructed, can improve the current situation.

For now at least, the legal system still considers you to be an emotional beast enrobed in rational thought. Throughout this book, we've systematically challenged this myth by evidence and observation, but there's one remaining assumption that we haven't questioned yet: are beasts even emotional? Are the brains of our close primate cousins, such as chimps, capable of constructing emotion? What about dogs: do they have concepts and social reality as we do? Just how unique in the animal kingdom are our emotional abilities? We'll explore these topics in the next chapter.

12

Is a Growling Dog Angry?

I don't have a dog, but several friends' dogs are part of my extended family. One of my favorites is Rowdy, part Golden Retriever and part Bernese Mountain Dog, who is an energetic, playful mutt, always ready for action. True to his name, Rowdy is a barker and a jumper, and he's known to growl when other dogs or strangers come near. In other words, he's a dog.

Sometimes Rowdy can barely contain himself, and once this nearly proved to be his undoing. Rowdy was out for a walk with his owner, my friend Angie, when a teenage boy approached to pet him. Rowdy did not know the boy and proceeded to bark and jump up on him. The boy was not visibly hurt, so it was a surprise when a few hours later, his mother (who had not been present) had Rowdy arrested and registered as a "potentially dangerous dog." Poor Rowdy had to be muzzled on walks for several years afterward. And if Rowdy ever again jumps up on someone, he will be registered as vicious and maybe even put down.

The boy was afraid of Rowdy and perceived him as angry and dangerous. When you encounter a dog who barks and growls, does he actually feel anger? Or is this merely territorial behavior, or an overly boisterous attempt to be friendly? In short, can dogs experience emotion?

Common sense seems to say yes, *of course*, Rowdy feels emotion when he growls. Numerous popular books explore the issue, like *The Emotional Lives of Animals* by Marc Bekoff, *Animal Wise* by Virginia Morell, and *How Dogs Love Us* by Gregory Berns, to name just a few. Dozens of news stories inform us of scientific discoveries in animal emotion: dogs get jealous, rats experience regret, crayfish feel anxiety, and even flies fear the incoming

Figure 12-1: Rowdy

flyswatter. And of course, if you live with pets, you've certainly seen them behave in ways that seem emotional: running around in fear, jumping up in joy, whining in sadness, purring with love. It seems so *obvious* that animals experience emotions just the way we do.* Carl Safina, author of *Beyond Words: What Animals Think and Feel,* puts it succinctly: "So, do other animals have human emotions? Yes, they do. Do humans have animal emotions? Yes, they're largely the same."[1]

Some scientists are not so sure. They suggest that emotions in animals are just illusions: that Rowdy has brain circuits that trigger behaviors for survival but not for emotion. From their perspective, Rowdy can approach or withdraw in dominance or submission, to defend his territory or to avoid a threat. In these instances, the argument goes, Rowdy might experience pleasure, pain, arousal, or other varieties of affect, but he does not have the mental machinery to experience more than that. This latter explanation is deeply unsatisfying because it denies our own experiences. Millions of pet owners would bet money that their dogs growl in anger, droop in sadness, and hide their heads in shame. It's hard to conceive that these perceptions are illusions built around some general affective responses.[2]

---

* For simplicity, I'll use the words "animal," "mammal," "primate," and "ape" strictly to mean the non-human kind. Of course we humans also belong to these categories.

I myself have succumbed to the allure of animal emotions. For years, my daughter has maintained a herd of guinea pigs in her bedroom. One day, we acquired a small baby, Cupcake. Every night for the first week, all by herself in a strange pen, Cupcake sounded like she was crying. I'd carry her around in my sweater pocket, all warm and cozy, which made her chirp with happiness. Whenever I approached the cage, the other pigs would squeal and run away, but little Cupcake would sit still as if waiting for me to pick her up, and then immediately crawl into the crook of my neck for a nuzzle. In those moments, it was very hard to resist the belief that she loved me. For many months, Cupcake was my late-night companion. She would nestle in my lap, purring, as I worked at my desk. Everyone in our house suspected that Cupcake was actually a puppy trapped in a guinea pig's body. And yet, as a scientist, I knew that my perceptions did not necessarily reveal what little Cupcake was actually feeling.

In this chapter, we'll systematically explore what animals are capable of feeling, based on their brain circuitry and on experimental research. We'll have to set aside our fond feelings for our pets, as well as the essentialist theory of human nature, to look carefully at the evidence. Scientists pretty much agree that many of the earth's animals, from insects to worms to humans, share the same basic nervous system plan. They even agree, more or less, that animal brains were built according to the same general blueprint. But as anyone who has renovated a house has learned, the devil is in the details when translating a blueprint into reality. When it comes to comparing brains of different species, even if they have the same networks of regions, microscopic differences in wiring are sometimes as important as these large-scale similarities.[3]

The theory of constructed emotion prompts us to ask whether animals have three necessary ingredients for making emotion. The first ingredient is interoception: do animals have the neural equipment to create interoceptive sensations and experience them as affect? The second is emotion concepts: can animals learn purely mental concepts like "Fear" and "Happiness," and if so, can they predict with these concepts to categorize their sensations and make emotions like ours? Finally, there's social reality: can animals share emotion concepts with each other so they are passed down to the next generation?

To see what animals are capable of feeling, we'll focus primarily on monkeys and great apes because they're our closest evolutionary cousins. In the process, we'll discover whether animals share the kinds of emotions that we feel . . . and the answer has an unexpected twist.

• • •

All animals regulate their body budget to stay alive, so they all must have an interoceptive network of some sort. My lab, together with neuroscientists Wim Vanduffel and Dante Mantini, set out to verify this network in macaque monkeys and were successful. (Macaques and humans shared their last common ancestor about 25 million years ago.) The macaque interoceptive network has some of the same parts as the human interoceptive network we discovered, as well as some differences. The macaque network is structured to function by prediction in the same way that the human network does.[4]

Macaques also likely experience affect. They can't tell us verbally how they feel, of course, but one of my former doctoral students, Eliza Bliss-Moreau, has evidence that they show the same bodily changes in the same situations that we humans do when we feel affect. Eliza studies macaques at the California National Primate Research Center at the University of California, Davis. Her monkeys watched three hundred videos of other monkeys playing, fighting, sleeping, and so on, while Eliza tracked their eye movements and cardiovascular responses. She found that the activity in the monkeys' autonomic nervous system mirrored what a human's would do when viewing these videos. In humans, this nervous system activity is related to the affect they feel, suggesting that macaques experience pleasant affect when watching positive behaviors like foraging and grooming, and unpleasant affect when watching negative behaviors like cowering.[5]

Based on these and other clues from biology, macaques pretty definitely process interoception and feel affect, and if that's the case, then great apes such as chimpanzees, bonobos, gorillas, and orangutans surely feel affect as well. As for mammals in general, it's harder to say for sure. They undoubtedly feel pleasure and pain, as well as alertness and fatigue. Many mammals have circuitry that looks similar to ours but has different functions, so we can't answer the question just by examining the wiring. No one, to my knowledge, has specifically studied the interoceptive circuitry of dogs, but it seems pretty clear from their behavior that they have an affective life. And how about birds, fish, or reptiles? We don't know for sure. I have to admit that these questions preoccupy me as a civilian (as my husband calls me in non-scientist moments). I can't shop for meat or eggs in a supermarket or attempt to rid my kitchen of bloody irritating fruit flies without asking my-self . . . what do these creatures feel?

I think it's best to assume all animals can experience affect. I realize this discussion has the potential to transport us from the land of science to the land of ethics, coming perilously close to moral issues such as pain and suf-fering in laboratory animals, creatures who are factory-farmed for food, and whether fish feel pain when a hook enters their mouth. The natural chemicals that relieve suffering within our own nervous systems, opioids, are found in fish, nematodes, snails, shrimps, crabs, and some insects. Even tiny flies might feel pain; we know that they can learn to avoid odors that are paired with electric shock.[6]

The eighteenth-century philosopher Jeremy Bentham thought that an animal belongs in the human moral circle only if we can prove the animal can feel pleasure or pain. I disagree. An animal is worthy of inclusion in our moral circle if there is *any possibility at all* that it can feel pain. Does that keep me from killing a fly? No, but I'll make it quick.[7]

Macaques do have an important difference from humans where affect is concerned. Many, many objects and events in your world, from the tini-est insect to the largest mountain, cause fluctuations in your body budget and change your affective feelings. That is, you have a large affective niche. Macaques, however, don't care about as many things as you and I do. Their affective niche is much smaller than ours; the sight of a majestic mountain rising in the distance doesn't impact their body budget in the least. Simply put, more things matter to us.[8]

An affective niche is one area of life where size truly matters. In the lab, if we present a human toddler with a collection of toys, they are usually within her affective niche. My daughter, Sophia, would sort her toys by shape, by color, by size, for the sheer fun of it, over and over, statistically honing the various concepts involved. Not so with macaques. The toys alone are unin-teresting and don't impact the macaque body budget or prompt the mon-keys to form concepts. We must offer the macaque a reward of some kind, like a tasty drink or treat, to bring the toys into the macaque's affective niche so statistical learning can proceed. (Eliza tells me that favorite monkey treats include white grape juice, dried fruit, Honey Nut Cheerios, grapes, cucumbers, clementines, and popcorn.) Repeat the reward enough times and the macaque will learn similarities among the toys.

A human infant also receives rewards from his human caregivers: not just tasty treats like breast milk or formula but also the day-to-day effects of tending to his body budget. His caregivers become part of his affective niche because they feed him, keep him warm, and so on. He is born with rudimentary concepts for his mother's scent and voice, learned in utero. In the first few weeks of life, he learns to integrate his mother's other perceptual regularities, such as the feel of her touch and eventually the sight of her face, because she is regulating his body budget. She and other caregivers also guide the infant's attention to things of interest in the world. He follows their gaze to an object (say, a lamp), then they look at him, then at the lamp again, and talk about what he is looking at. They say the word "lamp" to him with intent, alerting and orienting him with a "baby talk" tone of voice.[9]

Other primates do not share attention like this, and so they cannot use it to regulate each other's body budgets the way that humans do. A mother macaque may follow her infant's gaze, but she will not look back and forth from the object to the infant's face, as if inviting her baby to wonder what is in her mind. Baby primates do learn concepts without the explicit reward of their mother's presence, but not with the range and diversity that baby humans do.[10]

Why do humans and macaques have such differently sized affective niches? For starters, a macaque's interoceptive network is less developed than a human's, particularly the circuitry that helps control prediction error. This means a macaque is not as nimble in directing attention to stuff in the world based on past experience. More importantly, a human brain is almost five times as large as a macaque brain. We have much greater connectivity in our control network and in parts of our interoceptive network. The human brain employs this heavy-duty machinery to compress and summarize prediction error in the way we discussed in chapter 6. This allows us to integrate and process more sensory information from more sources more efficiently than a macaque can, to learn purely mental concepts. That's why you can have majestic mountains in your affective niche and a macaque cannot.[11]

• • •

An interoceptive network, along with the affective niche it helps create, is not sufficient for feeling and perceiving emotions. For that, a brain must also be equipped to build a conceptual system, to construct emotion concepts, and to make sensations meaningful as emotions in themselves and others. A hypothetical macaque with the capacity for emotions must be able to look at another macaque swinging in a tree and see not only the physical movement but an instance of "Joy."

Animals can definitely learn concepts. Monkeys, sheep, goats, cows, raccoons, hamsters, pandas, harbor seals, bottlenose dolphins, and plenty of other animals learn concepts by smell. You might not think of smell as conceptual knowledge, but each time you smell the same aroma, such as popcorn in a movie theater, you're categorizing. The mix of chemicals in the air differs each time, and yet you perceive buttered popcorn. Similarly, most mammals use olfactory concepts to recognize friends, foes, and offspring. Many other animals learn concepts by sight or sound as well. Sheep apparently recognize one another by face (!), and goats by vocal bleats.[12]

In the lab, animals can learn additional concepts if you reward them with food or drink, widening their affective niche. Baboons can learn to distinguish a "B" from a "3" regardless of font, and macaques can distinguish animal images from food images. Rhesus macaques can learn the concept "Rhesus macaque" as distinct from "Japanese macaque," even though they are the same species and differ only by color. (Does this remind you of something that humans do?) Macaques can even learn concepts to distinguish painting styles by Claude Monet, Vincent van Gogh, and Salvador Dalí.[13]

The concepts that animals learn will not be the same as human concepts, however. Humans construct goal-based concepts, and a macaque brain simply lacks the necessary wiring to do so. It's the same lack of wiring that accounts for their smaller affective niche.

What about apes — can they construct goal-based concepts? Chimpanzees, our genetically closest cousins, have larger brains than macaques do, with more of the wiring necessary for integrating sensory information. A human brain is still three times as large as a chimp brain, though, with more of this critical wiring. That doesn't rule out goal-based concepts for chimps. It's just likely that your brain is better equipped to create purely mental concepts, such as "Wealth," whereas a chimpanzee brain is better equipped to create concepts for actions and concrete objects, like "Eating" and "Gathering" and "Banana."[14]

Apes almost certainly have concepts for physical behaviors, such as swinging from branch to branch. The big question is, can one chimp watch another chimp swinging in a tree and perceive an instance of "Joy"? That would require the observing chimp to have a purely mental concept and infer the swinging chimp's intention, making a mental inference. Most scientists assume that mental inference is a core ability of the human mind. So a lot is at stake if apes can do it. We know that monkeys cannot; they can understand what a human is doing but not what he is thinking, desiring, or feeling.[15]

Where apes are concerned, it's conceivable they could make mental inferences and construct goal-based concepts, but the scientific jury is still out. Chimps might have the prerequisites because they can create some mental similarities amid perceptual differences. For instance, they know that leopards climb trees, snakes climb trees, and monkeys climb trees. It's conceivable that chimps could extend this concept to a new animal who can perform a similar action, such as a housecat, and predict that the cat will climb a tree. But a human concept "To Climb" is more than just an action; it's a goal. So the real test would be whether chimps would understand that a person running up a flight of stairs, ambling up a ladder, and crawling up a rock face all share the goal "To Climb." That mental feat would show us that chimps really can go beyond physical similarities, grouping together instances of climbing that look very different but have a shared mental goal. And if chimps could comprehend that moving up a social hierarchy is also climbing, then their concepts would be identical to our own. Human infants can accomplish such feats, as we learned in chapter 5, if they have a word to represent the concept. The next question, then, is whether great apes have the capacity to learn words and use them for learning concepts in the way that human infants do.[16]

Scientists have been trying to teach language to apes since the 1960s, usually with a visual symbol system such as American Sign Language because their vocal machinery is not well-adapted for human speech. Apes can learn to use hundreds of words or other symbols to refer to particular features of the world if there is a reward along the way. They can even combine symbols to communicate complex requests for food, such as "cheese eat — wanting to" and "gum hurry — wanting to have some." Scientists still debate whether these apes understand the meanings of the symbols or are just mimicking their trainers in order to request rewards. For our purposes, the most important questions are whether great apes can learn and use words or symbols under their own steam, without an explicit reward, and whether they can build purely mental concepts like "Wealth" or "Sadness."[17]

So far, we have very little evidence that apes can learn and use symbols on their own. They appear to have only one such concept that they can map to a symbol without requiring an external reward: "Food." But when apes do learn to use a word, do they take the next step? Do they use a word as an invitation to go beyond what they see, hear, touch, and taste to infer the mental? We don't know yet. Words certainly don't prompt apes to search the minds of other creatures for concepts the way a human infant does. But there are intriguing possibilities. For example, it appears that chimps can categorize dissimilar-looking objects according to their function — tools, containers, food — if you reward them, and if they already have firsthand experience with the function. Moreover, if you teach and reward them to associate a symbol with a category like "Tool," they can match the symbol to unfamiliar tools.[18]

Do apes use words in this way *only* to request rewards? Skeptics point out that apes certainly don't use symbols or words to talk about the weather or their children; they can *refer* to something other than a reward, but only if a reward is waiting at the other end. (It would be interesting to observe what would happen to symbol-trained apes if their trainers stopped rewarding them. Would they continue to use the symbols?) The important point, I think, is that words don't seem to be intrinsically part of most apes' affective niche, as they are for typical human babies. To apes, words alone are not worth learning.[19]

One important exception to this story might be bonobos. They are very social creatures, far more egalitarian and cooperative than common chimps. They also have a larger social network and play longer before assuming adult roles. And some bonobos appear able to complete tasks without external rewards, whereas chimps seem to require them. Take the story of Kanzi, an infant bonobo who watched his stepmother and other adult bonobos earn food rewards for learning language-like symbols. By six months old, Kanzi appeared to be learning the symbols too, on his own, by watching other bonobos earn rewards. At a certain point, the scientists realized with careful testing that Kanzi appeared to understand some spoken English. So it is possible that a bonobo brain, immersed in a language-rich environment, can learn the meaning of concrete words.[20]

Chimps, in contrast to bonobos, have been characterized as charming, clever creatures with a dark side. They hunt and kill each other opportunistically to take over territory or get food. They also attack strangers for no reason, maintain a rigid dominance hierarchy, and beat females into sexual submission. Bonobos would rather work out their conflicts by having sex. That's a much better alternative than genocide.

Nevertheless, chimps may have been given a bad rap in the lab when it comes to concept learning. Chimps in language experiments were removed from their mothers in infancy and raised in a human-like environment vastly different from their natural habitat. These infants would normally live with their mother for up to ten years and nurse with them for five, so this premature separation could have changed the wiring of each chimp's interoceptive network and strongly influenced the results of the experiments. (Imagine separating a human infant from his mother like this!)[21]

When tested under more natural circumstances, a chimp's affective niche appears to be broader than many experiments suggest. For this insight, we have to thank the primatologist Tetsuro Matsuzawa at Kyoto University's Primate Research Institute. Matsuzawa has accomplished a truly impressive task. He has three generations of chimps who live in an outdoor compound built to look like a forest. Each day, chimps come to the lab *by choice* to do experiments. Sometimes they are rewarded, of course, but to emphasize this is to miss the point. These animals have a long-term, trusting relationship with Matsuzawa and the other human experimenters at the institute. A mother chimp will hold her baby on her lap and allow a human to run an experiment with her infant. For example, one study tested human and chimp infants as they learned concepts for mammals, furniture, and vehicles (using lifelike miniatures). This learning proceeded with *no rewards* as each infant was tested while sitting on his or her mother's lap. The infant's proximity to the mother, in relation to the trusting bond with the human experimenter, may have been enough to bring this situation into the chimp infant's affective niche. Incredibly, the chimp and human infants formed concepts equally well under these conditions. Still, the human infants spontaneously manipulated the objects, like moving toy trucks around, making concept formation more likely, whereas the chimps did not.[22]

Matsuzawa's troupe would be ideal for learning the limits of a chimp's conceptual abilities. We could test infant chimps, whose conceptual systems are still malleable, in a natural environment on their mothers' laps, perhaps conducting concept-building experiments like those in chapter 5. Would chimp infants be able to use a nonsense word like "toma" to group together objects or images that share little perceptual similarity, as human infants can?

At present, however, we have no firm evidence that chimps can form goal-based concepts. They cannot imagine something completely novel, like a flying leopard, even though they and macaques have a network that's analogous to the human default mode network (part of the interoceptive network). They cannot consider the same situation from different points of view. They can't imagine a future that is different from the present. They also do not realize that goal-based information resides inside the heads of other creatures. That's why chimps and other great apes most likely cannot create goal-based concepts. When rewarded, apes can learn a word, but they cannot spontaneously use the word to form a mental concept with a goal, like "Things That Taste Good with Termites."[23]

Any concept can be goal-based — recall that "Fish" can be a pet or a dinner — but emotion concepts are *only* goal-based, so it seems very likely that chimps cannot learn emotion concepts like "Happiness" and "Anger." Even if they can learn an emotion word like "angry," it's not clear that they can understand it or use it in a goal-based way, like categorizing another creature's actions as anger.

Sometimes apes appear to understand a purely mental concept when they do not. In one experiment, chimps earned tokens for completing tasks and could exchange them for food. They spontaneously learned to save up their tokens to exchange them for a desired treat. When you watch chimps engage in this transaction, it is tempting to infer that chimps understand the concept "Money." But here, the token was just a tool for obtaining food, rather than a form of currency that's exchangeable for goods in general. The chimps did not understand, as many humans do, that money comes to have value for its own sake.[24]

If chimps cannot form goal-based concepts, then necessarily, chimps are not naturally equipped to teach concepts to one another; that is, they don't have social reality. Even if they could learn a concept like "Anger" from a human trainer, one generation doesn't create the context for the next generation to bootstrap concepts into their brains. Chimps and other primates do have shared practices, like cracking nuts with rocks, but chimp mothers don't spontaneously instruct their infants on the finer culinary points; the children learn by observation. For example, in a troop of macaques in Japan, one member began washing her food before eating it, and within ten years, three-quarters of the adults in her troop had picked up the practice. This sort of collective intentionality is very limited compared to what we humans do with words and the mental concepts that we invent.[25]

The human capacity for social reality appears unique in the animal kingdom. Only we can create and share purely mental concepts using words. Only we can use these concepts to more effectively regulate our own body budgets and each other's, while we cooperate and compete with one another. Only we have concepts for mental states, such as emotion concepts, for predicting and making sense of sensations. Social reality is a human superpower.[26]

Which brings us back to Matsuzawa and his chimps. It is remarkable how he nestled a chimp troupe, preserving its family relationships, into human culture in an intimate way. I wonder whether, over time, Matsuzawa's very human cultural context will influence the brain development of the infant chimps, as they are raised by mothers who are acculturated by a group of trusting, loving humans.

One example that I find particularly striking, relayed by Virginia Morell in her book *Animal Wise*, describes two human experimenters who provide social support to a nursing mother chimp. The mother is reluctant to nurse her infant, but the experimenters gently encourage her to be brave. In Morell's words, "A researcher gently picks up the baby and places it in the mother's arms. The infant's hands latch on to her fur. The mother then attempts to nurse but cries when the baby takes her nipple; she seems about to drop her infant to the floor. But then the soft voice of the scientist is heard again. Yes, yes, he says soothingly, it may hurt at first, but soon it will not. And slowly the mother settles down, cradling her baby against her breast and letting the infant nurse." Thousands of human mothers each day experience nursing for the first time, and I can tell you from experience that it hurts like hell. But someone else (a nurse, an older female relative, or a friend) offers supportive encouragement and shows you what to do, and eventually all is well.[27]

To the mother chimp, these helpful humans are not merely her caretakers: they are affectively salient to her, regulating her body budget. She and her infant and their relationship are being bathed in human culture. Will this social contact make a difference to the language and conceptual abilities of these chimps long-term? If their offspring eventually become able to form goal-based concepts, it's a whole new ballgame.

• • •

Okay, so chimps and other primates don't appear to have emotion concepts or social reality. How about dogs like Rowdy? After all, we have bred dogs to be human companions, so they, like us, are truly social creatures. If any non-human animals were to be capable of emotion, dogs would seem to be prime candidates.

Just a couple of decades ago, it took the Russian scientist Dimitri Belyaev only about forty generations to transform wild foxes into something that approximated domesticated dogs. Each time female foxes gave birth,

Belyaev chose the fox pups who were most interested in and least aggressive toward humans and selectively bred them. The experimentally bred beasts looked like dogs; their skulls were shorter and they had wider muzzles, curly tails, and floppy ears, even though Belyaev did not select for these features. Their chemical makeup was closer to dogs than foxes. And they had a strong motivation to interact with humans. Modern dogs also have long been bred for certain desirable characteristics, like attaching to a human caregiver, and other characteristics surely have come along for the ride, perhaps even something like human emotion concepts.[28]

One of those inadvertently bred characteristics, I speculate, is a certain kind of dog nervous system. We can regulate a dog's body budget, and dogs can regulate ours in turn. (I wouldn't be surprised if dogs and their human owners even synchronize their heart rates, the way close humans do for each other.) We also probably selected for dogs with eyes that we perceive as expressive and facial muscles that move easily to serve as a canvas upon which we can paint complex mental states. We love dogs so much that we bred them to love us back, or at least to see them as loving us. We treat them as little almost-humans with four legs and a fur coat. But do dogs experience or perceive human emotion?[29]

Dogs, like other mammals, feel affect. No big surprise here. One way they appear to express affect is by wagging their tails. They apparently make larger tail-wagging movements to the right during pleasant events, such as seeing their owner, and to the left for unpleasant events, such as seeing an unfamiliar dog. The choice of side has been associated with brain activity: wagging to the right is said to mean relatively greater activity on the left side of the brain, and vice versa.[30]

Dogs also appear to look at each other's tails to perceive affect. They're more relaxed when they view movies of right-wagging tails and more stressed for left-wagging tails, as measured by heart rate and other factors. Dogs also appear to perceive affect in the faces and voices of humans. I haven't come across any relevant brain-imaging experiments on dogs, but if they have affect, it stands to reason that they have some sort of interoceptive network. Just how large their affective niche is no one knows, but given their social nature, I'll bet it is yoked to their owners in some way.[31]

Dogs can learn concepts too. Again, not surprising. They can distinguish dogs from other animals in photographs, for example, if trained to do so. It takes them a thousand or more trials to get the knack of it, compared to human infants who need only a few dozen trials. But dogs can learn to be accurate over 80 percent of the time, even if the dog in the photo is completely new or embedded in a complex scene. Not bad for a dog brain.[32]

Dogs also form olfactory concepts. They can distinguish the smell of an individual human, grouping together different smells from different parts of the body to treat as equivalent, and yet distinct from the smells of other humans. And of course, we know that dogs can be trained to track categories of objects by smell. Anyone who's been caught in an airport with food or drugs in their suitcase can tell you so.[33]

I will gingerly concede that dogs appear to infer intentions of some sort. Dogs are better than chimps at perceiving human gestures and following human gaze. When Sophia was younger and would play in the sand with her favorite beach dog, Harold, the two of them often looked to a human adult for permission to run farther away: Sophia to me and Harold to his owner. Dogs use our gaze to tell them what to attend to, and their skill is so great that they *seem* to read our mind in our eyes. Even more remarkably, dogs follow *each other's* gaze to get information about the world. When Rowdy wants to know what's going on, he spontaneously looks to his "sister," Biscuit, a Golden Retriever, and follows her gaze. The two of them freeze as they reference each other, and then ... they both suddenly leap into action. It's like watching a silent movie.[34]

But being the skeptic that I am, I have my doubts that dogs are making goal-based mental inferences. They could be just really good at perceiving human actions, because, let's be honest, we've bred them to be sensitive to our every whim.

Dogs do appear to understand that humans use symbols to communicate intent. For example, in one study, an experimenter put dog toys in different rooms and then used miniature replicas of the toys as symbols. Her test subjects (Border Collies) understood she was asking them, via the miniature, to retrieve the matching toy from the other room. This is rather more sophisticated than playing fetch. Studies also show that dogs use different growls and barks to communicate with each other, although they might just be communicating arousal (affect) in the acoustic signal. One study even shows that a dog named Sofia, like our chimp friends, could be trained to press symbols on a keyboard to communicate a few basic concepts: a walk, toy, water, play, food, and her crate.[35]

Clearly, dogs have something nontrivial going on upstairs, but even so, scientists have no indication yet that dogs have emotion concepts. In fact, there's pretty good evidence that they don't, though many dog behaviors look emotional. Dog owners, for example, infer guilt when they believe their dog is hiding something (for example, avoiding eye contact) or is being submissive (such as drooping the ears, lying down and showing the belly, or holding the tail low). But do dogs have a concept of guilt?

A clever study investigated this question. In each trial, a dog owner offered his or her dog a desirable biscuit, then explicitly instructed the dog not to eat it and promptly left the room. Unbeknownst to the owner, however, an experimenter then entered the room and influenced the dog's behavior, either handing the treat to the dog (who ate it) or removing the treat from the room. Afterward, the experimenter either told the owner the truth or lied. Half the owners were told that their dog had obeyed and to greet their dog in a warm and friendly manner; the rest heard that the dog had eaten the biscuit and should be scolded. This created four different scenarios: obedient dog with a friendly owner, obedient dog being scolded, disobedient dog with a friendly owner, and disobedient dog being scolded. What happened? The scolded dogs performed more behaviors that people perceive as stereotypically guilty, regardless of whether or not the dogs had disobeyed. This is evidence that dogs were not experiencing guilt at performing a forbidden act; rather, their owners were perceiving guilt when they believed the dog had eaten the biscuit.[36]

Another study looked at jealousy in dogs, asking owners to interact with a toy dog while the real dog watched. The toy barked, whined, and wagged its tail. The study found that dogs in this situation would snap, whine, push at the owner and the toy, and insert themselves between the owner and the toy, more often than when the owner interacted with a different toy (a jack-o-lantern) or read a book. The authors interpreted these findings to mean that the dogs were jealous, particularly because many of the dogs tested sniffed the anus of the toy dog. Unfortunately, the experimenters did not test to see if the owners were behaving differently in the three conditions (toy dog, jack-o-lantern, and reading) in any way that could account for the dogs' behavior. They assumed that the owner's behavior was identical, and that the dog understood that jealously was called for in only one condition. So even though many pet owners are confident that their dogs experience jealousy, we have no scientific evidence to support this belief.[37]

Scientists are still exploring the limits of what dogs can do, emotionally speaking. Their affective niche is broader than ours in some respects, because their senses of smell and hearing are superior; but their affective niche is narrower in other respects, because they can't travel into the future to imagine a world other than the present one. My view, from evaluating the evidence, is that dogs don't have human emotion concepts like anger, guilt, and jealousy. It's conceivable that one individual dog could develop some emotion-like concept of its own, different from any human emotion concept, in relation to its owner. Without language, however, the dog's emotion concept would necessarily be narrower than a human's, and it couldn't teach the concept to other dogs. So the possibility of a common "Anger" (or similar concept) experienced by dogs is vanishingly remote.

Even if dogs don't share human emotions, it's remarkable just how much dogs and other animals can accomplish through affect alone. Many animals can experience unpleasant affect when another animal nearby is suffering. The first animal's body budget is taxed by the second animal's discomfort, so the first animal tries to fix the situation.* Even a rat will help another rat who is in distress, for example. Human infants can comfort another infant who is in distress. You don't need emotion concepts for this ability, just a nervous system with interoception that produces affect.[38]

Amid the accumulating evidence that dogs have some truly remarkable skills, we still severely misunderstand dogs. We see them relative to ourselves, using the outmoded essentialist theory of human nature, instead of seeing them on their own terms. John Bradshaw, the author of *Dog Sense,* explains that we view dogs wrongly as having a dominance-seeking "inner wolf" that needs to be tamed by a civilizing force, their owners (an intriguing parallel to our own mythical inner beast that must be tamed by rationality). Dogs are extremely social creatures, continues Bradshaw, as are wolves in the wild when you don't toss them into zoos with a bunch of strangers. Put a few dogs together in a park and in a few moments they're playing together. What looks like dominance in dogs is what Bradshaw calls "anxiety," and what we'd say is a body budget that's out of balance. Think about it: we take an affiliative, affectionate creature whose body budget we regulate, and we abandon it for most of every day. (Can you imagine doing that to a human child?) Of course their body budget will get out of whack and they'll feel high-arousal, unpleasant affect. We've bred them to be affectively

---

* I am studiously avoiding the word "empathy" here. For some scientists, empathy means simple synchrony of affect. For others, empathy is a complex, purely mental concept rooted in social reality. These two completely different ideas, unfortunately, are named by the same word in English.

dependent on us. So owners must take care with their dogs' body budget. Dogs might not feel fear, anger, and other human emotions, but they do experience pleasure, distress, attachment, and other affective feelings. But for dogs to be successful as a species, living cooperatively with their human companions, affect may be enough.[39]

• • •

Let's recap where we are. Do animals regulate their body budgets by interoception? I cannot speak for the entire animal kingdom here but for mammals — rats, monkeys, apes, dogs — I think we are on pretty safe ground answering yes. Do animals experience affect? Again, I think we can give a pretty confident yes, based on some biological and behavioral clues. Can animals learn concepts and can they categorize predictively with those concepts? Definitely. Can they learn action-based concepts? Unquestionably yes. Can they learn the meaning of words? Under some circumstances, some animals can learn words or other symbol systems, in the sense that the symbols become part of the statistical patterns that a brain can capture and store for later use.

But can animals use words to go beyond the statistical regularities in the world, to create goal-based similarities that unite actions or objects that look, sound, or feel different? Can they use words as invitations to form mental concepts? Do they realize that part of the information they need about the world resides in the minds of other creatures around them? Can they categorize actions and make them meaningful as mental events?

Probably not. At least not in the way that we humans do. Apes can construct categorizations that are more similar to our own than we might have imagined. But right now, there is no clear evidence that any non-human animals on the planet have the sorts of emotion concepts that humans do. We alone have all the ingredients necessary to create and transmit social reality, including emotion concepts. This holds true even for Man's Best Friend.

So, let's return to Rowdy: was he angry when he growled and jumped up on the boy? Based on our discussion so far, Rowdy lacks emotion concepts, so you might guess that my answer is no.

Well, not exactly. (Get ready for that twist I mentioned at the beginning of the chapter.)

From the perspective of the theory of constructed emotion, the question "Is a growling dog angry?" is *the wrong question to ask* in the first place, or at least incomplete. It assumes that a dog is measurably angry or not angry in some objective sense. But as you've learned, emotion categories have no consistent, biological fingerprints. Emotions are always constructed from some perceiver's point of view. So the question "Was Rowdy angry?" is actually two separate scientific questions:

- "Was Rowdy angry from the boy's perspective?"
- "Was Rowdy angry from his own perspective?"

These questions have substantially different answers.

The first question asks, "Could the boy construct a perception of anger from Rowdy's actions?" Absolutely. When we observe a dog's behavior, we use our own emotion concepts to make predictions and construct perceptions. Rowdy was angry, from a human perspective, if the boy constructed a perception of anger.

Was the boy correct in his assessment? Accuracy for categories of social reality, you may remember, is a matter of consensus. Let's say that you and I are walking past Rowdy's house and he growls loudly. You experience him as angry. I don't. Accuracy could be: Do we agree? Do our experiences of Rowdy agree with his owner Angie's experience, as she knows him best? Do our experiences of Rowdy match the social norms of the situation, because this is social reality after all? If we agree, then our constructions are in sync.

Now let's consider the second question, regarding Rowdy's experience. Did he feel anger when he growled? Was he able to construct an experience of anger from his sensory predictions? The answer is almost certainly no. Dogs do not have the human emotion concepts necessary to construct an instance of anger. Lacking a Western concept of "Anger," dogs cannot categorize their interoceptive and other sensory information to create an instance of emotion. Nor can they perceive emotion in other dogs or in humans. Dogs do perceive distress and pleasure and a handful of other states, a feat that requires only affect.

Dogs may well have some emotion-like concepts. For example, a number of scientists now suspect that very social animals, such as dogs and elephants, have some concept of death and can experience some kind of grief. This grief need not have exactly the same features as human grief, but both could be rooted in something similar: the neurochemical basis of attachment, body budgeting, and affect. In humans, the loss of a parent, lover, or close friend can wreak havoc with your budget and cause much distress that operates similarly to drug withdrawal. When one creature loses another who helped to keep its body budget on track, the first creature will feel miserable from the budget imbalance. So Brian Ferry of the rock band Roxy Music was right — love *is* a drug.[40]

Rowdy's misadventure has a backstory that may have affected his behavior on that fateful day. Earlier that week, before his arrest, Rowdy lost his "sister" Sadie, a Golden Retriever who died of old age. Their owner Angie believes this is why Rowdy jumped up on the boy that day. She said Rowdy was grieving, which in canine terms means he lost a creature who helped to regulate his body budget, and he temporarily forgot his training. Rowdy knows he is not supposed to jump, but maybe he just wasn't himself that day — whatever self a dog can have.

There are anecdotal reports of dogs who stop eating or become apathetic after the death of another dog in the family. Some people see these cases as evidence of grief in dogs, but they also could be understood more simply as an effect of body-budget imbalance, accompanied by unpleasant affect. After all, Angie was probably grieving Sadie's death, and Rowdy, being very sensitive to her behavior, could have detected some affective change in her, throwing off his own budget even more.

Dividing our growling dog question into two questions, reflecting human and canine perceptions separately, is not a parlor trick. I'll admit, the distinctions I'm making here are subtle. Construction views of emotion are frequently misinterpreted as saying "dogs don't have emotions" (and sometimes even "people don't have emotions"). Such simplistic statements are meaningless because they assume emotions have essences so that they can exist, or not, independent of any perceiver. But emotions are perceptions, and every perception requires a perceiver. And therefore every question about an instance of emotion must be asked from a particular point of view.

• • •

If apes, dogs, and other animals don't have the capacity to experience human emotions, why are there so many news stories about emotions being discovered in animals, even in insects? It all comes down to a subtle mistake that's repeated over and over in science, and which is very difficult to detect and overcome.

Picture this: a rat is placed into a small box with an electrical grid on the floor. Scientists play a loud tone and then a moment later give the rat an electrical shock. The shock causes the rat to freeze and its heart rate and blood pressure to rise, as it stimulates a circuit that involves key neurons in the amygdala. The scientists repeat this process many times, pairing the tone and the shock, with the same results. Eventually, they play the tone without the shock, and the rat, having learned that the tone foreshadows the shock, again freezes and has increased heart rate and blood pressure. The rat's brain and body respond as if expecting the shock.

Scientists who adhere to the classical view say that the rat has learned to be afraid of the tone, calling this phenomenon "fear learning." (This is the same type of experiment performed on SM, the woman with no amygdala who allegedly couldn't learn fear, as described in chapter 1.) All over the world, for decades, scientists have been shocking rats, flies, and other animals to map how neurons in the amygdala allow them to learn to freeze. Having identified this freezing circuit, scientists then infer that the amygdala contains a fear circuit — the essence of fear — and the increased heart rate, blood pressure, and freezing is said to represent a consistent, biological fingerprint for fear. (I've never been sure why they decided it's fear. Couldn't the rat be learning surprise, or vigilance, or maybe just pain? If I were the rat, I'd be pretty pissed off about the shocks, so why isn't it "anger learning"?)[41]

Anyway, these scientists go on to say that their fear learning analysis extends from rats to humans, because the relevant fear circuitry in the amygdala has been passed to us through mammalian evolution à la the "triune brain." These fear learning studies helped to establish the amygdala as the supposed brain location of fear.[42]

In psychology and neuroscience, so-called fear learning has become an industry. Scientists use it to explain anxiety disorders like post-traumatic stress disorder (PTSD). It's employed to aid with drug discovery in the pharmaceutical industry and to understand sleep disturbance. With over 100,000 hits on Google, "fear learning" is one of the most commonly used phrases in psychology and neuroscience. And yet, under the hood, fear learning is just a fancy name for another well-known phenomenon: classical conditioning or Pavlovian conditioning, named after the physiologist Ivan Pavlov, who discovered it with his famous experiments on salivating dogs.* The classic fear learning experiment demonstrates that a benign stimulus, such as a tone, can acquire the ability to trigger certain amygdala

---

* Feed a dog and it salivates. Ring a bell before feeding the dog, repeat this sequence enough times, and the dog will salivate when it hears the bell. Pavlov was awarded a Nobel Prize in 1904 for this discovery.

circuitry in anticipation of uncertain danger. Scientists have spent years mapping this circuitry in elegant detail.[43]

Now comes the subtle mistake I alluded to. Freezing is a behavior, whereas fear is a much more complex mental state. The scientists who believe they study fear learning are categorizing a freezing behavior as "Fear" and the underlying circuit for freezing as a fear circuit. Just as I categorized Cupcake the guinea pig as happy, when she herself couldn't construct an experience of happiness, these scientists unknowingly apply their own emotion concepts, construct perceptions of fear, and attribute fear to the freezing rat. I call this general scientific mistake the *mental inference fallacy.*

Mental inference is normal; we all do it every day, automatically and effortlessly. When you see a friend smile, you might instantly infer that she is happy. When you see a man drinking a glass of water, you might infer that he's thirsty. Alternatively, you might infer that he's feeling dry-mouthed anxiety or pausing dramatically before making a point. When you're on a lunch date and you feel hot and flushed, you might infer that it's caused by romantic feelings or by a case of the flu.[44]

Children of course perceive emotions in their toys and their security blankets and have fascinating two-way conversations with them, but adults are also experts in this regard. In a famous experiment from the 1940s, Fritz Heider and Mary-Ann Simmel created a simple animation of geometric shapes to see if viewers would infer mental states. The video features two triangles and a circle moving around a large rectangle. The video contains no sound and no explanation for the movements. Even so, viewers readily assigned emotions and other mental states to the shapes. The large triangle, some said, was bullying the small, innocent triangle until the brave circle came to the rescue.

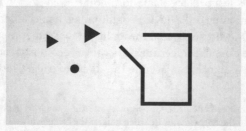

Figure 12-2: Still image from a Heider-Simmel
video available at heam.info/heider-simmel

Scientists, as members of the human species, make mental inferences when interpreting the findings of their own experiments. In fact, every time scientists record a physical measurement and assign it a mental cause, they commit the mental inference fallacy. "That change in heartbeat was caused by excitement." "That scowl is expressing anger." "That activity in the anterior insula was caused by disgust." "That test subject pressed the computer key slightly faster because of anxiety." Emotions do not cause these actions in any objective, perceiver-independent sense. These actions, on their own, are surely evidence that something psychological has occurred, but the scientists are guessing what it is. That is what scientists do: we measure stuff, and then we transform the pattern of numbers into something meaningful by making an inference. But when scientific explanation is your goal, some inferences are better than others.[45]

The fear learning phenomenon is the most dramatic example of the mental inference fallacy in the science of emotion.* Its practitioners blur the important distinction among movement, behavior, and experience. Contracting a muscle is a movement. Freezing is a behavior because it involves multiple, coordinated muscle movements. The feeling of fear is an experience that may or may not occur together with behaviors like freezing. Circuitry that controls freezing is not circuitry for fear. This egregious scientific misunderstanding, along with the phrase "fear learning," has sown confusion for decades and turned what's effectively an experiment on classical conditioning into an industry of fear.[46]

The whole notion of fear learning is fraught with other problems. Rats in threatening situations do not always freeze. When you put them into a small box with tones and shocks arriving together at unpredictable times, rats indeed freeze, but in a larger enclosure, rats run away, unless they're cornered, in which case they attack. If you restrain the rat during the tone (which shouldn't matter, because the rat is going to freeze anyway), its heart rate goes *down* instead of up. Additionally, not all of these varied behaviors require the amygdala. To date, scientists have identified at least three alleged fear pathways in the rat brain, each associated with a specific behavior, all of them products of the mental inference fallacy. Finally, a simple behavior

---

* If we scanned the brains of scientists as they write papers about "fear learning," we'd probably see evidence of mental inference as activity in nodes of the interoceptive and control networks as they describe their freezing rats as afraid.

like freezing is supported by multiple circuits within a distributed network that is not specific to freezing or fear.[47]

In a nutshell, you can't study fear by shocking rats unless at the outset you have defined "fear" circularly as "the freezing response of a shocked rat."

Humans, like rats, act in various ways when threatened. We might freeze, flee, or attack. We might also crack jokes, faint, or ignore what's going on. Such behaviors might be evoked by distinct circuitry in the brain that is shared among mammals, but they are not inherently emotional, and they're not evidence that emotions have biological essences.

Nevertheless, some scientists continue to write that they've isolated highly complex mental states in animals. Baby rats, for example, when separated from their mother after birth, make a high-pitched noise that sounds like crying. Some scientists inferred that the brain circuitry responsible for the crying must be the circuitry for distress. But these baby rats aren't sad. They're *cold*. The sound is just a byproduct as the baby rats try to regulate their body temperature — part of their body budget — a task normally done by their absent mothers. It has nothing to do with emotion. The mental inference fallacy strikes again.[48]

From now on, any time that you read an article about animal emotion, watch for this pattern. If a scientist labels a behavior like freezing using a mental state word like "fear," you should think, "Aha, the mental inference fallacy!"

To be fair, it's extremely hard for scientists to avoid the trap of mental inference. Grant agencies prefer to fund research that is directly relevant to humans. Scientists must also recognize that they are performing a mental inference in the first place, which is a nontrivial feat of introspection. And then they must be brave enough to face the criticism and scorn of their colleagues for swimming against the tide. But it can be done.

The neuroscientist Joseph E. LeDoux, who popularized the idea of fear learning in his acclaimed book *The Emotional Brain*, now argues against using the term "fear" altogether when referring to a rat. In taking this stand, he is a scientist of rare intellectual courage. He had published hundreds of papers on so-called fear learning, and a popular book on the brain basis of fear in the amygdala, yet he carefully considered the contrary evidence and revised his position. In his revised view, freezing helps keep an animal safe when facing threat; it is a survival behavior. His classic experiments reveal what he now calls a survival circuit that controls freezing behavior, not a mental state like fear. LeDoux's theoretical shift is just another example of the new scientific revolution of the mind and brain, steering the field toward a more scientifically defensible theory of emotion.[49]

Although LeDoux and other like-minded scientists have made the shift, you can still easily find the mental inference fallacy in YouTube videos and TED talks by other researchers who study emotion in animals. The speaker shows you a compelling movie or a picture of an animal engaging in some behavior. See how the rat is happy when you tickle it; see how sad the dog is when he whimpers; see how afraid the rat is when she freezes. But remember, emotions are not observed, they are constructed. When you watch the video, you have no awareness that you're using conceptual knowledge to make an inference, any more than you were aware of the processes that turned random blobs into a bee in chapter 2. So to you, it seems like the animal is emotional.

In chapter 4, I explained that every so-called emotionally reactive brain region is issuing predictions to regulate the body budget. Add the mental inference fallacy, mix well, and you have a recipe for a grand mythology of how emotions work in the brain. It's one thing to observe that a rodent's anterior cingulate cortex increases its activity when a neighbor is in pain. It's quite another to say the rodent is feeling empathy. A simpler explanation is that the two animals are just influencing each other's body budgets, as so many creatures do.[50]

You're more likely to engage in mental inference when the animal in question is similar to yourself. It's easier to perceive joy in a scampering dog than in a scampering cockroach. It's easier to see love in a mother bunny sleeping with her young than in a mother caecilian, a worm-like amphibian, feeding her little babies on her own flesh. The Oscar-nominated science-fiction film *District 9* provides a fantastic example of this phenomenon. Its alien creatures seem at first like disgusting, human-sized insects, but once we glimpse that they have families and loved ones, we feel empathy for them. Even Heider's and Simmel's shapes seem human-like, because their speed and trajectories are reminiscent of people chasing one another. We start to perceive their actions in terms of mental causes, and they enter our moral circle.[51]

Mental inference toward animals is not a bad thing in itself—it's completely normal. Every day, I drive by a billboard featuring an adorable baby orangutan. I beam every time I approach it, no matter what else I am brooding about, even though I know the orangutan is not really smiling toward me and does not share a mind like my own. Frankly, if everyone engaged in the mental inference fallacy with animals, and in the process we admitted those animals into our moral circle, maybe we'd have fewer poachers who slaughter elephants and rhinos for their ivory or hunt gorillas and bonobos as food. If people engaged in more mental inference when observing their fellow humans, perhaps we'd have less cruelty and fewer wars. When we have our scientist hats on, however, we must resist the lure of mental inference.[52]

We are accustomed to thinking about animals in terms of ourselves: how similar they are to us, what they teach us about ourselves, how they might be useful to us, how we are superior to them. It's okay for us to anthropomorphize animals if it's going to protect them. But when we see animals through the lens of our own identity, we can harm them in ways that we often don't think about. We treat anxiously attached dogs as "too dominant" and punish them when we should be offering them predictable care and affection. We rip baby chimps from their mothers when in the wild they would nurse until they are five years old, secure in the warmth and smell of their mother's fur.

Our challenge is to understand animal minds for their own sake, not as inferior human minds. The latter idea comes from the classical view of human nature, which implies that chimps and other primates are less evolved, diminished versions of ourselves. They're not. They're adapted to the ecological niche that they live in. Chimps have to forage for food and modern humans largely do not, so a chimp brain is wired to identify and remember details, not to build mental similarities.[53]

In the end, if we learn about animals on their own terms, we will benefit because our relationship with them will be better. We humans will do less damage to them and to the world that we all inhabit.

• • •

Animals are emotional creatures, at least as far as human perceivers are concerned. This is part of the social reality that we create. We grant emotions to our cars, our houseplants, and even little circles and triangles in a movie. We also grant emotions to animals. However, this does not mean that animals *experience* emotion. Animals with a small affective niche cannot form emotion concepts. A lion cannot hate a zebra when she hunts and kills it as prey. That is why we don't find the lion's actions immoral. Anytime you read a book or news story about animals experiencing human emotions ("News

Flash: Cats Feel Schadenfreude toward Mice"), keep this mindset and you'll quickly see the mental inference fallacy materialize before your eyes.

Some scientists still presume that all vertebrates share preserved, core emotion circuits to justify the claim that animals feel as humans do. One prominent neuroscientist, Jaak Panksepp, routinely invites his audiences to see evidence of such circuits in his photos of growling dogs and hissing cats, and in videos of baby birds "crying for their mothers." It is doubtful, however, that these proposed emotion circuits exist in any animal brain. You do have survival circuits for behaviors like the famous "four F's" (fighting, fleeing, feeding, and mating); they're controlled by body-budgeting regions in your interoceptive network, and they cause bodily changes that you experience as affect, but they are not dedicated to emotion. For emotion, you also need emotion concepts for categorization.[54]

The search for emotional capacities in animal minds is ongoing. Bonobos and perhaps chimpanzees, our close cousins, might have the hot-wiring in their brain circuitry to form their own sort of emotion concepts. Elephants are another intriguing possibility; they are long-lived, social animals who form strong bonds in close-knit herds. Ditto for dolphins. Even dogs like Rowdy are good candidates, having been bred alongside humans for thousands of years. Something more may be going on in these animals, even if it is not *human* emotion. As for laboratory rats, Cupcake the guinea pig, and most other animals that we experience as having emotion, they cannot construct emotion because they don't have the necessary emotion concepts. Non-human animals feel affect, but the reality of their emotion is, for the moment, only within ourselves.

13

From Brain to Mind:
The New Frontier

The human brain is a master of deception. It creates experiences and directs actions with a magician's skill, never revealing how it does so, all the while giving us a false sense of confidence that its products — our day-to-day experiences — reveal its inner workings. Joy, sadness, surprise, fear, and other emotions seem so distinct and feel so built-in that we assume they have separate causes inside us. When you have a brain that essentializes, it's easy to come up with a wrong theory of the mind. We are, after all, a bunch of brains trying to figure out how brains work.

For millennia, the deception has been largely a success. Oh, the essences of the mind received a makeover every century or two, but for the most part, the idea of mental organs has pretty much stuck around.* Casting away those essences remains a challenge today because the brain is wired to categorize, and categories breed essentialism. Every noun we utter is an opportunity to invent an essence without intending to do so.

Little by little, the science of the mind is finally removing its training wheels. The skull is no longer the force field that it was, now that brain-imaging technology can peer harmlessly into a human head. New wearable measurement devices are moving psychology and neuroscience out of the

---

* In a nutshell, the idea that concepts depend on experience (empiricism) keeps being soundly trounced by beliefs that concepts are built-in, either because you are endowed with them (nativism) or because they come from intuition or logic (rationalism). Every attempt at empiricism has failed in one way or another, from the associationist philosophers of the seventeenth century to the behaviorists of the twentieth century.

lab and into the real world. As we amass petabytes of brain data with our twenty-first-century tech toys, however, the media, venture capitalists, most textbooks, and some scientists are still interpreting that data with a seventeenth-century theory of the mind (having upgraded to a fancy version of phrenology from Plato 1.0). Neuroscience has delivered a far better understanding of the brain and its function than our own experiences ever could, not just for emotion but for all mental events.

If I have done my job correctly, you now realize that many seeming facts about emotions in textbooks and in the popular media are highly doubtful and must be reconsidered. In these pages, you've learned that emotions are part of the biological makeup of the human brain and body, but not because you have dedicated circuits for each one. Emotions are a result of evolution, but not as essences passed down from ancestral animals. You experience emotions without conscious effort, but that does not mean you're a passive recipient of these experiences. You perceive emotions without formal instruction, but that does not mean that emotions are innate or independent of learning. What's innate is that humans use concepts to build social reality, and social reality, in turn, wires the brain. Emotions are very real creations of social reality, made possible by human brains in concert with other human brains.

In this final chapter, we will employ the theory of constructed emotion as a flashlight to focus on larger issues of the mind and brain. We'll take a hard look at the predicting brain and everything we've learned about it, such as degeneracy, core systems, and the wiring for concept development, to illuminate the *kind of mind* most likely to emerge from this kind of brain. We'll see which aspects of the mind are universal or inevitable, which are not, and what this means for your broader understanding of other people and yourself.

• • •

For as long as people have been writing about humanity, there's been a pervasive assumption that the human mind is created by some all-powerful force. For the Ancient Greeks, that force was nature, embodied as gods. Christianity wrenched human nature away from Mother Nature and placed it in the hands of a single, omnipotent God. Darwin yanked it back and attributed it to a specific feature of nature called evolution. Suddenly you were no longer an immortal soul, and your mind was no longer a battleground of good and evil, righteousness and sin. You were instead a collection of specialized inner forces, sculpted by evolution, that struggle to control your actions. Your brain allegedly battles your body, rationality battles emotionality, cortex battles subcortex, and forces outside of you battle forces within you. With your animal brain wrapped in rational cortex, you are supposedly distinct from other animals in nature, not because you have a soul while they are soulless but because you are the pinnacle of evolution, endowed with insight and reason. You therefore came into the world preformed to respond to what it has to offer in a specific way, not in God's image but by your genes. Experiences like emotions are heralded as evidence that you are an animal through and through. But you are considered special in the animal kingdom because you can overcome your inner beast.

As you have learned in this book, however, new discoveries about the brain have revolutionized our understanding of what it means to be human.

Your mind is definitely a product of evolution, but it is not sculpted by genes alone. Sure, your brain is made of networked neurons, but that's just one factor in growing a human mind. Your brain also developed inside of a body, nestled among other human brains in bodies, who balanced your body budget and expanded your affective niche through actions and words.

Your mind is not a battleground between opposing inner forces — passion and reason — that determine how responsible you are for your behavior. Rather, your mind is a computational moment within your constantly predicting brain.

Your brain predicts with its concepts, and while scientists debate whether certain concepts are innate or learned, it's unquestionable that you learned a slew of them as your brain wired itself to its physical and social surroundings. Those concepts come from your culture and help negotiate the quintessential dilemma of living in groups — getting ahead versus getting along — a tug-of-war that has more than one solution. On balance, some cultures favor getting along, while others favor getting ahead.

All these discoveries reveal a crucial insight: The human brain evolved, in the context of human cultures, to create *more than one kind of mind*. People in Western cultures, for example, experience thoughts and emotions as fundamentally different and sometimes in conflict. At the same time, Balinese and Ilongot cultures, and to a certain extent cultures guided by Buddhist philosophy, do not make hard distinctions between thinking and feeling.[1]

How do different kinds of minds emerge from one kind of brain with the same set of networks? How can one type of brain create *your* mind, full of emotion concepts and experiences, and *my* mind, which has different instances of the same concepts or maybe some different emotion concepts, and a Balinese mind that has no separate concepts or experiences for thoughts and feelings, each of which is adapted to its physical and social environment?

On the surface, all normally developing human brains look pretty similar, particularly if you take off your glasses and squint. They all have two hemispheres. Every cortex has five lobes, with up to six layers. The neurons within every cortex are wired to compress information into efficient summaries, creating a conceptual system that shapes action and experience. Many of these features are present in other mammals, and some truly ancient aspects of your nervous system are even shared with insects. (One example is Hox genes, which organize a vertebrate's nervous system from head to tail.)

Nevertheless, brains vary significantly from person to person: in the placement of every cortical groove and ridge, in the number of neurons within particular layers of the cortex or in subcortical regions, in the microwiring between neurons, and in the strength of connectivity within brain networks. When you take into account these fine details, no two brains from the same species are structured completely alike.[2]

Also, within a single brain such as your own, the wiring is not static. Just as the arbor of a tree grows in the spring and shrinks in the fall, interconnections between your axons and dendrites increase and decrease as you age. You even grow new neurons in certain brain regions. This kind of anatomical change, called plasticity, also occurs with experience. Your experiences become encoded in your brain's wiring and can eventually change the wiring, increasing the chances that you'll have the same experience again, or use a previous experience to create a new one.[3]

And from one moment to another, your billions of neurons continually reconfigure themselves from one pattern into another. Chemicals called neurotransmitters make this possible. They enable signals to pass between neurons, and they dial up or dial down neural connections in a split second, so information flows along different paths. Neurotransmitters empower a single brain with a single set of networks to construct diverse mental events, creating something greater than the sum of the parts.[4]

Then, of course, we have degeneracy: different sets of neurons produce the same outcomes. Plus, no matter how finely or coarsely you look at brain tissue — as networks, regions, or individual neurons — that tissue contributes to more than one category of mental event, such as anger, attention, or even vision or hearing.[5]

Microwiring. Neurotransmitters. Plasticity. Degeneracy. Multipurpose circuitry. Neuroscientists sum up this incredible well of variation by calling the brain a "complex system." I don't mean complexity colloquially, as in "gosh, that brain sure is complicated," but something more formal. *Complexity* is a metric to describe any structure that efficiently creates and transmits information. A system with high complexity can create many new patterns by combining bits and pieces of old patterns. You can find complex systems in neuroscience, physics, mathematics, economics, and other scholarly disciplines.[6]

The human brain is a high-complexity system because, within one physical structure, it can reconfigure its billions of neurons to construct a huge repertoire of experiences, perceptions, and behaviors. It achieves high complexity via an ultra-efficient arrangement for communication centered on the critical "hubs" mentioned in chapter 6. This organization permits the brain to integrate so much information from multiple sources so efficiently that it can support consciousness. In contrast, the model of the brain posited by the classical view — independent blobs with distinct functions — would be a low-complexity system because each blob would accomplish its single function by itself.[7]

A brain with high complexity and degeneracy brings distinct advantages. It can create and carry more information. It's more robust and reliable, with multiple paths to get to the same end. It's more resistant to injury and illness; you've seen living examples in the twins with amygdala damage (chapter 1) and Roger with his ravaged predictive brain circuitry (chapter 4). Such a brain therefore makes you more likely to survive and pass your genes to the next generation.[8]

Natural selection favors a complex brain. Complexity, not rationality, makes it possible for you to be an architect of your experience. Your genes allow you, and others, to remodel your brain and therefore your mind.[9]

Complexity implies that the wiring diagram of a brain is not a set of instructions for a single kind of mind with universal mental organs. But the human brain has few preset mental concepts, such as perhaps pleasantness and unpleasantness (valence), agitation and calmness (arousal), loudness and softness, brightness and darkness, and other properties of consciousness. Instead, variation is the norm. The human brain is structured to learn many different concepts and to invent many social realities, depending on

the contingencies it is exposed to. This variability is not infinite or arbitrary; it is constrained by the brain's need for efficiency and speed, by the
outside world, and by the human dilemma of getting along versus getting
ahead. Your culture handed you one particular system of concepts, values,
and practices to address that dilemma.[10]

We don't need one universal mind, with one set of universal concepts, to
claim that we are all one species. All we need is an exceptionally complex
human brain that wires itself to its social and physical surroundings, ultimately producing different kinds of minds.

• • •

A human brain can create many kinds of minds, yet all human minds do
have some common ingredients. For millennia, scholars believed that the
inevitable bits of the mind were essences, but they are not. The ingredients
are three aspects of the mind that we've encountered in this book: affective
realism, concepts, and social reality. They (and perhaps others) are inevitable and therefore universal, barring illness, based on the anatomy and function of the brain.

Affective realism, the phenomenon that you experience what you believe, is inevitable because of your wiring. The body-budgeting regions in
your interoceptive network — your inner loudmouthed, mostly deaf scientist with a megaphone — are the most powerful predictors in your brain,
and your primary sensory regions are eager listeners. Body-budget predictions laden with affect, not logic and reason, are the main drivers of your experience and behavior. We all think a food "is delicious" as if the flavor were
embedded in the food, when flavor is a construction and the deliciousness
is our own affect. When a soldier in a warzone perceives a gun in someone's
hands when no gun is present, he might actually *see* that gun; it's not a mistake but a genuine perception. Judges who are hungry during parole hearings render more negative decisions.

Nobody can completely escape affective realism. Your own perceptions
are not like a photograph of the world. They are not even a painting of photographic quality, like a Vermeer. They are more like a Van Gogh or Monet.
(Or on a very bad day, perhaps a Jackson Pollock.)[11]

But you *can* recognize affective realism by its effects. Anytime you have
a gut feeling that you know something to be true, that's affective realism.
When you hear some news or read a story that you immediately believe,
that's affective realism too. Or if you are immediately dismissive of a message, or even dislike the messenger, that is also affective realism. We all like

things that support our beliefs, and usually dislike things that violate those beliefs.

Affective realism keeps you believing something even when the evidence puts it highly in doubt. It's not because of ignorance or malevolence — it is simply a matter of how the brain is wired and operates. Everything you believe, and everything you see, is colored by your brain's budget-balancing act.

Affective realism, when left unchecked, leads people to be dead certain and inflexible. When two opposing groups believe deeply that they are right, they engage in political skirmishes, ideological battles, even wars. The two views of human nature you've seen in this book, from the classical view and construction, have been duking it out for several thousand years.[12]

In this ongoing battle, affective realism has led each side to stereotype the other's point of view. The classical view is caricatured as biological determinism, that culture is completely irrelevant and genes are absolute destiny, justifying the present social order of who is wealthy and who struggles. That caricature depicts an extreme version of favoring "getting ahead" over "getting along." Construction, on the other hand, is criticized as absolute collectivism at the expense of the individual, or as the mistaken view that humans are one big superorganism like the Borg from *Star Trek,* and that the brain is "a uniform meatloaf" in which every neuron has exactly the same function. It's an exaggerated version of "getting along" trumping "getting ahead." Each side in this battle ignores the subtleties and variations that necessarily arise in scientific communities. If you've read this far, you've seen that the evidence points to a more nuanced conclusion: the dividing line between biology and culture is porous. Culture arose from natural selection, and as culture gets under the skin and into the brain, it helps to shape the next generation of humans.[13]

Affective realism is an inevitability, and yet you are not helpless against it. The best defense against affective realism is curiosity. I tell my students to be particularly mindful when you love or hate something you read. These feelings probably mean that the ideas you've read are firmly in your affective niche, so keep an open mind about them. Your affect is not evidence that the science is good or bad. The biologist Stuart Firestein in his lovely book *Ignorance* encourages curiosity as a way to learn about the world. Try to become comfortable with uncertainty, he suggests, finding pleasure in mystery, and being mindful enough to cultivate doubt. These practices will help you take a calm look at evidence that violates your own deeply held beliefs and experience the pleasure of the hunt for knowledge.[14]

The second inevitability of the mind is that you have concepts, because the human brain is wired to construct a conceptual system. You build concepts for the smallest physical details, like fleeting bits of light and sound, and for incredibly complex ideas like "Impressionism" and "Things Not to Bring on Airplane Rides." (The latter includes loaded guns, herds of elephants, and your boring Aunt Edna.) Your brain's concepts are a model of the world that keeps you alive, serves to meet your body's energy needs, and ultimately determines how well you propagate your genes.

What is not inevitable, however, is that you have *particular* concepts. Sure, everyone may have some basic concepts as a function of their wiring, such as "Positive" versus "Negative," but not every mind has distinct concepts for "Feeling" and "Thinking." Any set of concepts that helps you regulate your body budget and stay alive, as far as your brain is concerned, will do just fine. The emotion concepts that you learned in childhood are just one salient example.

Concepts are not just "in your head." Suppose you and I are chatting over coffee, and when I make some witty remark, you smile and nod. If my brain predicted your smile and your nod, and the visual input to my brain confirms these movements, then my own prediction — say, to nod back at you — becomes my behavior. You in turn might have predicted my nod, along with a host of other possibilities, which causes a change in your sensory input, which interacts with your predictions. In other words, your neurons influence one another not only through direct connections but indirectly through the outside environment, in an interaction with me. We are performing a synchronized dance of prediction and action, regulating each other's body budgets. This same synchrony is the basis of social connection and empathy; it makes people trust and like each other, and it's crucial for parent-infant bonding.[15]

Your personal experience, therefore, is actively constructed by your actions. You tweak the world, and the world tweaks you back. You are, in a very real sense, an architect of your environment as well as your experience. Your movements, and other people's movements in turn, influence your own incoming sensory input. These incoming sensations, like any experience, can rewire your brain. So you're not only an architect of your experience, you're also an electrician.

Concepts are vital to human survival, but we must also be careful with them because concepts open the door to essentialism. They encourage us to see things that aren't present. Firestein opens *Ignorance* with an old proverb, "It is very difficult to find a black cat in a dark room, especially when there is no cat." This statement beautifully sums up the search for essences. History has many examples of scientists who searched fruitlessly for an essence because they used the wrong concept to guide their hypotheses. Firestein gives the example of luminiferous ether, a mysterious substance that was thought to fill the universe so that light would have a medium to move through. The ether was a black cat, writes Firestein, and physicists had been theorizing in a dark room, and then experimenting in it, looking for evidence of a cat that did not exist. The same applies to the classical view of emotion, whose mental organs are a human invention that mistakes the question for the answer.

Concepts also encourage us *not* to see things that *are* present. One illusory stripe of a rainbow contains an infinite number of frequencies, but your concepts for "Red," "Blue," and other colors cause your brain to ignore the variability. Likewise, the frowny-faced stereotype of "Sadness" is a concept that downplays the great variation in that emotion category.

The third inevitability of the mind that we've discussed is social reality. When you are born, you can't regulate your body budget by yourself — somebody else has to do it. In the process, your brain learns statistically, creates concepts, and wires itself to its environment, which is filled with other people who have structured their social world in particular ways. That social world becomes real to you as well. Social reality is the human superpower; we're the only animal that can communicate purely mental concepts among ourselves. No *particular* social reality is inevitable, just one that works for the group (and is constrained by physical reality).

Social reality is in some ways a Faustian bargain. For some crucial human activities, such as building civilizations, social reality confers distinct advantages. Culture works most smoothly if we believe in our own mental creations, such as money and laws, without realizing that we're doing so. We don't suspect the involvement of our own hand (or neurons, as it were) in these constructions, so we just treat them as reality.

And yet, this same superpower that makes us effective civilization-builders also impedes our own understanding of how we do it. We constantly mistake perceiver-dependent concepts — flowers, weeds, colors, money, race, facial expressions, and so on — for perceiver-independent reality.

Many concepts that people consider to be purely physical are in fact beliefs about the physical, such as emotions, and many that appear to be biological are actually social. Even something that seems obviously biological, such as blindness, is not objective in biology. Some sightless people do not think of themselves as blind, because they get around in the world just fine.[16]

When you create social reality but fail to realize it, the result is a mess. Many psychologists, for example, do not realize that every psychological concept is social reality. We debate the differences between "will power" and "tenacity" and "grit" as if they were each distinct in nature, rather than constructions shared through collective intentionality. We separate "emotion," "emotion regulation," "self-regulation," "memory," "imagination," "perception," and scores of other mental categories, all of which can be explained as emerging from interoception and sensory input from the world, made meaningful by categorization, with assistance from the control network. These concepts are clearly social reality because not all cultures have them, whereas the brain is the brain is the brain. So, as a field, psychology keeps rediscovering the same phenomena and giving them new names and searching for them in new places in the brain. That's why we have a hundred concepts for "the self." Even brain networks themselves go by multiple names. The default mode network, which is part of the interoceptive network, has more aliases than Sherlock Holmes.[17]

When we misconstrue the social as the physical, we misunderstand our world and ourselves. In this regard, social reality is a superpower only if we know that we have it.

• • •

From these three inevitabilities of the mind, we see that construction teaches us to be skeptical. Your experiences are not a window into reality. Rather, your brain is wired to model your world, driven by what is relevant for your body budget, and then you experience that model as reality. Your moment-to-moment experience may feel like one discrete mental state followed by another, like beads on a string, but as you have learned in this book, your brain activity is continuous throughout intrinsic, core networks. Your experiences might seem to be triggered by the world outside the skull, but they're formed in a storm of prediction and correction. Ironically, each of us has a brain that creates a mind that misunderstands itself.

Where construction advocates skepticism, essentialism is deeply committed to certainty. It says, "Your brain is as your mind appears to be." You have thoughts, therefore you must have a blob in the brain for thoughts.

You experience emotions, therefore you must have blobs in the brain for emotions. You see evidence of thoughts, emotions, and perceptions in other people around the world, so the corresponding brain blobs must be universal and everyone must have the same mental essences. Genes have allegedly produced a mind that is common to all humans. You also see emotions in this animal and that—Darwin even saw emotion in flies—and so these creatures by implication must have the same universal emotion blobs that you do. Neural activity passes from one blob to another like runners in a relay race pass a baton.

Essentialism lays out not just a view of human nature but a worldview. It implies that your place in society is shaped by your genes. Therefore, if you are smarter, faster, or more powerful than others, you can justifiably succeed where others cannot. People get what they deserve and they deserve what they get. This view is a belief in a genetically just world, backed by a scientific-sounding ideology.

What we experience as "certainty"—the feeling of knowing what is true about ourselves, each other, and the world around us—is an illusion that the brain manufactures to help us make it through each day. Giving up a bit of that certainty now and then is a good idea. For instance, we all think about ourselves and other people in terms of characteristics. He is "generous." She is "loyal." Your boss is "an asshole." Our own sense of certainty tempts us to treat generosity, loyalty, and asshole-ness as if their essences actually live in those people, and as if they are detectable and measurable in objective terms. This not only determines our behavior toward them; we also feel justified in that behavior, even if the "generous" guy is just trying to suck up to you, the "loyal" woman is secretly self-serving, and your "asshole" boss has his mind on his sick kid at home. Certainty leads us to miss other explanations. I'm not saying that we are dumb or ill-equipped to grasp reality. I'm saying there is no single reality to grasp. Your brain can create more than one explanation for the sensory input around you—not an infinite number of realities, but definitely more than one.

A healthy dose of skepticism yields a worldview that is different from the genetically just world of the classical view. Your place in society is not random but neither is it inevitable. Consider an African American child born into poverty. She is less likely to receive proper nutrition during her early years of brain development—circumstances that will, in particular, negatively impact the development of her prefrontal cortex (PFC). These neurons are particularly important for learning (i.e., processing prediction error) and control; not surprisingly, the size and performance of PFC regions is linked to many skills that are required for doing well in school. Poorer nutrition equals a thinner PFC, which is linked to poorer performance in school, and less education, like not completing high school, leads back to poverty. In this cyclic manner, society's stereotypes about race, which are social reality, can *become the physical reality of brain wiring,* thereby making it seem as if the cause of poverty were simply genes all along.[18]

Some research seems to show that such stereotypes are more accurate than we might think. Steven Pinker writes in *The Blank Slate*, for example, that "people who believe that African Americans are more likely to be on welfare than whites . . . are not being irrational or bigoted. Those beliefs are correct" when compared to census figures. He and others argue that many scientists dismiss stereotypes as inaccurate because we are bullied into political correctness, are condescending toward ordinary people, or are biased by our own muddled assumptions about human nature. But as you've just seen, there is another possibility: the official welfare statistics are true because we, as a society, made them so.[19]

By virtue of our values and practices, we restrict options and narrow possibilities for some people while widening them for others, and then we say that stereotypes are accurate. They are accurate only in relation to a shared social reality that our collective concepts created in the first place. People aren't a bunch of billiard balls knocking one another around. We are a bunch of brains regulating each other's body budgets, building concepts and social reality together, and thereby helping to construct each other's minds and determine each other's outcomes.

Some readers might dismiss this sort of constructionist worldview as a stereotypically bleeding-heart liberal ivory tower academic viewpoint from the Land Where Everything Is Relative. In fact, this view cuts across traditional political lines. The idea that you're molded by your culture is stereotypically liberal. At the same time, as we discussed in chapter 6, you are responsible in a broad sense for the concepts you have, which ultimately influence your behavior. Individual responsibility is a deeply conservative idea. You are also somewhat responsible to others, not only the less fortunate but also future generations, for how you influence their wiring. It matters how you treat other people. That is a fundamentally religious idea. The American Dream traditionally says, "If you work hard, anything is possible." Construction agrees that you're indeed the agent of your own destiny, but you are bounded by your surroundings. Your wiring, determined in part by your culture, influences your later options.

I don't know about you, but I find some comfort in a bit of uncertainty. It's refreshing to question the concepts that have been given to us, and to be curious about which are physical and which are social. There is a kind of freedom in realizing that we categorize to create meaning, and therefore it is possible to change meaning by recategorizing. Uncertainty means that things can be other than they appear. This realization brings hope in difficult times and can prompt gratitude in good times.

• • •

Now it's time for me to drink my own Kool-Aid. Prediction, interoception, categorization, and the roles I've described for your various brain networks are not objective facts. They are concepts invented by scientists to describe the physical activity within a brain. I claim these concepts are the best way to understand certain computations being performed by neurons. However, there are many other ways to read the brain's wiring diagram (some of which wouldn't call it a wiring diagram at all). The theory of constructed emotion maps to the brain more closely than do so-called psychological essences or mental organs. In the future, I wouldn't be surprised to see more useful and functional concepts for the brain's structure emerge. As Firestein observes in *Ignorance,* no fact is "safe from the next generation of scientists with the next generation of tools."[20]

The history of science, however, has been a slow but steady march in the direction of construction. Physics, chemistry, and biology began with intuitive, essentialist theories, rooted in naive realism and certainty. We progressed beyond these ideas because we noticed that the old observations held true only under certain conditions. So, we had to replace our concepts. A scientific revolution swaps out one social reality for another, just like a political revolution does with its new government and social order. Again and again in science, our new sets of concepts have led us away from essentialism toward variation, and from naive realism to construction.[21]

The theory of constructed emotion predicts and matches the latest scientific evidence about emotion, the mind, and the brain, and yet so much about the brain is still a mystery. We're finding that neurons aren't the only important cells in the brain; glial cells, long ignored, turn out to do a hell of a lot, possibly even communicating with each other without synapses. The enteric nervous system, which controls your stomach and intestines, is looking more and more important for understanding the mind, but it's extremely difficult to measure and therefore largely unexplored. We're even finding that microbes in your stomach have a huge effect on mental states, and nobody knows how or why. There's so much innovative research going on that in ten years, today's experts might feel like Plato in the presence of a brain-scanning machine.

As our tools improve and our knowledge grows, I am confident that we'll discover the brain to be even more steeped in construction than we now know it to be. Perhaps our core ingredients like interoception and concepts will one day be seen as too essentialist, as we discover something even more finely constructed going on behind the scenes. Our scientific story is still evolving, but that's not surprising. Progress in science isn't always about finding the answers; it's about asking better questions. Today, those questions have forced a paradigm shift in the science of emotion, and more broadly in the science of mind and brain.

In the coming years, I hope we'll all see fewer and fewer news stories about brain blobs for emotion in people or rats or fruit flies, and more about how brains and bodies construct emotion. In the meantime, whenever you see an essentialism-steeped news story about emotion, if you even feel a twinge of doubt, then you're playing a role in this scientific revolution.

Like most important paradigm shifts in science, this one has the potential to transform our health, our laws, and who we are. To forge a new reality. If you've learned within these pages that you are an architect of your experience — and the experiences of those around you — then we're building that new reality together.

Acknowledgments

They say that it takes a village to raise a child, and this book, which my daughter took to calling her "baby brother," was no exception. The sheer number of people who contributed their comments, criticism, science, and support over the past three and a half years is a testament to both the richness of the subject area and the wonderful friends, family, and colleagues that I am so fortunate to know.

This book had a nontraditional family with more than the usual number of parents. It began life with the editors Courtney Young and Andrea Schulz at Houghton Mifflin Harcourt, and eighteen months later, both had been wooed away by compelling job offers. For a few months, I was a single parent with support from Bruce Nichols, publisher at HMH and effectively the book's great-great-grandfather. Houghton Mifflin Harcourt then hired Alex Littlefield as the new editor, who had a strikingly different vision of child-rearing from mine (leading to a stormy adolescence), but as is often the case, the best ideas come from vigorous debate, and I thank Alex for the way we ultimately shepherded a leaner and stronger book to its graduation day and released it into the world.

I'm extremely grateful to the book's adopted uncle, Jamie Ryerson at the *New York Times,* who helped at the last minute to trim three chapters that had become too lengthy and overwhelmingly technical. I am in awe of Jamie's skill to pare down material to the absolute essentials while retaining its style and voice. He may look like a mild-mannered editor, but when he stands in just the right light, you can see his knightly armor glinting in the sun.

Max Brockman, who is my agent and the village wizard, played an absolutely essential role in bringing this book to life. Not only did he navigate me through the ins and outs of the business, but each time we hit a hurdle during the long writing process, he was always ready with wise council. Thank you, thank you, thank you.

Yes, it takes a village to write a book, but my village is not the only one on the planet of emotion research. The other major village, which I've called "the classical view," is home to many creative and accomplished scientists, some of whom are my close colleagues. Our villages share territory, so we necessarily have conflicts and rivalries, but at the end of the day, we continue the debates over drinks and dinner. For two decades of animated discussion and close friendship, I thank James Gross and George Bonanno. Likewise, I am grateful to Paula Niedenthal, who introduced me to embodied cognition in general and to Larry Barsalou's work in particular. For informative conversations, I also thank Andrea Scarantino, Disa Sauter (for details on her study of the Himba), Ralph Adolphs, and Steven Pinker. I'd also like to thank Jaak Panksepp, who a number of years ago graciously accepted Jim Russell's and my invitation to come to Boston and teach a month-long graduate seminar on his theoretical views.

In a similar vein, I owe a very special debt of gratitude to my distinguished colleague Bob Levenson. It is a gift when someone with a different point of view engages you in honest conversation, and Bob truly embodies this spirit of scientific exploration every time we meet. His curiosity and insightful observations consistently challenge me, and I consider him one of my most valued colleagues. I also have a deep appreciation and respect for Paul Ekman, who helped to chart the course of research on emotion for the past five decades. We may not agree on the scientific details, but I admire his courageous path. When Paul began presenting his findings in the 1960s, he was shouted down at meetings, called a fascist and a racist, and generally disrespected due to prevailing attitudes of the time.* He showed formidable tenacity to pursue his vision of the classical view, and ultimately he brought the science of emotion into the public eye.

Back in the village of constructed emotion, I offer my heartfelt thanks to the Interdisciplinary Affective Science Laboratory at Northeastern University and Massachusetts General Hospital, which I direct with Karen Quig-

---

* As related by Steven Pinker in *The Blank Slate*.

ley. Our lab is one of the enduring pleasures and sources of pride in my career as a scientist. The community of hard-working, talented research assistants, graduate students, postdoctoral fellows, and research scientists contributed immeasurably to the body of knowledge that made this book possible. All the members (past and present) can be found at affective-science.org/people.shtml. Those whose valuable contributions are specifically cited in this book include Kristen Lindquist, Eliza Bliss-Moreau, Maria Gendron, Alexandra Touroutoglou, Christy Wilson-Mendenhall, Ajay Satpute, Erika Siegel, Elizabeth Clark-Polner, Jennifer Fugate, Kevin Bickart, Mariann Weierich, Suzanne Oosterwijk, Yoshiya Moriguchi, Lorena Chanes, Eric Anderson, Jiahe Zhang, and Myeong-Gu Seo. In addition to their important scientific contributions, I am grateful to the lab members for their endless patience and encouragement. They never once complained about my periodic absences (at least when I was in earshot) and occasionally endured long delays in their own progress as I raced to complete this book.

I am especially grateful to my collaborators for their friendship, commitment, and rompingly insightful discussions as we pursued some of the research you've just read about. First and foremost, my deepest thanks to Larry Barsalou for his foundational work on concepts; Larry is one of the most creative, rigorous thinkers of his generation, and I will be forever grateful for the opportunity to work with him. Nothing can convey the gratitude that I feel toward Jim Russell, who, when I was a young assistant professor, took my ideas seriously when many of our colleagues thought I was nuts. His seminal work on the affective circumplex is so well-accepted in the field that people rarely cite him for it anymore! Larry and Jim maximize discovery and explanation in their scientific pursuits, rather than fame and fortune, and I find this particularly inspiring (because sometimes in science, the latter interferes with the former). In this way, they remind me of my dissertation advisors, Mike Ross and Eric Woody, to whom I will be forever grateful.

I also owe a very big thank-you to Brad Dickerson for helping me to chip away at the false boundaries between emotion and cognition, to Moshe Bar for our work on how affect influences vision (and many other projects), to Tor Wager for our meta-analysis collaboration, and to Paula Pietromonaco for our longstanding collaboration on emotion in relationships. I am particularly grateful to Debi Roberson for making it possible, by our collaboration, for my lab to study the Himba of Namibia, and Alyssa Crittenden for likewise making it possible to study emotion perception in the Hadza of Tanzania.

The influence of my newer collaborations can also be seen in this book, and so I send enthusiastic thanks to Kyle Simmons, who works with me on the architecture and function of the predictive brain; Martijn van den Heuvel for listening to my far-out ideas about network connectivity and brain hubs that often turn out to be not so crazy; Wim Vanduffel and Dante Mantini for our work on brain networks in macaques; Talma Hendler for our collaboration on network dynamics while watching emotional films; Wei Gao for allowing me to join the adventure of studying the developing newborn brain; Tim Johnson for his partnership in showing that pattern classification does not provide evidence for neural fingerprints; Stacy Marcella for opening my eyes to the possibilities for studying simulation and prediction with computational models in virtual reality; and Dana Brooks, Deniz Erdogmus, Jennifer Dy, Sarah Brown, Jaume Coll-Font, and the rest of the B/SPIRAL group at Northeastern University for their patience and interest in immigrating to my village, and for crafting a computational framework to test the theory of constructed emotion.

This book would not have been possible without the support of the larger village of colleagues who generously shared their expertise on my journey from the land of clinical psychology to the land of neuroscience, with stops in social psychology, psychophysiology, and cognitive science along the way. My friends Jim Blascovich and Karen Quigley mentored me in the basics of the peripheral nervous system, and Karen taught me facial EMG. My neuroscience education began with the incomparable Michael Numan, who was encouraging and constantly available for questions, and Richard Lane, who encouraged me when I was first interested in the brain basis of emotion and introduced me to Scott Rauch at Massachusetts General Hospital. Scott enthusiastically gave me the opportunity to learn brain imaging, although I had no clue what I was doing at the time. I am also indebted to Chris Wright, who helped me conduct my first brain-imaging study, and with whom I secured my first large imaging grant from the National Institute on Aging. And my heartfelt thanks go out to the generous and thoughtful colleagues who spent time answering my questions, including Howard Fields, who was always available for enticing and enlightening discussions about the relation between nociception, reward, and interoceptive processing; Vijay Balasubramanian, who provided extremely useful explanations in response to my extensive questioning about the visual system; Thom Cleland, who enthusiastically shared his insights on the olfactory system; Moran Cerf, who gave me the inside scoop on intracranial electrical recording in live humans; and Karl Friston, who rewarded my out-of-the-blue email on predictive coding with an insightful email discussion wrapped in encouragement. Several others provided helpful answers to my questions via email or Skype, including Dayu Lin, who provided a detailed discussion of her research using optogenetics; Mark Bouton, who taught me the basics of contextual learning in mammals; Earl Miller for explaining the implications of his single-cell recording research on category learning in macaques; and Matthew Rushworth, who offered additional details about his mapping of the anterior cingulate cortex.

I also offer my enduring thanks to some of my neuroanatomy colleagues who responded quickly, and in good cheer, to my incessant questions, no matter how arcane: Barb Finlay for knowing everything about everything, off the top of her head, and sharing generously; Helen Barbas for her model of information flow in the cortex, which is the cornerstone of my approach to the predictive brain; Miguel Ángel García Cabezas for his detailed explanations of neuroanatomy at the cellular level; Bud Craig, who knows more about the insula than perhaps anyone else on the planet; Larry Swanson for his rapid and informative answers and for connecting me with other neuroscientists, such as Murray Sherman, who answered my questions about the thalamus; and Georg Striedter for his expertise on brain evolution.

For sharing their expertise in developmental psychology, I offer warm thanks to Linda Camras and Harriet Oster, who were my guides to the emotional capacities of infants and young children. I am also indebted to Fei Xu, Susan Gelman, and Sandy Waxman for reviewing chapter 5, and for their willingness to trample the traditional scientific boundary between cognitive and emotional development, to help me explore the idea that words scaffold the development of emotion concepts in infancy. I am also grateful to Susan Carey for discussions of innate concepts.

Chapter 11 on emotion and the legal system would not have been possible without my dear friends Judy Edersheim and Amanda Pustilnik and their insights and encouragement during our long discussions about psychology, neuroscience, and the law; that chapter is best viewed as a collaboration between the three of us. I am grateful to former U.S. federal judge Nancy Gertner for inviting me to contribute to her course on the law and neuroscience at Harvard Law School. I'd also like to thank the many others from the Center for Law, Brain, and Behavior at Massachusetts General Hospital for inviting me into their village. Thanks also to Nita Farahany for the DNA example in chapter 11.

This book was made possible by many generous colleagues across diverse fields who offered me their insights. On primate cognition: Eliza Bliss-Moreau, Herb Terrace, and Tetsuro Matsuzawa. On topics related to culture: Aneta Pavlenko, Batja Mesquita, Jeanne Tsai, Michele Gelfand, and Rick Shweder. On the history of smiling: Colin Jones and Mary Beard. On autism: Jillian Sullivan, Matthew Goodwin, and Oliver Wilde-Smith. On essentialism: Susan Gelman, John Coley, and Marjorie Rhodes. On affective realism and economics: Marshall Sonenshine. On contemplative philosophy and practice: Christy Wilson Mendenhall, John Dunne, Larry Barsalou, Paul Condon, Wendy Hasenkamp, Arthur Zajonc, and Tony Back. More generally, an enthusiastic thank-you goes to Jerry Clore for being consistently thoughtful, curious, and supportive; to Helen Mayberg for our multi-year conversation about the puzzle of depression; and to Joe LeDoux, whom I greatly admire for many reasons, not least for his incredible open-mindedness. My discussions with other insightful colleagues also shaped this book, including Amitai Shenhav, Dagmar Sternad, Dave DeSteno, David Borsook, Derek Isaacowitz, Elissa Epel, Emre Demiralp, Iris Berent, Jo-Anne Bachorowski, the late Michael Owren, Jordan Smoller, Philippe Schyns, Rachael Jack, José-Miguel Fernández-Dols, Kevin Ochsner, Kurt Gray, Linda Bartoshuk, Matt Lieberman, Maya Tamir, Naomi Eisenberger, Paul Bloom, Paul Whalen, Margaret Clark, Peter Salovey, Phil Rubin, Steve Cole, Tania Singer, Wendy Mendes, Will Cunningham, Beatrice de Gelder, Leah Summerville, and Joshua Buckholtz.

I benefited greatly from valuable comments and criticisms offered by early readers: Aaron Scott (who is also the extraordinary graphic designer who created most of the figures), Ann Kring (my most faithful reader, who provided valuable insights on every draft), Ajay Satpute, Aleza Wallace, Amanda Pustilnik, Anita Nevyas-Wallace, Anna Neumann, Christy Wilson-Mendenhall, Dana Brooks, Daniel Renfro, Deborah Barrett, Eliza Bliss-Moreau, Emil Moldovan, Eric Anderson, Erika Siegel, Fei Xu, Florin Luca, Gibb Backlund, Herbert Terrace, Ian Kleckner, Jiahe Zhang, Jolie Wormwood, Judy Edersheim, Karen Quigley, Kristen Lindquist, Larry Barsalou, Lorena Chanes, Nicole Betz, Paul Condon, Paul Gade, Sandy Waxman, Shir Atzil, Stephen Barrett, Susan Gelman, Tonya LeBel, Victor Danilchenko, and Zac Rodrigo.

I am also especially grateful to Joanne Miller, chair of the psychology department at Northeastern University, and to the rest of my colleagues in the department, for their support and patience as I completed this book.

I am indebted to the funding agencies and fellowships that made it feasible for me to write this book. These include fellowships from the American Philosophical Society and the James McKeen Cattell Fund from the Association for Psychological Science, as well as generous support from the U.S. Army Research Institute for the Behavioral and Social Sciences; in particular, I am most grateful to Paul Gade, who was my program officer at ARI at the time, and who has continued to offer me encouragement and moral support. The research reported in this book was additionally funded by the generous support of granting agencies under the helpful guidance of their program officers. This includes the National Science Foundation, particularly Steve Breckler, who gave me my first neuroscience grant; the National Institute of Mental Health, particularly Susan Brandon, who oversaw my K02 Independent Scientist Award, Kevin Quinn, and Janine Simmons; the National Institute on Aging, particularly Lis Nielsen; the National Cancer Institute, particularly Paige Green and Becky Ferrer; the National Institutes of Health Director's Pioneer Award; the National Institute of Child Health and Development; the U.S. Army Research Institute for the Behavioral and Social Sciences, particularly Paul Gade, Jay Goodwin, and Greg Ruark; and the Mind and Life Institute, particularly Wendy Hasenkamp and Arthur Zajonc.

I owe a very special debt of gratitude to the people who handled the legal, administrative, and logistical aspects of the book: Fred Polner (my attorney) and Michael Healy (attorney at Brockman, Inc.); Emma Hitchcock and Jiahe Zhang for creating some of the brain images contained in this book; Rosemary Marrow at Redux Pictures; Chris Martin and Elyna Anderson at the Paul Ekman Group; Beverly Ornstein, Rona Menashe, and Dick Guttman for permission to use Martin Landau's photograph; Nicole Betz, Anna Neumann, Kirsten Ebanks, and Sam Lyons for ultra-fast search and retrieval of research papers on request; and Jeffrey Eugenides for his wonderful conceptual combinations for much-needed emotion concepts.

I'm also grateful to Ronda Heilig, an agent with the Federal Bureau of Investigation, and Peter DiDomenica, who developed the TSA's Screening Passengers by Observation Techniques (SPOT) program while director of security policy at Boston Logan International Airport, for speaking with me about the ways that the classical view informed training at their respective agencies.

Thanks also to the rest of the team at Houghton Mifflin Harcourt: Naomi Gibbs, Taryn Roeder, Ayesha Mirza, Leila Meglio, Lori Glazer, Pilar Garcia-Brown, Margaret Hogan, and Rachael DeShano.

I realize that it might sound odd, but I also want to acknowledge the Internet for playing an important part in the writing of this book, which required integrating and synthesizing copious material from diverse fields rapidly. When I had an idea, I was able to investigate it instantly by downloading relevant research papers in minutes, or buying virtually any book with overnight shipping. So a hearty thank-you to the engineers who brought us Google, Amazon (though for the amount I spent, they should be thanking me), and the many scientific journal websites that make their papers available online. This book was created in part with open-source software, including Subversion and a suite of Linux-based tools.

And let's not forget those who kept my body budget solvent during the writing of this book. I am truly, deeply grateful for their love and encouragement to Ann Kring, Batja Mesquita, Barb Fredrickson, James Gross, Judy Edersheim, Karen Quigley, Angie Hawk, and Jeanne Tsai. They provided both intellectual challenge and comfort during the long months of writing, not to mention the continual influx of chocolates, coffee, and other treats to keep me going. Special thanks for vital social support also goes to Florin and Magdalena Luca, and Carmen Valencia. I am deeply thankful for my extended family's support. This includes my sisters-in-law, Louise Greenspan and Deborah Barrett; my goddaughter, Olivia Allison; and my nephew, Zac Rodrigo; and of course the incomparable (Uncle) Kevin Allison, whom you met virtually in chapters 6 and 7. And my deepest thanks to Mike Alves, trainer extraordinaire, and Barry Meklir, my miracle-working physical therapist, who together kept me walking and typing after sitting for sixteen hours a day; and to Victoria Krutan, who embodies the best of what massage therapy has to offer.

My daughter, Sophia, with grace and forbearance beyond her years, tolerated three years of my late-night, early-morning, and weekend preoccupations with her "little brother" (not to mention my occasional bouts of bad temper). If there was ever a justified case for sibling rivalry, this is it. Sophia, you are my girl. I wrote this book for you. I want you to understand the power of your own mind. When you were little, you would sometimes wake from a nightmare. We'd position your stuffed animals in a protective circle around your bed, and I would sprinkle some "fairy dust" and you'd get back to sleep. What's remarkable is not that you believed in magic, but that you *didn't.* We both knew it was pretend, and yet it worked. Your exuberant little four-year-old self had the superpower to create social reality with me, just as your courageous, funny, and insightful teenage self does now. You are an architect of your experience, even in times when you feel buffeted by the world.

If Sophia was the reason that I began this book, then my husband, Dan, is the reason I completed it. Dan is often the calm behind my storm. For as long as I have known him, he has had an unshakeable belief in my ability to do the extraordinary. Dan read every word of every book draft, often several times, and made this book better than anything I could have managed on my own. My brain will never be free from his oft-asked question, "Is this for the 1 percent?" (by which he meant my scientific colleagues, as opposed to a general audience), although now I am more likely to smile when my brain is simulating it. Among his many superpowers is the ability to simultaneously edit this book, soothe my worries, rub my back, cook dinner, suspend our entire social life without a trace of bitterness, and collect enough takeout menus to sustain us during my final months of writing. He never flinched, not once, even after it became clear that I had gotten us into something much more challenging than either of us knew at the outset. Dan's other superpower (beyond his uncanny ability to choose the right-sized Tupperware every time) is that he can make me laugh when no one else can, because he knows me in a way that no one else does. I awaken every day of my life filled with gratitude and awe that he is beside me.

Appendix A

Brain Basics

Every Halloween, I create a life-sized model of the brain out of gelatin. I pour boiling water into peach-flavored gelatin, add condensed milk to make the mixture opaque, and dribble in some green food coloring to turn the brain a jiggly gray. The brain is a prop for an elaborate haunted house that my family and lab have designed and run since 2004 as a charity event. Visitors who make it through the haunted house always exclaim (once they can speak normally again) how realistic the brain looks, which is interesting because a real brain is nothing like a uniform blob of gelatin. It is a massive network composed of billions of brain cells wired together to pass information back and forth.[1]

To get the most out of this book, you'll need a few basic facts about the human brain. The most important type of brain cell for our discussion is the *neuron*. There are a wide variety of neurons, but in general, each one consists of a cell body, some branch-like structures on the top called dendrites, and one root-like structure on the bottom called an axon, which has axon terminals at its end, as in figure AA-1.

The axon terminals of one neuron are close to the dendrites of other neurons — usually thousands — forming connections called synapses. A neuron "fires" by sending an electrical signal down its axon to its axon terminals, which release chemicals called neurotransmitters into the synapses, where they are picked up by receptors on the dendrites of other neurons. The neurotransmitters excite or inhibit each neuron on the other end of a synapse, changing its rate of firing. Through this process, one individual

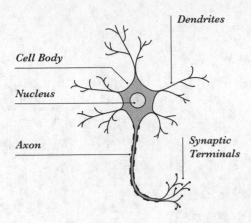

Figure AA-1: Neurons come in different shapes but they each
have a cell body, one long axon, and dendrites.

neuron influences thousands of others, and thousands of neurons can influence one, all simultaneously. This is the brain in action.[2]

At a more macro level, the human brain can be divided, more or less, into three major parts based on how the neurons are arranged.* The *cortex* is a sheet of neurons arranged in layers, anywhere from four to six (see figure AA-2), wired into circuits and networks. A cross section of this sheet reveals that neurons are organized into columns; neurons within the same column of cortex form synapses with each other, and with neurons in other columns.[3]

The cortex is folded around the *subcortical regions* that, in contrast to the layered cortex, are organized as clumps of neurons, as depicted in figure AA-3. The ever-popular amygdala, for example, is a subcortical region.[4]

The third part of the brain, the *cerebellum,* is toward the bottom of the brain, at the back. The cerebellum is important for coordinating physical movements and making that information available to the rest of the brain.[5]

Scientists must point to different collections of neurons, that is, "brain

---

* People divide the brain in many different ways, depending on their needs. Divisions may be spatial (top to bottom, back to front, outer to inner), anatomical (by lobe, by region, by network), chemical (by neurotransmitter), functional (which parts do which tasks), and more. Since the division between cortex and the subcortical regions is so important in the history of emotion, I'll talk about the brain in those simplified terms.

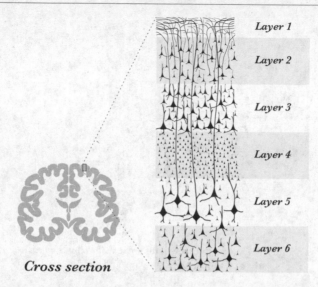

Figure AA-2: Cross section of six-layered cortex

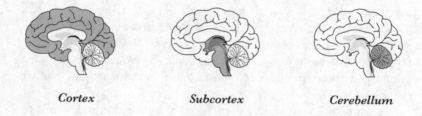

Figure AA-3: Three major parts of the brain

areas," so they have devised some terminology to help.* The cortex, which comes up repeatedly in this book, is divided into discrete areas called lobes, which are rather like continents in the brain (figure AA-4).

For navigating the entire brain, instead of using compass directions like east or northwest, scientists uses phrases like "dorsal anterior" (upper front)

---

* Different neuroscientists slice and dice the brain in different ways, using different terms to suit their goals and preferences. I'm presenting only a selection of the most conventional distinctions.

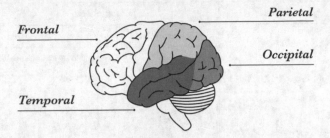

Figure AA-4: Lobes of the cortex

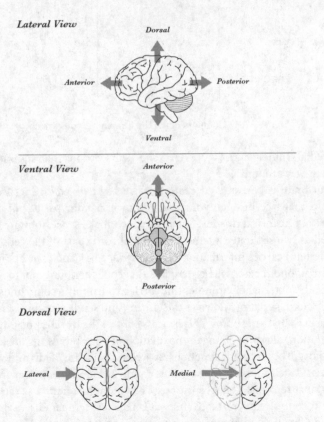

Figure AA-5: Road signs for the brain. *Anterior* = toward the front; *posterior* = toward the back; *dorsal* = toward the top; *ventral* = toward the bottom; *medial* = toward the midline or middle; and *lateral* = away from the midline toward the outside

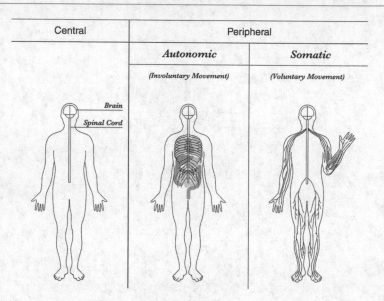

Figure AA-6: Components of the human nervous system

or "medial" (inner wall). Figure AA-5 shows the various road signs for finding your way around.

Your brain is part of your *central nervous system,* as distinct from the neurons that lace through your body, known as your *peripheral nervous system.* For historical reasons, not all of which make sense, they are usually studied as two separate systems. Your spinal cord (part of the central nervous system) carries information between your body and your brain.

Your autonomic nervous system is one avenue for your brain to regulate your body's internal environment. It carries your brain's commands to the body's internal organs, known as the viscera, and sends sensations from the viscera back up to the brain. This process controls heart rate, breathing rate, perspiration, digestion, hunger, the dilation of your pupils, sexual arousal, and a host of other bodily functions. It is responsible for "fight or flight" responses that tell your body to spend its energy resources, as well as "rest and digest" that replenish those resources. The autonomic nervous system also helps to control your metabolism, water balance, temperature, salt, heart and lung function, inflammation, and other resources across all systems of the body, like a budget. The somatic nervous system gives the brain access to muscles, joints, tendons, and ligaments.[6]

Appendix B

Supplement for Chapter 2

Stop! Read the beginning of chapter 2 before turning the page.

Figure AB-1: The mystery picture revealed

Appendix C

Supplement for Chapter 3

Stop! Read the introduction of chapter 3 before turning the page.

Figure AC-1: An ecstatic Serena Williams after she beat
her sister Venus in the 2008 U.S. Open tennis finals

Appendix D

Evidence for the Concept Cascade

I've described the brain in two ways that look like hierarchies. (They are metaphors to help understand brain activity; neurons are not wired in a strict hierarchy.) The first hierarchy in chapter 6 illustrates how the brain uses sensory input to form concepts, as a hierarchy of similarities and differences. This hierarchy is bottom-up and should be familiar to neuroscientists. Your primary sensory regions are at the bottom; their neurons fire to represent the different sensory details of bodily sensations, changing wavelengths of light, changes in air pressure, and so on, that make up a particular instance. The neurons at the top of the hierarchy represent the highest-level, efficient, multisensory summary of the instance.

The second hierarchy in chapter 4 illustrates how concepts unpack as predictions, based on the structure of the cortex. This hierarchy is top-down and incorporates some of my own discoveries. Body-budgeting circuitry (more commonly called visceromotor limbic circuitry), the loudmouth of the brain, is at the top, and it issues but does not receive predictions. Primary sensory regions are at the bottom, as they receive predictions but don't issue them to other cortical regions. In this manner, body-budgeting regions drive predictions throughout the brain and down to the primary sensory regions, in progressively finer detail.

The two hierarchies represent the same circuitry but operate in reverse. The former hierarchy is for learning concepts and the latter — which I call the *concept cascade* — is for applying those concepts to construct your perceptions and actions. In this manner, categorization is a whole-brain activity, with predictions flowing from simulated similarities to simulated differences, and prediction errors flowing in the other direction.

The concept cascade involves some reasoned speculation but is consistent with evidence from neuroscience. At present, we have scientific evidence that all the external sensory systems (vision, hearing, etc.) operate by prediction. Along with my colleague neuroscientist W. Kyle Simmons, I discovered that the interoceptive network is also structured to function this way.[1]

Right now, scientists have specific details of the conceptual cascade within the visual system. The broader conceptual cascade that I've outlined in this book is based on three very solid pieces of evidence: (1) the anatomical evidence in chapter 4 about how predictions and prediction errors flow across the structure of cortex, (2) the anatomical evidence showing that the cortex is structured to compress sensory differences into multisensory summaries in chapter 6, and (3) scientific evidence on the functions of several brain networks, which we'll discuss now.[2]

A prediction originates as a multisensory summary, representing the goal of the concept, in a portion of the interoceptive network known as the *default mode network*. Notice I did not say that concepts are "stored" in the default mode network. I specifically use the word "originate." Concepts do not live wholesale in the default mode network, or anywhere else, as if they were entities. This network simulates only part of the concept, namely, the efficient, multisensory summaries of the concept instances with none of their sensory details. When your brain constructs a concept of "Happiness" on the fly, for use in a specific situation, degeneracy is in play. Each instance is created with its own pattern of neurons. The more conceptually similar the instances are, the more the neural patterns will be close to one another in the default mode network, and some will even overlap, using some of the same neurons. Different representations need not be separate in the brain, just separable.[3]

The default mode network is an intrinsic network. In fact, it was the first intrinsic network to be discovered. Scientists noticed a set of brain regions that increased their activity when subjects were lying at rest. They named these regions the "default mode" because they are spontaneously active while the brain isn't being probed or stimulated by an experimental procedure. When I first learned about this network, I thought the choice of name was unfortunate, because numerous other intrinsic networks have since been discovered. But the name is ironic: Scientists originally believed

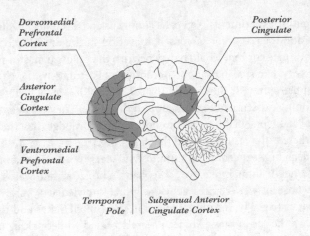

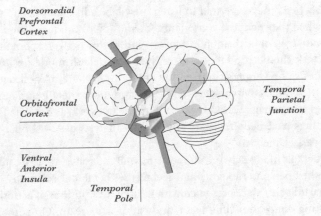

Figure AD-1: The default mode network, which lies within the interoceptive network. Body-budgeting regions, which launch predictions, are in dark gray. They send commands to the subcortical nuclei that control the body's tissues and organs, metabolism, and immune function. Top is medial view, bottom is lateral.

the brain's "default" activity was aimless mind-wandering between tasks, when in fact this network is at the core of every prediction in the brain. Your brain's "default mode" by which it interprets and navigates the world, namely, *prediction using concepts,* makes the name fit this network nicely.[4]

Neuroscientists have demonstrated pretty definitively that the default mode network represents key *portions* of concepts. This discovery required clever scientific experiments. You can't simply ask a test subject to simulate a concept and then look for increased activation in the default mode network. That single concept would barely perturb the brain's maelstrom of intrinsic activity, like spitting into an ocean wave. Fortunately, the cognitive neuroscientist Jeffrey R. Binder and his colleagues designed an ingenious brain-scanning experiment to work around this issue. They created two experimental tasks, one that used more conceptual knowledge than the other, and "subtracted" the results to yield the difference.

In Binder's first experimental task, test subjects in a brain scanner listened to names of animals like "fox," "elephant," and "cow," and were asked a question whose answer requires rich conceptual knowledge of purely mental similarities (e.g., "Is the animal found in the United States and used by people?"). In the second task, subjects were scanned while making a decision that requires more limited conceptual knowledge based on perceptual similarity (e.g., they were told to listen to syllables like "pa-da-su" and respond when they hear the consonants "b" and "d"). Both tasks should produce an increase in activation in sensory and motor networks, but only the former task should produce an increase in the default mode network. By "subtracting" one brain scan from the other, Binder and his colleagues removed the brain activity related to sensory and motor details and observed an increase in activity within the default mode network, as predicted. Binder's findings have been replicated by a meta-analysis of 120 similar brain-imaging experiments.[5]

The default mode network supports mental inference, that is, categorizing another person's thoughts and feelings with mental concepts. In one study, participants were presented with written descriptions of actions such as drinking coffee, brushing teeth, and eating ice cream. On some trials, participants were asked *how* people performed these actions: drinking coffee from a mug, brushing teeth with a toothbrush, eating ice cream with a spoon. Participants appeared to simulate these actions in motor regions of the brain. On other trials, participants were asked *why* people performed such actions: drinking coffee to stay awake, brushing teeth to avoid cavities, eating ice cream because it tastes good. These judgments require purely mental concepts, and they were more associated with activity in the default mode network.[6]

A growing number of cognitive neuroscientists, social psychologists, and neurologists speculate that the default mode network has a general function: it allows you to simulate how the world might be different from the way it is right now. This includes remembering the past and imagining the future from different points of view. This remarkable ability provides you with a leg up when negotiating the two big challenges of human life: getting along with others and getting ahead to benefit yourself. The social psychologist Daniel T. Gilbert, author of *Stumbling on Happiness,* who is famous for his humorous eloquence, calls the default mode network an "experience simulator," akin to flight simulators for training pilots. By simulating a future world, you are better equipped to reach your future goals.[7]

The default mode network unites past, present, and future. Information from the past, constructed as concepts, forms predictions about the present, which make you better equipped to reach your future goals.

I find it useful to think about the default mode network as playing a key role in categorization. The network initiates predictions to create simulations, thereby allowing the brain to work its magic of modeling the world. The "world" in this case includes the outside world, the minds of other people, and the body that holds the brain. Sometimes these simulations are corrected by the outside world, like when you construct emotions, and other times they aren't, like when you imagine or dream.[8]

Of course, the default mode network is not working alone. It contains only part of the pattern required for making a concept, namely, the mental, goal-based, multisensory knowledge that initiates a cascade. Anytime you imagine things, or your mind wanders, or your brain performs other intrinsic activity, you also simulate sights, sounds, changes in your body budget, and other sensations that are the domain of sensory and motor networks. Thus, it stands to reason that the default mode network should be interacting with these other networks to construct instances of concepts. (And they do, which you'll see shortly.)[9]

Newborns don't have a fully formed default mode network, hence their inability to predict and their diffuse "lantern" of attention; newborn brains spend a lot of time learning from prediction error. It very well may be that experience with the multisensory world, anchored in body budgeting, provides the needed inputs that help the default mode network to form. This occurs sometime during the first few years of life as the brain is bootstrapping concepts into its wiring. What begins as "outside" becomes "inside" as you become wired by your environment.[10]

My lab has been investigating the biology of concepts and categorization for some time, and we've uncovered considerable evidence about the roles of the default mode network, the rest of the interoceptive network, and the control network. When we peer into the brains of people who are experiencing emotion, or perceiving emotion in blinks, furrowed brows, muscle twitches, and the lilting voices of others, we see pretty clearly that key parts of these networks are hard at work. For starters, you might remember my lab's meta-analysis that examined every published neuroimaging study of emotion, which we saw in chapter 1. We divided the entire brain into tiny cubes called "voxels" (akin to "pixels" of the brain), and then identified voxels that consistently showed a significant increase in activity for any of the emotion categories we studied. We could not localize a single emotion category to any brain region. This same meta-analysis also provided evidence for the theory of constructed emotion. We identified groups of voxels that activated together with high probability, like a network would. These groups of voxels consistently fell within the interoceptive and control networks.[11]

When you consider that our meta-analysis, at the time it was conducted, covered over 150 diverse, independent studies by hundreds of scientists, in which subjects viewed faces, smelled scents, listened to music, watched movies, remembered past events, and performed many other emotion-evoking tasks, the emergence of these networks is particularly compelling. These findings are even more remarkable to me because the studies covered by the meta-analysis weren't designed to test the theory of constructed emotion. Most were inspired by classical view theories and designed to localize each emotion to a different region of the brain. And most of them studied only the most stereotyped examples of emotion categories and did not examine each emotion in all its real-life variations.

Our meta-analysis project is ongoing, and we have collected almost four hundred brain-imaging studies to date. From this data, my colleagues and I used pattern classification analysis (chapter 1) to produce five summaries of emotion categories, shown in figure AD-2. In all five, the interoceptive network played a significant role. The control network was also present for all five, but less clearly for happiness and sadness. Remember that you're not looking at neural fingerprints here, just abstract summaries. No single instance of anger, disgust, fear, happiness, or sadness looks exactly like its associated summary. Each instance can use diverse combinations of neurons, as we know from the principle of degeneracy. For each study of (say) anger in the meta-analysis, the brain activity was closer to the anger summary than to the other summaries, so it was identified as anger. So we can *diagnose* an instance of anger, but we cannot specify which neurons will be active. In other words, we have applied Darwin's principle of population thinking to the construction of anger. The same result follows for the other four emotion categories we studied.[12]

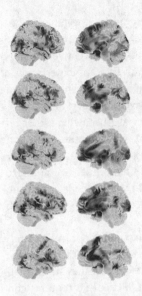

Figure AD-2: Statistical summaries of the concepts (top to bottom) "Anger," "Disgust," "Fear," "Happiness," and "Sadness." These are not neural fingerprints (see chapter 1). Left is lateral view, right is medial view.

When we specifically design experiments to test the theory of constructed emotion, we find similar results. In one study, my collaborators Christine D. Wilson-Mendenhall and Lawrence W. Barsalou, and I asked subjects to immerse themselves in imagined scenarios while we performed brain scans. We saw evidence of the resulting simulations as increased activity in sensory and motor regions. We could also see evidence that their body budgets were perturbed, associated with changes in the interoceptive network. In a second phase after each immersion, test subjects were shown a word and asked to categorize their interoceptive sensations as instances of either "Anger" or "Fear." As our subjects simulated these concepts, we saw even more increased activity in the interoceptive network. We also saw activations representing the low-level sensory and motor details, as well as increased activity in a key node in the control network.[13]

In a later study, we had subjects construct atypical, infrequent simulations, such as the pleasant fear of riding a rollercoaster and the unpleasant happiness of injuring yourself while winning a competition. We hypothesized that less typical simulations would require the interoceptive network to work harder to issue predictions, compared to simulating more typical instances like pleasant happiness and unpleasant fear, which are like mental habits. This is exactly what we observed.[14]

In a more recent set of experiments, our test subjects watched evocative movie scenes, and we saw the interoceptive network construct ongoing emotional experiences. Talma Hendler's lab at Tel Aviv University in Israel chose film clips that would create a variety of different experiences of sadness, fear, and anger. For example, some test subjects watched a scene from *Sophie's Choice* where the title character, played by Meryl Streep, must choose one of her children to be taken from her at Auschwitz. Other test subjects watched a clip from the film *Stepmom*, where Susan Sarandon's character reveals to her children that she is dying of cancer. In all cases, we observed that the default mode network and the remainder of the interoceptive network were firing more in synchrony in the moments when subjects reported more intense emotional experiences, and less so when subjects reported less intense experiences.[15]

Other studies make a similar case for emotion perception. In one study, subjects watched movies and explicitly categorized the characters' physical movements as emotional expressions. In other words, they made mental inferences about what the movements meant, a task that requires concepts. Their brains showed increased activity in the interoceptive network, in nodes of the control network, and in visual cortex where objects are represented.[16]

· · ·

When discussing concepts, we must be mindful not to essentialize because it's super easy to imagine concepts as "stored" in your brain. For example, you could think concepts live in the default mode network alone (as if the summaries exist apart from their sensory and motor details). There is abundant evidence (and very little doubt), however, that any instance of any concept is represented by the entire brain. As you look at the hammer in figure AD-3, neurons in your motor cortex that control your hand movements have increased their firing. (And if you are like me, the neurons that simulate pain in your thumb are also firing madly.) This increase even occurs when you read the name of the object ("hammer"). Viewing the hammer also makes it easier for you to make a gripping motion with your hand.[17]

Figure AD-3:
Tweaking your motor cortex

Likewise, as you read these words:

- Apple, Tomato, Strawberry, Heart, Lobster

neurons that process color sensations in early visual cortex also increase their firing rate, because all of the objects are typically red. So concepts have no mental core in the default mode network; they are represented throughout the entire brain.[18]

A second essentialist misconception is that your default mode network has a single set of neurons for each goal, like little essences, even if the rest of the concept, such as sensory and motor features, is distributed throughout the brain. This cannot be the case, however. If it were, then in brain scans we'd see this "essence" activate first, under all conditions, because it's at the top of their concept cascade, followed by the more variable sensory and motor differences depending on the situation, but we see nothing of the kind.[19]

Here again, essentialism yields to degeneracy. Each time you construct an instance of an emotion concept like "Happiness" with a particular goal, such as being with a close friend, the pattern of neural firing can be different. Even the highest-level, multisensory summary of "Happiness," represented by sets of neurons in the default mode network, can be different each time. None of these instances need be physically alike, and yet they are all instances of "Happiness." What is binding them together? Nothing. They are not "bound" together in any permanent way. But they are very likely initiated concurrently, as predictions. When you read the word "happy" or hear it spoken, or when you find yourself surrounded by your favorite people, your brain launches a variety of predictions, each with some prior probability of being likely in whatever the specific situation is. Words are powerful. This is reasoned speculation on my part because the brain operates on degeneracy, words are key to concept learning, and the default mode network and the language network share many brain regions.[20]

A third mistake of essentialism is thinking of concepts as "things." When I was an undergraduate student, I took a course in astronomy where I learned the universe was expanding. At first, I was baffled: expanding into *what*? I was confused because I harbored an incorrect intuition that the universe was expanding into space. After some reflection, I realized that I conceived of "space" in rather literal, physical terms, as a big, dark, empty bucket. Instead, "space" is a theoretical idea — a concept — not a concrete, fixed entity; space is always computed in relation to something else. ("Space and time are in the eye of the beholder.")[21]

Something similar happens when people think about concepts. A concept is not a "thing" that exists in the brain, any more than "space" is a physical thing that the universe expands into. "Concept" and "space" are ideas. It is a verbal convenience to talk about "a" concept. Really you have a conceptual system. When I write "you have a concept for awe," this translates as "you have many instances that you have categorized, or that have been categorized for you, as awe, and each can be reconstituted as a pattern in your brain." The "concept" refers to all the knowledge you construct about awe in your conceptual system in a given moment. Your brain is not a vessel that "contains" concepts. It enacts them as a computational moment over some period of time. When you "use a concept," you are really constructing an instance of that concept on the spot. You don't have little packets of knowledge called "concepts" stored in your brain, any more than you have little packets called "memories" stored in your brain. Concepts have no existence separate from the process that creates them.[22]

Bibliography

Abrams, Kathryn, and Hila Keren. 2009. "Who's Afraid of Law and the Emotions." *Minnesota Law Review* 94: 1997.

Adler, Nancy E., Thomas Boyce, Margaret A. Chesney, Sheldon Cohen, Susan Folkman, Robert L. Kahn, and S. Leonard Syme. 1994. "Socioeconomic Status and Health: The Challenge of the Gradient." *American Psychologist* 49 (1): 15–24.

Adolphs, Ralph, and Daniel Tranel. 1999. "Intact Recognition of Emotional Prosody Following Amygdala Damage." *Neuropsychologia* 37 (11): 1285–1292.

———. 2000. "Emotion Recognition and the Human Amygdala." In *The Amygdala. A Functional Analysis*, edited by J. P. Aggleton, 587–630. New York: Oxford University Press.

———. 2003. "Amygdala Damage Impairs Emotion Recognition from Scenes Only When They Contain Facial Expressions." *Neuropsychologia* 41 (10): 1281–1289.

Adolphs, Ralph, Daniel Tranel, Hanna Damasio, and Antonio Damasio. 1994. "Impaired Recognition of Emotion in Facial Expressions Following Bilateral Damage to the Human Amygdala." *Nature* 372 (6507): 669–672.

Aglioti, Salvatore M., Paola Cesari, Michela Romani, and Cosimo Urgesi. 2008. "Action Anticipation and Motor Resonance in Elite Basketball Players." *Nature Neuroscience* 11 (9): 1109–1116.

Akil, Huda. 2015. "The Depressed Brain: Sobering and Hopeful Lesson." National Institutes of Health Wednesday Afternoon Lectures, June 10. http://videocast.nih.gov/summary. asp?Live=16390.

Albright, Madeleine. 2003. *Madam Secretary: A Memoir.* New York: Miramax Books.

Allport, Floyd. 1924. *Social Psychology.* Boston: Houghton Mifflin.

Altschul, Drew, Greg Jensen, and Herbert S. Terrace. 2015. "Concept Learning of Ecological and Artificial Stimuli in Rhesus Macaques." *PeerJ Preprints* 3. doi:10.7287/peerj. preprints.967v1.

American Academy of Pain Medicine. 2012. "AAPM Facts and Figures on Pain." http:// www.painmed.org/patientcenter/facts_on_pain.aspx.

American Kennel Club. 2016. "The Golden Retriever." http://www.akc.org/dog-breeds/ golden-retriever/.

American Psychological Association. 2012. "Stress in America: Our Health at Risk." https:// www.apa.org/news/press/releases/stress/2011/final-2011.pdf.

American Society for Aesthetic Plastic Surgery. 2016. "Initial Data from the American Society for Aesthetic Plastic Surgery Points to 20% Increase in Procedures in 2015." http://www. surgery.org/media/news-releases/initial-data-from-the-american-society-for-aesthetic -plastic-surgery-points-to-20percent-increase-in-procedures-in-2015-300226241.html.

Amso, Dima, and Gaia Scerif. 2015. "The Attentive Brain: Insights from Developmental Cognitive Neuroscience." *Nature Reviews Neuroscience* 16 (10): 606–619.

Anderson, Craig A., Leonard Berkowitz, Edward Donnerstein, L. Rowell Huesmann, James D. Johnson, Daniel Linz, Neil M. Malamuth, and Ellen Wartella. 2003. "The Influence of Media Violence on Youth." *Psychological Science in the Public Interest* 4 (3): 81–110.

Anderson, Eric, Erika H. Siegel, Dominique White, and Lisa Feldman Barrett. 2012. "Out of Sight but Not Out of Mind: Unseen Affective Faces Influence Evaluations and Social Impressions." *Emotion* 12 (6): 1210–1221.

Anderson, Michael L. 2014. *After Phrenology: Neural Reuse and the Interactive Brain.* Cambridge, Mass.: MIT Press.

Anleu, Sharyn Roach, and Kathy Mack. 2005. "Magistrates' Everyday Work and Emotional Labour." *Journal of Law and Society* 32 (4): 590–614.

Ansell, Emily B., Kenneth Rando, Keri Tuit, Joseph Guarnaccia, and Rajita Sinha. 2012. "Cumulative Adversity and Smaller Gray Matter Volume in Medial Prefrontal, Anterior Cingulate, and Insula Regions." *Biological Psychiatry* 72 (1): 57–64.

Antoni, Michael H., Susan K. Lutgendorf, Steven W. Cole, Firdaus S. Dhabhar, Sandra E. Sephton, Paige Green McDonald, Michael Stefanek, and Anil K. Sood. 2006. "The Influence of Bio-Behavioural Factors on Tumour Biology: Pathways and Mechanisms." *Nature Reviews Cancer* 6 (3): 240–248.

Apkarian, A. Vania, Marwan N. Baliki, and Melissa A. Farmer. 2013. "Predicting Transition to Chronic Pain." *Current Opinion in Neurology* 26 (4): 360–367.

Arkowitz, Hal, and Scott O. Lilienfeld. 2010. "Why Science Tells Us Not to Rely on Eyewitness Accounts." *Scientific American Mind*, January 1. http://www.scientificamerican. com/article/do-the-eyes-have-it/.

Atkinson, Anthony P., Andrea S. Heberlein, and Ralph Adolphs. 2007. "Spared Ability to Recognise Fear from Static and Moving Whole-Body Cues Following Bilateral Amygdala Damage." *Neuropsychologia* 45 (12): 2772–2782.

Avena, Nicole M., Pedro Rada, and Bartley G. Hoebel. 2008. "Evidence for Sugar Addiction: Behavioral and Neurochemical Effects of Intermittent, Excessive Sugar Intake." *Neuroscience and Biobehavioral Reviews* 32 (1): 20–39.

Aviezer, Hillel, Ran R. Hassin, Jennifer Ryan, Cheryl Grady, Josh Susskind, Adam Anderson, Morris Moscovitch, and Shlomo Bentin. 2008. "Angry, Disgusted, or Afraid? Studies on the Malleability of Emotion Perception." *Psychological Science* 19 (7): 724–732.

Aviezer, Hillel, Yaacov Trope, and Alexander Todorov. 2012. "Body Cues, Not Facial Expressions, Discriminate Between Intense Positive and Negative Emotions." *Science* 338 (6111): 1225–1229.

Bachman, Jerald G., Lloyd D. Johnston, and Patrick M. O'Malley. 2006. "Monitoring the Future: Questionnaire Responses from the Nation's High School Seniors." Institute for Social Research Survey Research Center, University of Michigan. www.monitoring thefuture.org/datavolumes/2006/2006dv.pdf.

Balasubramanian, Vijay. 2015. "Heterogeneity and Efficiency in the Brain." *Proceedings of the IEEE* 103 (8): 1346–1358.

Bandes, Susan A. Forthcoming. "Share Your Grief but Not Your Anger: Victims and the Expression of Emotion in Criminal Justice." In *Emotional Expression: Philosophical, Psychological, and Legal Perspectives*, edited by Joel Smith and Catharine Abell. New York: Cambridge University Press.

Bandes, Susan A., and Jeremy A. Blumenthal. 2012. "Emotion and the Law." *Annual Review of Law and Social Science* 8: 161–181.

Banks, Siobhan, and David F. Dinges. 2007. "Behavioral and Physiological Consequences of Sleep Restriction." *Journal of Clinical Sleep Medicine* 3 (5): 519–528.

Bar, Moshe. 2007. "The Proactive Brain: Using Analogies and Associations to Generate Predictions." *Trends in Cognitive Sciences* 11 (7): 280–289.

———. 2009. "The Proactive Brain: Memory for Predictions." *Philosophical Transactions of the Royal Society B: Biological Sciences* 364 (1521): 1235–1243.

Barbas, Helen. 2015. "General Cortical and Special Prefrontal Connections: Principles from Structure to Function." *Annual Review of Neuroscience* 38: 269–289.

Barbas, Helen, and Nancy Rempel-Clower. 1997. "Cortical Structure Predicts the Pattern of Corticocortical Connections." *Cerebral Cortex* 7 (7): 635–646.

Bargmann, C. I. 2012. "Beyond the Connectome: How Neuromodulators Shape Neural Circuits." *Bioessays* 34 (6): 458–465.

Barrett, Deborah. 2012. *Paintracking: Your Personal Guide to Living Well with Chronic Pain*. New York: Prometheus Books.

Barrett, Lisa Feldman. 2006a. "Are Emotions Natural Kinds?" *Perspectives on Psychological Science* 1 (1): 28–58.

———. 2006b. "Solving the Emotion Paradox: Categorization and the Experience of Emotion." *Personality and Social Psychology Review* 10 (1): 20–46.

———. 2009. "The Future of Psychology: Connecting Mind to Brain." *Perspectives on Psychological Science* 4 (4): 326–339.

———. 2011a. "Bridging Token Identity Theory and Supervenience Theory Through Psychological Construction." *Psychological Inquiry* 22 (2): 115–127.

———. 2011b. "Was Darwin Wrong about Emotional Expressions?" *Current Directions in Psychological Science* 20 (6): 400–406.

———. 2012. "Emotions Are Real." *Emotion* 12 (3): 413–429.

———. 2013. "Psychological Construction: The Darwinian Approach to the Science of Emotion." *Emotion Review* 5: 379–389.

Barrett, Lisa Feldman, and Moshe Bar. 2009. "See It with Feeling: Affective Predictions During Object Perception." *Philosophical Transactions of the Royal Society B: Biological Sciences* 364 (1521): 1325–1334.

Barrett, Lisa Feldman, and Eliza Bliss-Moreau. 2009a. "Affect as a Psychological Primitive." *Advances in Experimental Social Psychology* 41: 167–218.

———. 2009b. "She's Emotional. He's Having a Bad Day: Attributional Explanations for Emotion Stereotypes." *Emotion* 9 (5): 649–658.

Barrett, Lisa Feldman, James Gross, Tamlin Conner Christensen, and Michael Benvenuto. 2001. "Knowing What You're Feeling and Knowing What To Do About It: Mapping the Relation Between Emotion Differentiation and Emotion Regulation." *Cognition and Emotion* 15 (6): 713–724.

Barrett, Lisa Feldman, Kristen A. Lindquist, Eliza Bliss-Moreau, Seth Duncan, Maria Gendron, Jennifer Mize, and Lauren Brennan. 2007. "Of Mice and Men: Natural Kinds of Emotions in the Mammalian Brain? A Response to Panksepp and Izard." *Perspectives on Psychological Science* 2 (3): 297–311.

Barrett, Lisa Feldman, Kristen A. Lindquist, and Maria Gendron. 2007. "Language as Context for the Perception of Emotion." *Trends in Cognitive Sciences* 11 (8): 327–332.

Barrett, Lisa Feldman, Batja Mesquita, and Maria Gendron. 2011. "Context in Emotion Perception." *Current Directions in Psychological Science* 20 (5): 286–290.

Barrett, Lisa Feldman, Lucy Robin, Paula R. Pietromonaco, and Kristen M. Eyssell. 1998. "Are Women the 'More Emotional' Sex? Evidence from Emotional Experiences in Social Context." *Cognition and Emotion* 12 (4): 555–578.

Barrett, Lisa Feldman, and James A. Russell. 1999. "Structure of Current Affect: Controversies and Emerging Consensus." *Current Directions in Psychological Science* 8 (1): 10–14.

———, eds. 2015. *The Psychological Construction of Emotion.* New York: Guilford Press.

Barrett, Lisa Feldman, and Ajay B. Satpute. 2013. "Large-Scale Brain Networks in Affective and Social Neuroscience: Towards an Integrative Functional Architecture of the Brain." *Current Opinion in Neurobiology* 23 (3): 361–372.

Barrett, Lisa Feldman, and W. Kyle Simmons. 2015. "Interoceptive Predictions in the Brain." *Nature Reviews Neuroscience* 16 (7): 419–429.

Barrett, Lisa Feldman, Michele M. Tugade, and Randall W. Engle. 2004. "Individual Differences in Working Memory Capacity and Dual-Process Theories of the Mind." *Psychological Bulletin* 130 (4): 553–573.

Barsalou, Lawrence W. 1985. "Ideals, Central Tendency, and Frequency of Instantiation as Determinants of Graded Structure in Categories." *Journal of Experimental Psychology: Learning, Memory, and Cognition* 11 (4): 629–654.

———. 1992. *Cognitive Psychology: An Overview for Cognitive Scientists.* Mawah, NJ: Lawrence Erlbaum.

———. 1999. "Perceptual Symbol Systems." *Behavioral and Brain Sciences* 22 (4): 577–609.

———. 2003. "Situated Simulation in the Human Conceptual System." *Language and Cognitive Processes* 18: 513–562.

———. 2008a. "Cognitive and Neural Contributions to Understanding the Conceptual System." *Current Directions in Psychological Science* 17 (2): 91–95.

———. 2008b. "Grounded Cognition." *Annual Review of Psychology* 59: 617–645.

———. 2009. "Simulation, Situated Conceptualization, and Prediction." *Philosophical Transactions of the Royal Society B: Biological Sciences* 364 (1521): 1281–1289.

Barsalou, Lawrence W., W. Kyle Simmons, Aron K. Barbey, and Christine D. Wilson. 2003. "Grounding Conceptual Knowledge in Modality-Specific Systems." *Trends in Cognitive Sciences* 7 (2): 84–91.

Bartal, Inbal Ben-Ami, Jean Decety, and Peggy Mason. 2011. "Empathy and Pro-Social Behavior in Rats." *Science* 334 (6061): 1427–1430.

Beard, Mary. 2014. *Laughter in Ancient Rome: On Joking, Tickling, and Cracking Up.* Berkeley: University of California Press.

Bechara, Antoine, Daniel Tranel, Hanna Damasio, Ralph Adolphs, Charles Rockland, and Antonio R. Damasio. 1995. "Double Dissociation of Conditioning and Declarative Knowledge Relative to the Amygdala and Hippocampus in Humans." *Science* 269 (5227): 1115–1118.

Becker, Benjamin, Yoan Mihov, Dirk Scheele, Keith M. Kendrick, Justin S. Feinstein, Andreas Matusch, Merve Aydin, Harald Reich, Horst Urbach, and Ana-Maria Oros-Peusquens. 2012. "Fear Processing and Social Networking in the Absence of a Functional Amygdala." *Biological Psychiatry* 72 (1): 70–77.

Becquet, Celine, Nick Patterson, Anne C. Stone, Molly Przeworski, and David Reich. 2007.

Genetic Structure of Chimpanzee Populations. *PLOS Genetics* 3 (4): e66. doi:10.1371/journal.pgen.0030066.

Beggs, Simon, Gillian Currie, Michael W. Salter, Maria Fitzgerald, and Suellen M. Walker. 2012. "Priming of Adult Pain Responses by Neonatal Pain Experience: Maintenance by Central Neuroimmune Activity." *Brain* 135 (2): 404–417.

Bekoff, Marc, and Jane Goodall. 2008. *The Emotional Lives of Animals: A Leading Scientist Explores Animal Joy, Sorrow, and Empathy—and Why They Matter.* Novato, CA: New World Library.

Benedetti, Fabrizio. 2014. "Placebo Effects: From the Neurobiological Paradigm to Translational Implications." *Neuron* 84 (3): 623–637.

Benedetti, Fabrizio, Martina Amanzio, Sergio Vighetti, and Giovanni Asteggiano. 2006. "The Biochemical and Neuroendocrine Bases of the Hyperalgesic Nocebo Effect." *Journal of Neuroscience* 26 (46): 12014–12022.

Berent, Iris. 2013. "The Phonological Mind." *Trends in Cognitive Sciences* 17 (7): 319–327.

Bergelson, Elika, and Daniel Swingley. 2012. "At 6–9 Months, Human Infants Know the Meanings of Many Common Nouns." *Proceedings of the National Academy of Sciences* 109 (9): 3253–3258.

Berlau, Daniel J., and James L. McGaugh. 2003. "Basolateral Amygdala Lesions Do Not Prevent Memory of Context-Footshock Training." *Learning and Memory* 10 (6): 495–502.

Berns, Walter. 1979. *For Capital Punishment: Crime and the Morality of the Death Penalty.* New York: Basic Books.

"Better Than English." 2016. http://betterthanenglish.com/.

Beukeboom, Camiel J., Dion Langeveld, and Karin Tanja-Dijkstra. 2012. "Stress-Reducing Effects of Real and Artificial Nature in a Hospital Waiting Room." *Journal of Alternative and Complementary Medicine* 18 (4): 329–333.

Binder, Jeffrey R., and Rutvik H. Desai. 2011. "The Neurobiology of Semantic Memory." *Trends in Cognitive Sciences* 15 (11): 527–536.

Binder, Jeffrey R., Rutvik H. Desai, William W. Graves, and Lisa L. Conant. 2009. "Where Is the Semantic System? A Critical Review and Meta-Analysis of 120 Functional Neuroimaging Studies." *Cerebral Cortex* 19 (12): 2767–2796.

Binder, Jeffrey R., Julia A. Frost, Thomas A. Hammeke, P. S. F. Bellgowan, Stephen M. Rao, and Robert W. Cox. 1999. "Conceptual Processing During the Conscious Resting State: A Functional MRI Study." *Journal of Cognitive Neuroscience* 11 (1): 80–93.

Birklein, Frank. 2005. "Complex Regional Pain Syndrome." *Journal of Neurology* 252 (2): 131–138.

Black, Ryan C., Sarah A. Treul, Timothy R. Johnson, and Jerry Goldman. 2011. "Emotions, Oral Arguments, and Supreme Court Decision Making." *Journal of Politics* 73 (2): 572–581.

Bliss-Moreau, Eliza, and David G. Amaral. Under review. "Associative Affective Learning Persists Following Early Amygdala Damage in Nonhuman Primates."

Bliss-Moreau, Eliza, Christopher J. Machado, and David G. Amaral. 2013. "Macaque Cardiac Physiology Is Sensitive to the Valence of Passively Viewed Sensory Stimuli." *PLOS One* 8 (8): e71170. doi:10.1371/journal.pone.0071170.

Blow, Charles M. 2015. "Has the N.R.A. Won?" *New York Times,* April 20. http://www.nytimes.com/2015/04/20/opinion/charles-blow-has-the-nra-won.html.

Blumberg, Mark S., and Greta Sokoloff. 2001. "Do Infant Rats Cry?" *Psychological Review* 108 (1): 83–95.

Blumberg, Mark S., Greta Sokoloff, Robert F. Kirby, and Kristen J. Kent. 2000. "Distress Vocalizations in Infant Rats: What's All the Fuss About?" *Psychological Science* 11 (1): 78–81.

Boghossian, Paul. 2006. *Fear of Knowledge: Against Relativism and Constructivism*. Oxford: Clarendon Press.

Borsook, David. 2012. "Neurological Diseases and Pain." *Brain* 135 (2): 320–344.

Bourassa-Perron, Cynthia. 2011. *The Brain and Emotional Intelligence: New Insights*. Florence, MA: More Than Sound.

Bourke, Joanna. 2000. *An Intimate History of Killing: Face-to-Face Killing in Twentieth-Century Warfare*. New York: Basic Books.

Boyd, Robert, Peter J. Richerson, and Joseph Henrich. 2011. "The Cultural Niche: Why Social Learning Is Essential for Human Adaptation." *Proceedings of the National Academy of Sciences* 108 (Supplement 2): 10918–10925.

Brackett, Marc A., Susan E. Rivers, Maria R. Reyes, and Peter Salovey. 2012. "Enhancing Academic Performance and Social and Emotional Competence with the RULER Feeling Words Curriculum." *Learning and Individual Differences* 22 (2): 218–224.

Bradshaw, John. 2014. *Dog Sense: How the New Science of Dog Behavior Can Make You a Better Friend to Your Pet*. New York: Basic Books.

Brandone, Amanda C., and Henry M. Wellman. 2009. "You Can't Always Get What You Want: Infants Understand Failed Goal-Directed Actions." *Psychological Science* 20 (1): 85–91.

Bratman, Gregory N., J. Paul Hamilton, Kevin S. Hahn, Gretchen C. Daily, and James J. Gross. 2015. "Nature Experience Reduces Rumination and Subgenual Prefrontal Cortex Activation." *Proceedings of the National Academy of Sciences* 112 (28): 8567–8572.

Breiter, Hans C., Nancy L. Etcoff, Paul J. Whalen, William A. Kennedy, Scott L. Rauch, Randy L. Buckner, Monica M. Strauss, Steven E. Hyman, and Bruce R. Rosen. 1996. "Response and Habituation of the Human Amygdala During Visual Processing of Facial Expression." *Neuron* 17 (5): 875–887.

Brennan, William J., Jr. 1988. "Reason, Passion, and the Progress of the Law." *Cardozo Law Review* 10: 3.

Brescoll, Victoria L., and Eric Luis Uhlmann. 2008. "Can an Angry Woman Get Ahead? Status Conferral, Gender, and Expression of Emotion in the Workplace." *Psychological Science* 19 (3): 268–275.

Briggs, Jean L. 1970. *Never in Anger: Portrait of an Eskimo Family*. Cambridge, MA: Harvard University Press.

Broly, Pierre, and Jean-Louis Deneubourg. 2015. "Behavioural Contagion Explains Group Cohesion in a Social Crustacean." *PLOS Computational Biology* 11 (6): e1004290. doi:10.1371/journal.pcbi.1004290.

Browning, Michael, Timothy E. Behrens, Gerhard Jocham, Jill X. O'Reilly, and Sonia J. Bishop. 2015. "Anxious Individuals Have Difficulty Learning the Causal Statistics of Aversive Environments." *Nature Neuroscience* 18 (4): 590–596.

Bruner, Jerome S. 1990. *Acts of Meaning*. Cambridge, MA: Harvard University Press.

Bryant, Richard A., Kim L. Felmingham, Derrick Silove, Mark Creamer, Meaghan O'Donnell, and Alexander C. McFarlane. 2011. "The Association Between Menstrual Cycle and Traumatic Memories." *Journal of Affective Disorders* 131 (1): 398–401.

Büchel, Christian, Stephan Geuter, Christian Sprenger, and Falk Eippert. 2014. "Placebo Analgesia: A Predictive Coding Perspective." *Neuron* 81 (6): 1223–1239.

Buckholtz, Joshua W., Christopher L. Asplund, Paul E. Dux, David H. Zald, John C. Gore,

Owen D. Jones, and Rene Marois. 2008. "The Neural Correlates of Third-Party Punishment." *Neuron* 60 (5): 930–940.

Buckner, Randy L. 2012. "The Serendipitous Discovery of the Brain's Default Network." *Neuroimage* 62 (2): 1137–1145.

Bullmore, Ed, and Olaf Sporns. 2012. "The Economy of Brain Network Organization." *Nature Reviews Neuroscience* 13 (5): 336–349.

Burkett, J. P., E. Andari, Z. V. Johnson, D. C. Curry, F. B. M. de Waal, and L. J. Young. 2016. "Oxytocin-Dependent Consolation Behavior in Rodents." *Science* 351 (6271): 375–378.

Burns, Jeffrey M., and Russell H. Swerdlow. 2003. "Right Orbitofrontal Tumor with Pedophilia Symptom and Constructional Apraxia Sign." *Archives of Neurology* 60 (3): 437–440.

Bushnell, M. Catherine, Marta Čeko, and Lucie A. Low. 2013. "Cognitive and Emotional Control of Pain and Its Disruption in Chronic Pain." *Nature Reviews Neuroscience* 14 (7): 502–511.

Cabanac, M., and J. Leblanc. 1983. "Physiological Conflict in Humans: Fatigue vs. Cold Discomfort." *American Journal of Physiology* 244 (5): R621–628.

Cacioppo, John T., Gary G. Berntson, Jeff H. Larsen, Kristen M. Poehlmann, and Tiffany A. Ito. 2000. "The Psychophysiology of Emotion." In *Handbook of Emotions,* 2nd edition, edited by Michael Lewis and Jeannette M. Haviland-Jones, 173–191. New York: Guilford Press.

Caldwell-Harris, Catherine L., Angela L. Wilson, Elizabeth LoTempio, and Benjamin Beit-Hallahmi. 2011. "Exploring the Atheist Personality: Well-Being, Awe, and Magical Thinking in Atheists, Buddhists, and Christians." *Mental Health, Religion and Culture* 14 (7): 659–672.

Calhoun, Cheshire. 1999. "Making Up Emotional People: The Case of Romantic Love." In *The Passions of Law,* edited by Susan A. Bandes, 217–240. New York: New York University Press.

Calvin, Catherine M., G. David Batty, Gordon Lowe, and Ian J. Deary. 2011. "Childhood Intelligence and Midlife Inflammatory and Hemostatic Biomarkers: The National Child Development Study (1958) Cohort." *Health Psychology* 30 (6): 710–718.

Cameron, C. Daryl, B. Keith Payne, and John M. Doris. 2013. "Morality in High Definition: Emotion Differentiation Calibrates the Influence of Incidental Disgust on Moral Judgments." *Journal of Experimental Social Psychology* 49 (4): 719–725.

Camras, Linda A., Harriet Oster, Tatsuo Ujiie, Joseph J. Campos, Roger Bakeman, and Zhaolan Meng. 2007. "Do Infants Show Distinct Negative Facial Expressions for Fear and Anger? Emotional Expression in 11-Month-Old European American, Chinese, and Japanese Infants." *Infancy* 11 (2): 131–155.

Carhart-Harris, Robin L., Suresh Muthukumaraswamy, Leor Rosemana, Mendel Kaelena, Wouter Droog, et al. 2016. "Neural Correlates of the LSD Experience Revealed by Multimodal Neuroimaging." *Proceedings of the National Academy of Sciences* 113 (7): 4853–4858.

Caron, Rose F., Albert J. Caron, and Rose S. Myers. 1985. "Do Infants See Emotional Expressions in Static Faces?" *Child Development* 56 (6): 1552–1560.

Casey, Caroline. 2010. "Looking Past Limits." TED.com. https://www.ted.com/talks/caroline_casey_looking_past_limits.

Cassoff, Jamie, Sabrina T. Wiebe, and Reut Gruber. 2012. "Sleep Patterns and the Risk for ADHD: A Review." *Nature and Science of Sleep* 4: 73–80.

Centers for Disease Control and Prevention. 2015. "Prescription Opioid Analgesic Use Among Adults: United States, 1999–2012." http://www.cdc.gov/nchs/products/data briefs/db189.htm.

Ceulemans, Eva, Peter Kuppens, and Iven Van Mechelen. 2012. "Capturing the Structure of Distinct Types of Individual Differences in the Situation-Specific Experience of Emotions: The Case of Anger." *European Journal of Personality* 26 (5): 484–495.

Chanes, Lorena, and Lisa Feldman Barrett. 2016. "Redefining the Role of Limbic Areas in Cortical Processing." *Trends in Cognitive Sciences* 20 (2): 96–106.

Chang, Anne-Marie, Daniel Aeschbach, Jeanne F. Duffy, and Charles A. Czeisler. 2015. "Evening Use of Light-Emitting eReaders Negatively Affects Sleep, Circadian Timing, and Next-Morning Alertness." *Proceedings of the National Academy of Sciences* 112 (4): 1232–1237.

Chang, Luke J., Peter J. Gianaros, Stephen B. Manuck, Anjali Krishnan, and Tor D. Wager. 2015. "A Sensitive and Specific Neural Signature for Picture-Induced Negative Affect." *PLOS Biology* 13 (6): e1002180.

Chao, Linda L., and Alex Martin. 2000. "Representation of Manipulable Man-Made Objects in the Dorsal Stream." *Neuroimage* 12 (4): 478–484.

Charney, Evan. 2012. "Behavior Genetics and Postgenomics." *Behavioral and Brain Sciences* 35 (5): 331–358.

Chen, Lucy L. 2014. "What Do We Know About Opioid-Induced Hyperalgesia?" *Journal of Clinical Outcomes Management* 21 (3): 169–175.

Choi, Ki Sueng, Patricio Riva-Posse, Robert E. Gross, and Helen S. Mayberg. 2015. "Mapping the 'Depression Switch' During Intraoperative Testing of Subcallosal Cingulate Deep Brain Stimulation." *JAMA Neurology* 72 (11): 1252–1260.

Chomsky, Noam. 1980. "Rules and Representations." *Behavioral and Brain Sciences* 3 (1): 1–15.

Cisek, P., and J. Kalaska. 2010. "Neural Mechanisms for Interacting with a World Full of Action Choices." *Annual Review of Neuroscience* 33: 269–298.

Clark, Andy. 2013. "Whatever Next? Predictive Brains, Situated Agents, and the Future of Cognitive Science." *Behavioral and Brain Sciences* 36: 281–253.

Clark-Polner, E., T. Johnson, and L. F. Barrett. In press. "Multivoxel Pattern Analysis Does Not Provide Evidence to Support the Existence of Basic Emotions." *Cerebral Cortex.*

Clark-Polner, Elizabeth, Tor D. Wager, Ajay B. Satpute, and Lisa Feldman Barrett. In press. "Neural Fingerprinting: Meta-Analysis, Variation, and the Search for Brain-Based Essences in the Science of Emotion." In *Handbook of Emotions,* 4th edition, edited by Lisa Feldman Barrett, Michael Lewis, and Jeannette M. Haviland-Jones, 146–165. New York: Guilford Press.

Clave-Brule, M., A. Mazloum, R. J. Park, E. J. Harbottle, and C. Laird Birmingham. 2009. "Managing Anxiety in Eating Disorders with Knitting." *Eating and Weight Disorders-Studies on Anorexia, Bulimia and Obesity* 14 (1): e1–e5.

Clore, Gerald L., and Andrew Ortony. 2008. "Appraisal Theories: How Cognition Shapes Affect into Emotion." In *Handbook of Emotions,* 3rd edition, edited by Michael Lewis, Jeannette M. Haviland-Jones, and Lisa Feldman Barrett, 628–642. New York: Guilford Press.

Coan, James A., Hillary S. Schaefer, and Richard J. Davidson. 2006. "Lending a Hand: Social Regulation of the Neural Response to Threat." *Psychological Science* 17 (12): 1032–1039.

Cohen, Sheldon, William J. Doyle, David P. Skoner, Bruce S. Rabin, and Jack M. Gwaltney. 1997. "Social Ties and Susceptibility to the Common Cold." *JAMA* 277 (24): 1940–1944.

Cohen, Sheldon, William J. Doyle, Ronald Turner, Cuneyt M. Alper, and David P. Skoner. 2003. "Sociability and Susceptibility to the Common Cold." *Psychological Science* 14 (5): 389–395.

Cohen, Sheldon, and Gail M. Williamson. 1991. "Stress and Infectious Disease in Humans." *Psychological Bulletin* 109 (1): 5–24.

Cole, Steven W., and Anil K. Sood. 2012. "Molecular Pathways: Beta-Adrenergic Signaling in Cancer." *Clinical Cancer Research* 18 (5): 1201–1206.

Consedine, Nathan S., Yulia E. Chentsova Dutton, and Yulia S. Krivoshekova. 2014. "Emotional Acculturation Predicts Better Somatic Health: Experiential and Expressive Acculturation Among Immigrant Women from Four Ethnic Groups." *Journal of Social and Clinical Psychology* 33 (10): 867–889.

Copeland, William E., Dieter Wolke, Adrian Angold, and E. Jane Costello. 2013. "Adult Psychiatric Outcomes of Bullying and Being Bullied by Peers in Childhood and Adolescence." *JAMA Psychiatry* 70 (4): 419–426.

Copeland, William E., Dieter Wolke, Suzet Tanya Lereya, Lilly Shanahan, Carol Worthman, and E. Jane Costello. 2014. "Childhood Bullying Involvement Predicts Low-Grade Systemic Inflammation into Adulthood." *Proceedings of the National Academy of Sciences* 111 (21): 7570–7575.

Cordaro, Daniel T., Dacher Keltner, Sumjay Tshering, Dorji Wangchuk, and Lisa M. Flynn. 2016. "The Voice Conveys Emotion in Ten Globalized Cultures and One Remote Village in Bhutan." *Emotion* 16 (1): 117–128.

Cosmides, Leda, and John Tooby. 2000. "Evolutionary Psychology and the Emotions." In *Handbook of Emotions,* 2nd edition, edited by Michael Lewis and Jeannette M. Haviland-Jones, 91–115. New York: Guilford Press.

Craig, A. D. 2015. *How Do You Feel? An Interoceptive Moment with Your Neurobiological Self.* Princeton, NJ: Princeton University Press.

Creswell, J. D., A. A. Taren, E. K. Lindsay, C. M. Greco, P. J. Gianaros, A. Fairgrieve, A. L. Marsland, K. W. Brown, B. M. Way, R. K. Rosen, and J. L. Ferris. In press. "Alterations in Resting State Functional Connectivity Link Mindfulness Meditation with Reduced Interleukin-6." *Biological Psychiatry.*

Crivelli, Carlos, Pilar Carrera, and José-Miguel Fernández-Dols. 2015. "Are Smiles a Sign of Happiness? Spontaneous Expressions of Judo Winners." *Evolution and Human Behavior* 36 (1): 52–58.

Crivelli, Carlos, Sergio Jarillo, James A. Russell, and José-Miguel Fernández-Dols. 2016. "Reading Emotions from Faces in Two Indigenous Societies." *Journal of Experimental Psychology* 145 (7): 830–843.

Crossley, Nicolas A., Andrea Mechelli, Jessica Scott, Francesco Carletti, Peter T. Fox, Philip McGuire, and Edward T. Bullmore. 2014. "The Hubs of the Human Connectome Are Generally Implicated in the Anatomy of Brain Disorders." *Brain* 137 (8): 2382–2395.

Crum, Alia J., William R. Corbin, Kelly D. Brownell, and Peter Salovey. 2011. "Mind over Milkshakes: Mindsets, Not Just Nutrients, Determine Ghrelin Response." *Health Psychology* 30 (4): 424–429.

Crum, Alia J., Peter Salovey, and Shawn Achor. 2013. "Rethinking Stress: The Role of Mindsets in Determining the Stress Response." *Journal of Personality and Social Psychology* 104 (4): 716–733.

Curry, John, Susan Silva, Paul Rohde, Golda Ginsburg, Christopher Kratochvil, Anne Simons, Jerry Kirchner, Diane May, Betsy Kennard, and Taryn Mayes. 2011. "Recovery and Recurrence Following Treatment for Adolescent Major Depression." *Archives of General Psychiatry* 68 (3): 263–269.

Damasio, Antonio. 1994. *Descartes' Error: Emotion, Reason and the Human Brain.* New York: Avon.

———. 1999. *The Feeling of What Happens: Body and Emotion in the Making of Consciousness.* New York: Harcourt Brace & Company.

Damasio, Antonio, and Gil B. Carvalho. 2013. "The Nature of Feelings: Evolutionary and Neurobiological Origins." *Nature Reviews Neuroscience* 14 (2): 143–152.

Danese, Andrea, and Bruce S. McEwen. 2012. "Adverse Childhood Experiences, Allostasis, Allostatic Load, and Age-Related Disease." *Physiology and Behavior* 106 (1): 29–39.

Dannlowski, Udo, Anja Stuhrmann, Victoria Beutelmann, Peter Zwanzger, Thomas Lenzen, Dominik Grotegerd, Katharina Domschke, Christa Hohoff, Patricia Ohrmann, and Jochen Bauer. 2012. "Limbic Scars: Long-Term Consequences of Childhood Maltreatment Revealed by Functional and Structural Magnetic Resonance Imaging." *Biological Psychiatry* 71 (4): 286–293.

Dantzer, Robert, Cobi Johanna Heijnen, Annemieke Kavelaars, Sophie Laye, and Lucile Capuron. 2014. "The Neuroimmune Basis of Fatigue." *Trends in Neurosciences* 37 (1): 39–46.

Dantzer, Robert, Jan-Pieter Konsman, Rose-Marie Bluthé, and Keith W. Kelley. 2000. "Neural and Humoral Pathways of Communication from the Immune System to the Brain: Parallel or Convergent?" *Autonomic Neuroscience* 85 (1): 60–65.

Danziger, Kurt. 1997. *Naming the Mind: How Psychology Found Its Language.* London: Sage.

Danziger, Shai, Jonathan Levav, and Liora Avnaim-Pesso. 2011. "Extraneous Factors in Judicial Decisions." *Proceedings of the National Academy of Sciences* 108 (17): 6889–6892.

Darwin, Charles. (1859) 2003. *On the Origin of Species.* Facsimile edition. Cambridge, MA: Harvard University Press.

———. (1871) 2004. *The Descent of Man, and Selection in Relation to Sex.* London: Penguin Classics.

———. (1872) 2005. *The Expression of the Emotions in Man and Animals.* Stilwell, KS: Digireads.com.

Dashiell, John F. 1927. "A New Method of Measuring Reactions to Facial Expression of Emotion." *Psychological Bulletin* 24: 174–175.

De Boer, Sietse F., and Jaap M. Koolhaas. 2003. "Defensive Burying in Rodents: Ethology, Neurobiology and Psychopharmacology." *European Journal of Pharmacology* 463 (1): 145–161.

Deffenbacher, Kenneth A., Brian H. Bornstein, Steven D. Penrod, and E. Kiernan McGorty. 2004. "A Meta-Analytic Review of the Effects of High Stress on Eyewitness Memory." *Law and Human Behavior* 28 (6): 687–706.

De Leersnyder, Jozefien, Batja Mesquita, and Heejung S. Kim. 2011. "Where Do My Emotions Belong? A Study of Immigrants' Emotional Acculturation." *Personality and Social Psychology Bulletin* 37 (4): 451–463.

Demiralp, Emre, Renee J. Thompson, Jutta Mata, Susanne M. Jaeggi, Martin Buschkuehl, Lisa Feldman Barrett, Phoebe C. Ellsworth, Metin Demiralp, Luis Hernandez-Garcia, and Patricia J. Deldin. 2012. "Feeling Blue or Turquoise? Emotional Differentiation in Major Depressive Disorder." *Psychological Science* 23 (11): 1410–1416.

Deneve, Sophie, and Renaud Jardri. 2016. "Circular Inference: Mistaken Belief, Misplaced Trust." *Current Opinion in Behavioral Sciences* 11: 40–48.

Denham, Joshua, Brendan J. O'Brien, and Fadi J. Charchar. 2016. "Telomere Length Maintenance and Cardio-Metabolic Disease Prevention Through Exercise Training." *Sports Medicine*, February 25, 1–25.

Denham, Susanne A. 1998. *Emotional Development in Young Children*. New York: Guilford Press.

Denison, Stephanie, Christie Reed, and Fei Xu. 2013. "The Emergence of Probabilistic Reasoning in Very Young Infants: Evidence from 4.5- and 6-Month-Olds." *Developmental Psychology* 49 (2): 243–249.

Denison, Stephanie, and Fei Xu. 2010. "Twelve- to 14-Month-Old Infants Can Predict Single-Event Probability with Large Set Sizes." *Developmental Science* 13 (5): 798–803.

———. 2014. "The Origins of Probabilistic Inference in Human Infants." *Cognition* 130 (3): 335–347.

"Developments in the Law: Legal Responses to Domestic Violence." 1993. *Harvard Law Review* 106 (7): 1498–1620.

Dixon-Gordon, Katherine L., Alexander L. Chapman, Nicole H. Weiss, and M. Zachary Rosenthal. 2014. "A Preliminary Examination of the Role of Emotion Differentiation in the Relationship Between Borderline Personality and Urges for Maladaptive Behaviors." *Journal of Psychopathology and Behavioral Assessment* 36 (4): 616–625.

Donoghue, Philip C. J., and Mark A. Purnell. 2005. "Genome Duplication, Extinction and Vertebrate Evolution." *Trends in Ecology and Evolution* 20 (6): 312–319.

Dowlati, Yekta, Nathan Herrmann, Walter Swardfager, Helena Liu, Lauren Sham, Elyse K. Reim, and Krista L. Lanctôt. 2010. "A Meta-Analysis of Cytokines in Major Depression." *Biological Psychiatry* 67 (5): 446–457.

Dreger, Alice Domurat. 1998. *Hermaphrodites and the Medical Invention of Sex*. Cambridge, MA: Harvard University Press.

———. 2015. *Galileo's Middle Finger: Heretics, Activists, and the Search for Justice in Science*, New York: Penguin.

Dreger, Alice D., Cheryl Chase, Aron Sousa, Philip A. Gruppuso, and Joel Frader. 2005. "Changing the Nomenclature/Taxonomy for Intersex: A Scientific and Clinical Rationale." *Journal of Pediatric Endocrinology and Metabolism* 18 (8): 729–734.

Dreyfus, Georges, and Evan Thompson. 2007. "Asian Perspectives: Indian Theories of Mind." In *The Cambridge Handbook of Consciousness*, edited by Philip David Zelazo, Morris Moscovitch, and Evan Thompson, 89–114. New York: Cambridge University Press.

Drnevich, J., et al. 2012. "Impact of Experience-Dependent and -Independent Factors on Gene Expression in Songbird Brain." *Proceedings of the National Academy of Sciences of the United States of America* 109: 17245–17252.

Dubois, Samuel, Bruno Rossion, Christine Schiltz, Jean-Michel Bodart, Christian Michel, Raymond Bruyer, and Marc Crommelinck. 1999. "Effect of Familiarity on the Processing of Human Faces." *Neuroimage* 9 (3): 278–289.

Duffy, Elizabeth. 1934. "Emotion: An Example of the Need for Reorientation in Psychology." *Psychological Review* 41 (2): 184–198.

———. 1941. "An Explanation of 'Emotional' Phenomena Without the Use of the Concept 'Emotion.'" *Journal of General Psychology* 25 (2): 283–293.

Dunfield, Kristen A. and Valerie A. Kuhlmeier. 2013. "Classifying Prosocial Behavior: Children's Responses to Instrumental Need, Emotional Distress, and Material Desire." *Child Development*, 84 (5): 1766–1776.

Dunn, Elizabeth W., Daniel T. Gilbert, and Timothy D. Wilson. 2011. "If Money Doesn't Make You Happy, Then You Probably Aren't Spending It Right." *Journal of Consumer Psychology* 21 (2): 115–125.

Dunn, Elizabeth, and Michael Norton. 2013. *Happy Money: The Science of Smarter Spending*. New York: Simon and Schuster.

Dunsmore, Julie C., Pa Her, Amy G. Halberstadt, and Marie B. Perez-Rivera. 2009. "Parents' Beliefs About Emotions and Children's Recognition of Parents' Emotions." *Journal of Nonverbal Behavior* 33 (2): 121–140.

Durham, William H. 1991. *Coevolution: Genes, Culture, and Human Diversity*. Stanford, CA: Stanford University Press.

Edelman, Gerald M. 1987. *Neural Darwinism: The Theory of Neuronal Group Selection*. New York: Basic Books.

———. 1990. *The Remembered Present: A Biological Theory of Consciousness*. New York: Basic Books.

Edelman, G. M., and J. A. Gally. 2001. "Degeneracy and Complexity in Biological Systems." *Proceedings of the National Academy of Sciences* 98: 13763–13768.

Edelman, Gerald M., and Giulio Tononi. 2000. *A Universe of Consciousness: How Matter Becomes Imagination*. New York: Basic Books.

Edersheim, Judith G., Rebecca Weintraub Brendel, and Bruce H. Price. 2012. "Neuroimaging, Diminished Capacity and Mitigation." In *Neuroimaging in Forensic Psychiatry: From the Clinic to the Courtroom*, edited by Joseph R. Simpson, 163–193. West Sussex, UK: Wiley-Blackwell.

Einstein, Albert and Leopold Infeld. 1938. *Evolution of Physics*. Cambridge, United Kingdom: Cambridge University Press.

Eisenberger, Naomi I. 2012. "The Pain of Social Disconnection: Examining the Shared Neural Underpinnings of Physical and Social Pain." *Nature Reviews Neuroscience* 13 (6): 421–434.

Eisenberger, Naomi I., and Steve W. Cole. 2012. "Social Neuroscience and Health: Neurophysiological Mechanisms Linking Social Ties with Physical Health." *Nature Neuroscience* 15 (5): 669–674.

Eisenberger, Naomi I., Tristen K. Inagaki, Nehjla M. Mashal, and Michael R. Irwin. 2010. "Inflammation and Social Experience: An Inflammatory Challenge Induces Feelings of Social Disconnection in Addition to Depressed Mood." *Brain, Behavior, and Immunity* 24 (4): 558–563.

Ekkekakis, Panteleimon, Elaine A. Hargreaves, and Gaynor Parfitt. 2013. "Invited Guest Editorial: Envisioning the Next Fifty Years of Research on the Exercise-Affect Relationship." *Psychology of Sport and Exercise* 14 (5): 751–758.

Ekman, Paul. 1992. "An Argument for Basic Emotions." *Cognition and Emotion* 6: 169–200.

———. 2007. *Emotions Revealed: Recognizing Faces and Feelings to Improve Communication and Emotional Life*. New York: Henry Holt.

Ekman, Paul, and Daniel Cordaro. 2011. "What Is Meant by Calling Emotions Basic." *Emotion Review* 3 (4): 364–370.

Ekman, Paul, and Wallace V. Friesen. 1971. "Constants Across Cultures in the Face and Emotion." *Journal of Personality and Social Psychology* 17 (2): 124–129.

———. 1984. *EM-FACS Coding Manual*. San Francisco: Consulting Psychologists Press.

Ekman, Paul, Wallace V. Friesen, Maureen O'Sullivan, Anthony Chan, Irene Diacoyanni-Tarlatzis, Karl Heider, Rainer Krause, William Ayhan LeCompte, Tom Pitcairn, and Pio E. Ricci-Bitti. 1987. "Universals and Cultural Differences in the Judgments of Facial Expressions of Emotion." *Journal of Personality and Social Psychology* 53 (4): 712–717.

Ekman, Paul, Robert W. Levenson, and Wallace V. Friesen. 1983. "Autonomic Nervous System Activity Distinguishes Among Emotions." *Science* 221 (4616): 1208–1210.

Ekman, Paul, E. Richard Sorenson, and Wallace V. Friesen. 1969. "Pan-Cultural Elements in Facial Displays of Emotion." *Science* 164 (3875): 86–88.

Elfenbein, Hillary Anger, and Nalini Ambady. 2002. "On the Universality and Cultural Specificity of Emotion Recognition: A Meta-Analysis." *Psychological Bulletin* 128 (2): 203–235.

Ellingsen, Dan-Mikael, Johan Wessberg, Marie Eikemo, Jaquette Liljencrantz, Tor Endestad, Håkan Olausson, and Siri Leknes. 2013. "Placebo Improves Pleasure and Pain Through Opposite Modulation of Sensory Processing." *Proceedings of the National Academy of Sciences* 110 (44): 17993–17998.

Ellis, Bruce J., and W. Thomas Boyce. 2008. "Biological Sensitivity to Context." *Current Directions in Psychological Science* 17 (3): 183–187.

Emmons, Robert A., and Michael E. McCullough. 2003. "Counting Blessings Versus Burdens: An Experimental Investigation of Gratitude and Subjective Well-Being in Daily Life." *Journal of Personality and Social Psychology* 84 (2): 377–389.

Emmons, Scott W. 2012. "The Mood of a Worm." *Science* 338 (6106): 475–476.

Ensor, Rosie, and Claire Hughes. 2008. "Content or Connectedness? Mother-Child Talk and Early Social Understanding." *Child Development* 79 (1): 201–216.

Epley, Nicholas, Adam Waytz, and John T. Cacioppo. 2007. "On Seeing Human: A Three-Factor Theory of Anthropomorphism." *Psychological Review* 114 (4): 864–886.

Erbas, Yasemin, Eva Ceulemans, Johanna Boonen, Ilse Noens, and Peter Kuppens. 2013. "Emotion Differentiation in Autism Spectrum Disorder." *Research in Autism Spectrum Disorders* 7 (10): 1221–1227.

Erbas, Yasemin, Eva Ceulemans, Madeline Lee Pe, Peter Koval, and Peter Kuppens. 2014. "Negative Emotion Differentiation: Its Personality and Well-Being Correlates and a Comparison of Different Assessment Methods." *Cognition and Emotion* 28 (7): 1196–1213.

Erickson, Kirk I., Michelle W. Voss, Ruchika Shaurya Prakash, Chandramallika Basak, Amanda Szabo, Laura Chaddock, Jennifer S. Kim, Susie Heo, Heloisa Alves, and Siobhan M. White. 2011. "Exercise Training Increases Size of Hippocampus and Improves Memory." *Proceedings of the National Academy of Sciences* 108 (7): 3017–3022.

Ernst, Aurélie, and Jonas Frisén. 2015. "Adult Neurogenesis in Humans-Common and Unique Traits in Mammals." *PLOS Biology* 13 (1): e1002045. doi:10.1371/journal.pbio.1002045.

ESPN. 2014. "Bucks Hire Facial Coding Expert." December 27. http://espn.go.com/nba/story/_/id/12080142/milwaukee-bucks-hire-facial-coding-expert-help-team-improve.

Etkin, Amit, and Tor D. Wager. 2007. "Functional Neuroimaging of Anxiety: A Meta-Analysis of Emotional Processing in PTSD, Social Anxiety Disorder, and Specific Phobia." *American Journal of Psychiatry* 164 (10): 1476–1488.

Fabre-Thorpe, Michèle. 2010. "Concepts in Monkeys." In *The Making of Human Concepts*, edited by Denis Mareschal, Paul C. Quinn, and Stephen E. G. Lea, 201–226. New York: Oxford University Press.

Fachner, George, Steven Carter, and Collaborative Reform Initiative. 2015. "An Assessment of Deadly Force in the Philadelphia Police Department." Washington, DC: Office of Community Oriented Policing Services.

Feigenson, Lisa, and Justin Halberda. 2008. "Conceptual Knowledge Increases Infants' Memory Capacity." *Proceedings of the National Academy of Sciences* 105 (29): 9926–9930.

Feinstein, Justin S., Ralph Adolphs, Antonio Damasio, and Daniel Tranel. 2011. "The Human Amygdala and the Induction and Experience of Fear." *Current Biology* 21 (1): 34–38.

Feinstein, Justin S., David Rudrauf, Sahib S. Khalsa, Martin D. Cassell, Joel Bruss, Thomas J. Grabowski, and Daniel Tranel. 2010. "Bilateral Limbic System Destruction in Man." *Journal of Clinical and Experimental Neuropsychology* 32 (1): 88–106.

Felitti, Vincent J., Robert F. Anda, Dale Nordenberg, David F. Williamson, Alison M. Spitz, Valerie Edwards, Mary P. Koss, and James S. Marks. 1998. "Relationship of Childhood Abuse and Household Dysfunction to Many of the Leading Causes of Death in Adults: The Adverse Childhood Experiences (ACE) Study." *American Journal of Preventive Medicine* 14 (4): 245–258.

Feresin, Emiliano. 2011. "Italian Court Reduces Murder Sentence Based on Neuroimaging Data." *Nature News Blog*, September 1. http://blogs.nature.com/news/2011/09/italian_court_reduces_murder_s.html.

Fernald, Anne, Virginia A. Marchman, and Adriana Weisleder. 2013. "SES Differences in Language Processing Skill and Vocabulary Are Evident at 18 Months." *Developmental Science* 16 (2): 234–248.

Fernández-Dols, José-Miguel, and María-Angeles Ruiz-Belda. 1995. "Are Smiles a Sign of Happiness? Gold Medal Winners at the Olympic Games." *Journal of Personality and Social Psychology* 69 (6): 1113–1119.

Fields, Howard L., and Elyssa B. Margolis. 2015. "Understanding Opioid Reward." *Trends in Neurosciences* 38 (4): 217–225.

Finger, Stanley. 2001. *Origins of Neuroscience: A History of Explorations into Brain Function.* New York: Oxford University Press.

Finlay, Barbara L., and Ryutaro Uchiyama. 2015. "Developmental Mechanisms Channeling Cortical Evolution." *Trends in Neurosciences* 38 (2): 69–76.

Finn, Emily S., Xilin Shen, Dustin Scheinost, Monica D. Rosenberg, Jessica Huang, Marvin M. Chun, Xenophon Papademetris, and R. Todd Constable. 2015. "Functional Connectome Fingerprinting: Identifying Individuals Using Patterns of Brain Connectivity." *Nature Neuroscience* 18 (11): 1664–1671.

Firestein, Stuart. 2012. *Ignorance: How It Drives Science.* New York: Oxford University Press.

Fischer, Håkan, Christopher I. Wright, Paul J. Whalen, Sean C. McInerney, Lisa M. Shin, and Scott L. Rauch. 2003. "Brain Habituation During Repeated Exposure to Fearful and Neutral Faces: A Functional MRI Study." *Brain Research Bulletin* 59 (5): 387–392.

Fischer, Shannon. 2013. "About Face." *Boston Magazine*, July. 68–73.

Fisher, Helen E., Lucy L. Brown, Arthur Aron, Greg Strong, and Debra Mashek. 2010. "Reward, Addiction, and Emotion Regulation Systems Associated with Rejection in Love." *Journal of Neurophysiology* 104 (1): 51–60.

Fodor, Jerry A. 1983. *The Modularity of Mind: An Essay on Faculty Psychology.* Cambridge, MA: MIT Press.

Ford, Brett Q., and Maya Tamir. 2012. "When Getting Angry Is Smart: Emotional Preferences and Emotional Intelligence." *Emotion* 12 (4): 685–689.

Ford, Earl S. 2002. "Does Exercise Reduce Inflammation? Physical Activity and C-Reactive Protein Among US Adults." *Epidemiology* 13 (5): 561–568.

Fossat, Pascal, Julien Bacqué-Cazenave, Philippe De Deurwaerdère, Jean-Paul Delbecque, and Daniel Cattaert. 2014. "Anxiety-Like Behavior in Crayfish Is Controlled by Serotonin." *Science* 344 (6189): 1293–1297.

Foulke, Emerson, and Thomas G. Sticht. 1969. "Review of Research on the Intelligibility and Comprehension of Accelerated Speech." *Psychological Bulletin* 72 (1): 50–62.

Franklin, David W., and Daniel M. Wolpert. 2011. "Computational Mechanisms of Sensorimotor Control." *Neuron* 72 (3): 425–442.

Freddolino, Peter L., and Saeed Tavazoie. 2012. "Beyond Homeostasis: A Predictive-Dynamic Framework for Understanding Cellular Behavior." *Annual Review of Cell and Developmental Biology* 28: 363–384.

Fridlund, Alan J. 1991. "Sociality of Solitary Smiling: Potentiation by an Implicit Audience." *Journal of Personality and Social Psychology* 60 (2): 229–240.

"Fright Night." 2012. *Science* 338 (6106): 450.

Frijda, Nico H. 1988. "The Laws of Emotion." *American Psychologist* 43 (5): 349–358.

Friston, Karl. 2010. "The Free-Energy Principle: A Unified Brain Theory?" *Nature Reviews Neuroscience* 11: 127–138.

Froh, Jeffrey J., William J. Sefick, and Robert A. Emmons. 2008. "Counting Blessings in Early Adolescents: An Experimental Study of Gratitude and Subjective Well-Being." *Journal of School Psychology* 46 (2): 213–233.

Frost, Ram, Blair C. Armstrong, Noam Siegelman, and Morten H. Christiansen. 2015. "Domain Generality Versus Modality Specificity: The Paradox of Statistical Learning." *Trends in Cognitive Sciences* 19 (3): 117–125.

Fu, Cynthia H. Y., Herbert Steiner, and Sergi G. Costafreda. 2013. "Predictive Neural Biomarkers of Clinical Response in Depression: A Meta-Analysis of Functional and Structural Neuroimaging Studies of Pharmacological and Psychological Therapies." *Neurobiology of Disease* 52: 75–83.

Fugate, Jennifer, Harold Gouzoules, and Lisa Feldman Barrett. 2010. "Reading Chimpanzee Faces: Evidence for the Role of Verbal Labels in Categorical Perception of Emotion." *Emotion* 10 (4): 544–554.

Ganzel, Barbara L., Pamela A. Morris, and Elaine Wethington. 2010. "Allostasis and the Human Brain: Integrating Models of Stress from the Social and Life Sciences." *Psychological Review* 117 (1): 134–174.

Gao, Wei, Sarael Alcauter, Amanda Elton, Carlos R. Hernandez-Castillo, J. Keith Smith, Juanita Ramirez, and Weili Lin. 2014. "Functional Network Development During the First Year: Relative Sequence and Socioeconomic Correlations." *Cerebral Cortex* 25 (9): 2919–2928.

Gao, Wei, Amanda Elton, Hongtu Zhu, Sarael Alcauter, J. Keith Smith, John H. Gilmore, and Weili Lin. 2014. "Intersubject Variability of and Genetic Effects on the Brain's Functional Connectivity During Infancy." *Journal of Neuroscience* 34 (34): 11288–11296.

Gao, Wei, Hongtu Zhu, Kelly S. Giovanello, J. Keith Smith, Dinggang Shen, John H. Gilmore, and Weili Lin. 2009. "Evidence on the Emergence of the Brain's Default Network from 2-Week-Old to 2-Year-Old Healthy Pediatric Subjects." *Proceedings of the National Academy of Sciences* 106 (16): 6790–6795.

Garber, Megan. 2013. "Tongue and Tech: The Many Emotions for Which English Has No Words." *Atlantic,* January 8. http://www.theatlantic.com/technology/archive/2013/01/tongue-and-tech-the-many-emotions-for-which-english-has-no-words/266956/.

Gardner, Howard. 1975. *The Shattered Mind: The Person After Brain Damage.* New York: Vintage.

Garland, Eric L., Brett Froeliger, and Matthew O. Howard. 2014. "Effects of Mindfulness-Oriented Recovery Enhancement on Reward Responsiveness and Opioid Cue-Reactivity." *Psychopharmacology* 231 (16): 3229–3238.

Gelman, Susan A. 2009. "Learning from Others: Children's Construction of Concepts." *Annual Review of Psychology* 60: 115–140.

Gendron, M., and L. F. Barrett. 2009. "Reconstructing the Past: A Century of Ideas About Emotion in Psychology." *Emotion Review* 1 (4): 316–339.

———. In press. "How and Why Are Emotions Communicated." In *The Nature of Emotion: Fundamental Questions,* 2nd edition, edited by A. S. Fox, R. C. Lapate, A. J. Shackman, and R. J. Davidson. Oxford: Oxford University Press.

Gendron, Maria, Kristen A. Lindquist, Lawrence W. Barsalou, and Lisa Feldman Barrett. 2012. "Emotion Words Shape Emotion Percepts." *Emotion* 12 (2): 314–325.

Gendron, Maria, Debi Roberson, Jacoba Marieta van der Vyver, and Lisa Feldman Barrett. 2014a. "Cultural Relativity in Perceiving Emotion from Vocalizations." *Psychological Science* 25 (4): 911–920.

———. 2014b. "Perceptions of Emotion from Facial Expressions Are Not Culturally Universal: Evidence from a Remote Culture." *Emotion* 14 (2): 251–262.

Gertner, Nancy. 2015. "Will We Ever Know Why Dzhokhar Tsarnaev Spoke After It Was Too Late?" *Boston Globe,* June 30. http://clbb.mgh.harvard.edu/will-we-ever-know -why-dzhokhar-tsarnaev-spoke-after-it-was-too-late.

Gibson, William T., Carlos R. Gonzalez, Conchi Fernandez, Lakshminarayanan Ramasamy, Tanya Tabachnik, Rebecca R. Du, Panna D. Felsen, Michael R. Maire, Pietro Perona, and David J. Anderson. 2015. "Behavioral Responses to a Repetitive Visual Threat Stimulus Express a Persistent State of Defensive Arousal in Drosophila." *Current Biology* 25 (11): 1401–1415.

Gilbert, Charles D., and Wu Li. 2013. "Top-Down Influences on Visual Processing." *Nature Reviews Neuroscience* 14 (5): 350–363.

Gilbert, D. T. 1998. "Ordinary Personology." In *The Handbook of Social Psychology,* edited by S. T. Fiske and L. Gardner, 89–150. New York: McGraw-Hill.

Giuliano, Ryan J., Elizabeth A. Skowron, and Elliot T. Berkman. 2015. "Growth Models of Dyadic Synchrony and Mother-Child Vagal Tone in the Context of Parenting At-Risk." *Biological Psychology* 105: 29–36.

Gleeson, Michael, Nicolette C. Bishop, David J. Stensel, Martin R. Lindley, Sarabjit S. Mastana, and Myra A. Nimmo. 2011. "The Anti-Inflammatory Effects of Exercise: Mechanisms and Implications for the Prevention and Treatment of Disease." *Nature Reviews Immunology* 11 (9): 607–615.

Goldapple, Kimberly, Zindel Segal, Carol Garson, Mark Lau, Peter Bieling, Sidney Kennedy, and Helen Mayberg. 2004. "Modulation of Cortical-Limbic Pathways in Major Depression: Treatment-Specific Effects of Cognitive Behavior Therapy." *Archives of General Psychiatry* 61 (1): 34–41.

Goldstein, Andrea N., and Matthew P. Walker. 2014. "The Role of Sleep in Emotional Brain Function." *Annual Review of Clinical Psychology* 10: 679–708.

Goldstone, Robert L. 1994. "The Role of Similarity in Categorization: Providing a Groundwork." *Cognition* 52 (2): 125–157.

Goleman, Daniel. 1998. *Working with Emotional Intelligence.* New York: Bantam.

———. 2006. *Emotional Intelligence.* New York: Random House.

Golinkoff, Roberta Michnick, Dilara Deniz Can, Melanie Soderstrom, and Kathy Hirsh-

Pasek. 2015. "(Baby) Talk to Me: The Social Context of Infant-Directed Speech and Its Effects on Early Language Acquisition." *Current Directions in Psychological Science* 24 (5): 339–344.

Goodkind, Madeleine, Simon B. Eickhoff, Desmond J. Oathes, Ying Jiang, Andrew Chang, Laura B. Jones-Hagata, Brissa N. Ortega, Yevgeniya V. Zaiko, Erika L. Roach, and Mayuresh S. Korgaonkar. 2015. "Identification of a Common Neurobiological Substrate for Mental Illness." *JAMA Psychiatry* 72 (4): 305–315.

Goodman, Morris. 1999. "The Genomic Record of Humankind's Evolutionary Roots." *American Journal of Human Genetics* 64 (1): 31–39.

Goodnough, Abby. 2009. "Harvard Professor Jailed; Officer Is Accused of Bias." *New York Times*, July 20. http://www.nytimes.com/2009/07/21/us/21gates.html.

Gopnik, Alison. 2009. *The Philosophical Baby: What Children's Minds Tell Us About Truth, Love and the Meaning of Life*. New York: Random House.

Gopnik, Alison, and David M. Sobel. 2000. "Detecting Blickets: How Young Children Use Information About Novel Causal Powers in Categorization and Induction." *Child Development* 71 (5): 1205–1222.

Gosselin, Frédéric, and Philippe G. Schyns. 2003. "Superstitious Perceptions Reveal Properties of Internal Representations." *Psychological Science* 14 (5): 505–509.

Gottman, John M., Lynn Fainsilber Katz, and Carole Hooven. 1996. "Parental Meta-Emotion Philosophy and the Emotional Life of Families: Theoretical Models and Preliminary Data." *Journal of Family Psychology* 10 (3): 243–268.

Government Accountability Office (GAO). 2013. "Aviation Security: TSA Should Limit Future Funding for Behavior Detection Activities (GAO-14-159)." http://www.gao.gov/products/GAO-14-159.

Grandin, Temple. 1991. "An Inside View of Autism." http://www.autism.com/advocacy_grandin.

———. 2009. "How Does Visual Thinking Work in the Mind of a Person with Autism? A Personal Account." *Philosophical Transactions of the Royal Society of London B: Biological Sciences* 364 (1522): 1437–1442.

Graziano, Michael S. A. 2013. *Consciousness and the Social Brain*. New York: Oxford University Press.

———. 2016. "Ethological Action Maps: A Paradigm Shift for the Motor Cortex." *Trends in Cognitive Sciences* 20 (2): 121–132.

Greene, Brian. 2007. *The Fabric of the Cosmos: Space, Time, and the Texture of Reality*. New York: Vintage.

Grill-Spector, Kalanit, and Kevin S. Weiner. 2014. "The Functional Architecture of the Ventral Temporal Cortex and Its Role in Categorization." *Nature Reviews Neuroscience* 15 (8): 536–548.

Gross, Cornelius T., and Newton Sabino Canteras. 2012. "The Many Paths to Fear." *Nature Reviews Neuroscience* 13 (9): 651–658.

Gross, James J. 2015. "Emotion Regulation: Current Status and Future Prospects." *Psychological Inquiry* 26 (1): 1–26.

Gross, James J., and Lisa Feldman Barrett. 2011. "Emotion Generation and Emotion Regulation: One or Two Depends on Your Point of View." *Emotion Review* 3 (1): 8–16.

Guarneri-White, Maria Elizabeth. 2014. *Biological Aging and Peer Victimization: The Role of Social Support in Telomere Length and Health Outcomes*. Master's thesis, University of Texas at Arlington, 1566471.

Guillory, Sean A., and Krzysztof A. Bujarski. 2014. "Exploring Emotions Using Invasive Methods: Review of 60 Years of Human Intracranial Electrophysiology." *Social Cognitive and Affective Neuroscience* 9 (12): 1880–1889.

Gweon, Hyowon, Joshua B. Tenenbaum, and Laura E. Schulz. 2010. "Infants Consider Both the Sample and the Sampling Process in Inductive Generalization." *Proceedings of the National Academy of Sciences* 107 (20): 9066–9071.

Hacking, Ian. 1999. *The Social Construction of What?* Cambridge, MA: Harvard University Press.

Hagelskamp, Carolin, Marc A. Brackett, Susan E. Rivers, and Peter Salovey. 2013. "Improving Classroom Quality with the Ruler Approach to Social and Emotional Learning: Proximal and Distal Outcomes." *American Journal of Community Psychology* 51 (3–4): 530–543.

Halperin, Eran, Roni Porat, Maya Tamir, and James J. Gross. 2013. "Can Emotion Regulation Change Political Attitudes in Intractable Conflicts? From the Laboratory to the Field." *Psychological Science* 24 (1): 106–111.

Halpern, Jake. 2008. *Fame Junkies: The Hidden Truths Behind America's Favorite Addiction.* Boston: Houghton Mifflin Harcourt.

Hamlin, J. Kiley, George E. Newman, and Karen Wynn. 2009. "Eight-Month-Old Infants Infer Unfulfilled Goals, Despite Ambiguous Physical Evidence." *Infancy* 14 (5): 579–590.

Haney, Craig. 2005. *Death by Design: Capital Punishment as a Social Psychological System.* New York: Oxford University Press.

Hanson, Jamie L., Nicole Hair, Dinggang G. Shen, Feng Shi, John H. Gilmore, Barbara L. Wolfe, and Seth D. Pollak. 2013. "Family Poverty Affects the Rate of Human Infant Brain Growth." *PLOS One* 8 (12): e80954. doi:10.1371/journal.pone.0080954.

Hare, Brian, and Vanessa Woods. 2013. *The Genius of Dogs: How Dogs Are Smarter than You Think.* New York: Penguin.

Harmon-Jones, Eddie, and Carly K. Peterson. 2009. "Supine Body Position Reduces Neural Response to Anger Evocation." *Psychological Science* 20 (10): 1209–1210.

Harré, Rom. 1986. *The Social Construction of Emotions.* New York: Blackwell.

Harris, Christine R., and Caroline Prouvost. 2014. "Jealousy in Dogs." *PLOS One* 9 (7): e94597. doi:10.1371/journal.pone.0094597.

Harris, Paul L., Marc de Rosnay, and Francisco Pons. In press. "Understanding Emotion." In *Handbook of Emotions,* 4th edition, edited by Lisa Feldman Barrett, Michael Lewis, and Jeannette M. Haviland-Jones, 293–306. New York: Guilford Press.

Harrison, Neil A., Lena Brydon, Cicely Walker, Marcus A. Gray, Andrew Steptoe, and Hugo D. Critchley. 2009. "Inflammation Causes Mood Changes Through Alterations in Subgenual Cingulate Activity and Mesolimbic Connectivity." *Biological Psychiatry* 66 (5): 407–414.

Harrison, Neil A., Lena Brydon, Cicely Walker, Marcus A. Gray, Andrew Steptoe, Raymond J. Dolan, and Hugo D. Critchley. 2009. "Neural Origins of Human Sickness in Interoceptive Responses to Inflammation." *Biological Psychiatry* 66 (5): 415–422.

Hart, Betty, and Todd R. Risley. 1995. *Meaningful Differences in the Everyday Experience of Young American Children.* Baltimore: Paul H. Brookes.

———. 2003. "The Early Catastrophe: The 30 Million Word Gap by Age 3." *American Educator* 27 (1): 4–9.

Hart, Heledd, and Katya Rubia. 2012. "Neuroimaging of Child Abuse: A Critical Review." *Frontiers in Human Neuroscience* 6 (52): 1–24.

Harvey, Allison G., Greg Murray, Rebecca A. Chandler, and Adriane Soehner. 2011. "Sleep

Disturbance as Transdiagnostic: Consideration of Neurobiological Mechanisms." *Clinical Psychology Review* 31 (2): 225–235.

Hassabis, Demis, and Eleanor A. Maguire. 2009. "The Construction System of the Brain." *Philosophical Transactions of the Royal Society B: Biological Sciences* 364 (1521): 1263–1271.

Hathaway, Bill. 2015. "Imaging Study Shows Brain Activity May Be as Unique as Fingerprints." *YaleNews,* October 12. http://news.yale.edu/2015/10/12/imaging-study-shows-brain-activity-may-be-unique-fingerprints.

Hawkins, Jeff, and Sandra Blakeslee. 2004. *On Intelligence.* New York: St. Martin's Griffin.

Hermann, Christiane, Johanna Hohmeister, Sueha Demirakça, Katrin Zohsel, and Herta Flor. 2006. "Long-Term Alteration of Pain Sensitivity in School-Aged Children with Early Pain Experiences." *Pain* 125 (3): 278–285.

Hertzman, Clyde, and Tom Boyce. 2010. "How Experience Gets Under the Skin to Create Gradients in Developmental Health." *Annual Review of Public Health* 31: 329–347.

Hey, Jody. 2010. "The Divergence of Chimpanzee Species and Subspecies as Revealed in Multipopulation Isolation-with-Migration Analyses." *Molecular Biology and Evolution* 27 (4): 921–933.

Higashida, Naoki. 2013. *The Reason I Jump: The Inner Voice of a Thirteen-Year-Old Boy with Autism.* New York: Random House.

Higgins, E. Tory. 1987. "Self-Discrepancy: A Theory Relating Self and Affect." *Psychological Review* 94 (3): 319–340.

Hill, Jason, Terrie Inder, Jeffrey Neil, Donna Dierker, John Harwell, and David Van Essen. 2010. "Similar Patterns of Cortical Expansion During Human Development and Evolution." *Proceedings of the National Academy of Sciences* 107 (29): 13135–13140.

Hillix, William A., and Duane M. Rumbaugh. 2004. "Language Research with Nonhuman Animals: Methods and Problems." In *Animal Bodies, Human Minds: Ape, Dolphin, and Parrot Language Skills,* 25–44. New York: Kluwer Academic.

Hirsh-Pasek, Kathy, Lauren B. Adamson, Roger Bakeman, Margaret Tresch Owen, Roberta Michnick Golinkoff, Amy Pace, Paula K. S. Yust, and Katharine Suma. 2015. "The Contribution of Early Communication Quality to Low-Income Children's Language Success." *Psychological Science* 26 (7): 1071–1083. doi:10.1177/0956797615581493.

Hochschild, Arlie R. 1983. *The Managed Heart: Commercialization of Human Feeling.* Berkeley: University of California Press.

Hofer, Myron A. 1984. "Relationships as Regulators: A Psychobiologic Perspective on Bereavement." *Psychosomatic Medicine* 46 (3): 183–197.

———. 2006. "Psychobiological Roots of Early Attachment." *Current Directions in Psychological Science* 15 (2): 84–88.

Hohwy, Jakob. 2013. *The Predictive Mind.* Oxford: Oxford University Press.

Holt-Lunstad, Julianne, Timothy B. Smith, and J. Bradley Layton. 2010. "Social Relationships and Mortality Risk: A Meta-Analytic Review." *PLOS Med* 7 (7): e1000316. doi:1.1371/journal.pmed.1316.

Holtzheimer, Paul E., Mary E. Kelley, Robert E. Gross, Megan M. Filkowski, Steven J. Garlow, Andrea Barrocas, Dylan Wint, Margaret C. Craighead, Julie Kozarsky, and Ronald Chismar. 2012. "Subcallosal Cingulate Deep Brain Stimulation for Treatment-Resistant Unipolar and Bipolar Depression." *Archives of General Psychiatry* 69 (2): 150–158.

Horowitz, Alexandra. 2009. "Disambiguating the 'Guilty Look': Salient Prompts to a Familiar Dog Behaviour." *Behavioural Processes* 81 (3): 447–452.

Hoyt, Michael A., Annette L. Stanton, Julienne E. Bower, KaMala S. Thomas, Mark S. Litwin, Elizabeth C. Breen, and Michael R. Irwin. 2013. "Inflammatory Biomarkers and Emotional Approach Coping in Men with Prostate Cancer." *Brain, Behavior, and Immunity* 32: 173–179.

Hunter, Richard G., and Bruce S. McEwen. 2013. "Stress and Anxiety Across the Lifespan: Structural Plasticity and Epigenetic Regulation." *Epigenomics* 5 (2): 177–194.

Huntsinger, Jeffrey R., Linda M. Isbell, and Gerald L. Clore. 2014. "The Affective Control of Thought: Malleable, Not Fixed." *Psychological Review* 121 (4): 600–618.

Innocence Project. 2015. "Eyewitness Misidentification." http://www.innocenceproject. org/causes-wrongful-conviction/eyewitness-misidentification.

International Association for the Study of Pain. 2012. "IASP Taxonomy." http://www.iasp -pain.org/Taxonomy.

Inzlicht, Michael, Bruce D. Bartholow, and Jacob B. Hirsh. 2015. "Emotional Foundations of Cognitive Control." *Trends in Cognitive Sciences* 19 (3): 126–132.

Irwin, Michael R., and Steven W. Cole. 2011. "Reciprocal Regulation of the Neural and Innate Immune Systems." *Nature Reviews Immunology* 11 (9): 625–632.

Iwata, Jiro, and Joseph E. LeDoux. 1988. "Dissociation of Associative and Nonassociative Concomitants of Classical Fear Conditioning in the Freely Behaving Rat." *Behavioral Neuroscience* 102 (1): 66–76.

Izard, Carroll E. 1971. *The Face of Emotion.* East Norwalk, CT: Appleton-Century-Crofts.

———. 1994. "Innate and Universal Facial Expressions: Evidence from Developmental and Cross-Cultural Research." *Psychological Bulletin* 115 (2): 288–299.

Jablonka, Eva, Marion J. Lamb, and Anna Zeligowski. 2014. *Evolution in Four Dimensions: Genetic, Epigenetic, Behavioral, and Symbolic Variation in the History of Life.* Revised edition. Cambridge, MA: MIT Press.

James, William. 1884. "What Is an Emotion?" *Mind* 34: 188–205.

———. (1890) 2007. *The Principles of Psychology.* Vol. 1. New York: Dover.

———. 1894. "The Physical Basis of Emotion." *Psychological Review* 1: 516–529.

Jamieson, J. P., M. K. Nock, and W. B. Mendes. 2012. "Mind over Matter: Reappraising Arousal Improves Cardiovascular and Cognitive Responses to Stress." *Journal of Experimental Psychology: General* 141 (3): 417–422.

Jamieson, Jeremy P., Brett J. Peters, Emily Greenwood, and Aaron Altose. 2016. "Reappraising Stress Arousal Improves Performance and Reduces Evaluation Anxiety in Classroom Exam Situations." *Social Psychological and Personality Science* 7 (6): 579–587.

Jamieson, Jeremy P., Wendy Berry Mendes, Erin Blackstock, and Toni Schmader. 2010. "Turning the Knots in Your Stomach into Bows: Reappraising Arousal Improves Performance on the GRE." *Journal of Experimental Social Psychology* 46 (1): 208–212.

Jamieson, Jeremy P., Wendy Berry Mendes, and Matthew K. Nock. 2013. "Improving Acute Stress Responses: The Power of Reappraisal." *Current Directions in Psychological Science* 22 (1): 51–56.

Jamieson, Jeremy P., Matthew K. Nock, and Wendy Berry Mendes. 2013. "Changing the Conceptualization of Stress in Social Anxiety Disorder Affective and Physiological Consequences." *Clinical Psychological Science* 1: 363–374.

Jamison, Kay R. 2005. *Exuberance: The Passion for Life.* New York: Vintage Books.

Jeste, Shafali S., and Daniel H. Geschwind. 2014. "Disentangling the Heterogeneity of Autism Spectrum Disorder Through Genetic Findings." *Nature Reviews Neurology* 10 (2): 74–81.

Ji, Ru-Rong, Temugin Berta, and Maiken Nedergaard. 2013. "Glia and Pain: Is Chronic Pain a Gliopathy?" *Pain* 154: S10–S28.

Job, Veronika, Gregory M. Walton, Katharina Bernecker, and Carol S. Dweck. 2013. "Beliefs About Willpower Determine the Impact of Glucose on Self-Control." *Proceedings of the National Academy of Sciences* 110 (37): 14837–14842.

———. 2015. "Implicit Theories About Willpower Predict Self-Regulation and Grades in Everyday Life." *Journal of Personality and Social Psychology* 108 (4): 637–647.

Johansen, Joshua P., and Howard L. Fields. 2004. "Glutamatergic Activation of Anterior Cingulate Cortex Produces an Aversive Teaching Signal." *Nature Neuroscience* 7 (4): 398–403.

John-Henderson, Neha A., Michelle L. Rheinschmidt, and Rodolfo Mendoza-Denton. 2015. "Cytokine Responses and Math Performance: The Role of Stereotype Threat and Anxiety Reappraisals." *Journal of Experimental Social Psychology* 56: 203–206.

John-Henderson, Neha A., Jennifer E. Stellar, Rodolfo Mendoza-Denton, and Darlene D. Francis. 2015. "Socioeconomic Status and Social Support: Social Support Reduces Inflammatory Reactivity for Individuals Whose Early-Life Socioeconomic Status Was Low." *Psychological Science* 26 (10): 1620–1629.

Jones, Colin. 2014. *The Smile Revolution in Eighteenth Century Paris*. New York: Oxford University Press.

Josefsson, Torbjörn, Magnus Lindwall, and Trevor Archer. 2014. "Physical Exercise Intervention in Depressive Disorders: Meta-Analysis and Systematic Review." *Scandinavian Journal of Medicine and Science in Sports* 24 (2): 259–272.

Jussim, L., J. T. Crawford, S. M. Anglin, J. Chambers, S. T. Stevens, and F. Cohen. 2009. "Stereotype Accuracy: One of the Largest Relationships in All of Social Psychology." In *Handbook of Prejudice, Stereotyping, and Discrimination*, 2nd edition, edited by Todd D. Nelson, 31–64. New York: Psychology Press.

Jussim, Lee. 2012. *Social Perception and Social Reality: Why Accuracy Dominates Bias and Self-Fulfilling Prophecy*. New York: Oxford University Press.

Jussim, Lee, Thomas R. Cain, Jarret T. Crawford, Kent Harber, and Florette Cohen. 2009. "The Unbearable Accuracy of Stereotypes." *Handbook of Prejudice, Stereotyping, and Discrimination*, 2nd edition, edited by Todd D. Nelson, 199–227. New York: Psychology Press.

Kagan, Jerome. 2007. *What Is Emotion?: History, Measures, and Meanings*. New Haven, CT: Yale University Press.

Kahan, Dan M., David A. Hoffman, Donald Braman, and Danieli Evans. 2012. "They Saw a Protest: Cognitive Illiberalism and the Speech-Conduct Distinction." *Stanford Law Review* 64: 851.

Kahan, Dan M., and Martha C. Nussbaum. 1996. "Two Conceptions of Emotion in Criminal Law." *Columbia Law Review* 96 (2): 269–374.

Kahneman, Daniel. 2011. *Thinking, Fast and Slow*. New York: Macmillan.

Kaiser, Roselinde H., Jessica R. Andrews-Hanna, Tor D. Wager, and Diego A. Pizzagalli. 2015. "Large-Scale Network Dysfunction in Major Depressive Disorder: A Meta-Analysis of Resting-State Functional Connectivity." *JAMA Psychiatry* 72 (6): 603–611.

Kaminski, Juliane, Juliane Bräuer, Josep Call, and Michael Tomasello. 2009. "Domestic Dogs Are Sensitive to a Human's Perspective." *Behaviour* 146 (7): 979–998.

Karlsson, Håkan, Björn Ahlborg, Christina Dalman, and Tomas Hemmingsson. 2010. "Association Between Erythrocyte Sedimentation Rate and IQ in Swedish Males Aged 18–20." *Brain, Behavior, and Immunity* 24 (6): 868–873.

Karmiloff-Smith, Annette. 2009. "Nativism Versus Neuroconstructivism: Rethinking the Study of Developmental Disorders." *Developmental Psychology* 45 (1): 56–63.

Kashdan, Todd B., Lisa Feldman Barrett, and Patrick E. McKnight. 2015. "Unpacking Emotion Differentiation Transforming Unpleasant Experience by Perceiving Distinctions in Negativity." *Current Directions in Psychological Science* 24 (1): 10–16.

Kashdan, Todd B., and Antonina S. Farmer. 2014. "Differentiating Emotions Across Contexts: Comparing Adults With and Without Social Anxiety Disorder Using Random, Social Interaction, and Daily Experience Sampling." *Emotion* 14 (3): 629–638.

Kashdan, Todd B., Patty Ferssizidis, R. Lorraine Collins, and Mark Muraven. 2010. "Emotion Differentiation as Resilience Against Excessive Alcohol Use an Ecological Momentary Assessment in Underage Social Drinkers." *Psychological Science* 21 (9): 1341–1347.

Kassam, Karim S., and Wendy Berry Mendes. 2013. "The Effects of Measuring Emotion: Physiological Reactions to Emotional Situations Depend on Whether Someone Is Asking." *PLOS One* 8 (6): e64959. doi:10.1371/journal.pone.0064959.

Kassin, Saul M., V. Anne Tubb, Harmon M. Hosch, and Amina Memon. 2001. "On the 'General Acceptance' of Eyewitness Testimony Research: A New Survey of the Experts." *American Psychologist* 56 (5): 405–416.

Katz, Lynn Fainsilber, Ashley C. Maliken, and Nicole M. Stettler. 2012. "Parental Meta-Emotion Philosophy: A Review of Research and Theoretical Framework." *Child Development Perspectives* 6 (4): 417–422.

Keefe, P. R. 2015. "The Worst of the Worst." *New Yorker,* September 14. http://www.new yorker.com/magazine/2015/09/14/the-worst-of-the-worst.

Keil, Frank C., and George E. Newman. 2010. "Darwin and Development: Why Ontogeny Does Not Recapitulate Phylogeny for Human Concepts." In *The Making of Human Concepts,* edited by Denis Mareschal, Paul Quinn, and Stephen E. G. Lea, 317–334. New York: Oxford University Press.

Kelly, Megan M., John P. Forsyth, and Maria Karekla. 2006. "Sex Differences in Response to a Panicogenic Challenge Procedure: An Experimental Evaluation of Panic Vulnerability in a Non-Clinical Sample." *Behaviour Research and Therapy* 44 (10): 1421–1430.

Keltner, Dacher, and Jonathan Haidt. 2003. "Approaching Awe, a Moral, Spiritual, and Aesthetic Emotion." *Cognition and Emotion* 17 (2): 297–314.

Khandaker, Golam M., Rebecca M. Pearson, Stanley Zammit, Glyn Lewis, and Peter B. Jones. 2014. "Association of Serum Interleukin 6 and C-Reactive Protein in Childhood with Depression and Psychosis in Young Adult Life: A Population-Based Longitudinal Study." *JAMA Psychiatry* 71 (10): 1121–1128.

Kiecolt-Glaser, Janice K. 2010. "Stress, Food, and Inflammation: Psychoneuroimmunology and Nutrition at the Cutting Edge." *Psychosomatic Medicine* 72 (4): 365–369.

Kiecolt-Glaser, Janice K., Jeanette M. Bennett, Rebecca Andridge, Juan Peng, Charles L. Shapiro, William B. Malarkey, Charles F. Emery, Rachel Layman, Ewa E. Mrozek, and Ronald Glaser. 2014. "Yoga's Impact on Inflammation, Mood, and Fatigue in Breast Cancer Survivors: A Randomized Controlled Trial." *Journal of Clinical Oncology* 32 (10): 1040–1051.

Kiecolt-Glaser, Janice K., Lisa Christian, Heather Preston, Carrie R. Houts, William B. Malarkey, Charles F. Emery, and Ronald Glaser. 2010. "Stress, Inflammation, and Yoga Practice." *Psychosomatic Medicine* 72 (2): 113–134.

Kiecolt-Glaser, Janice K., Jean-Philippe Gouin, Nan-ping Weng, William B. Malarkey, David Q. Beversdorf, and Ronald Glaser. 2011. "Childhood Adversity Heightens the Impact of Later-Life Caregiving Stress on Telomere Length and Inflammation." *Psychosomatic Medicine* 73 (1): 16–22.

Killingsworth, M. A., and D. T. Gilbert. 2010. "A Wandering Mind Is an Unhappy Mind." *Science* 330 (6006): 932.

Kim, Min Y., Brett Q. Ford, Iris Mauss, and Maya Tamir. 2015. "Knowing When to Seek Anger: Psychological Health and Context-Sensitive Emotional Preferences." *Cognition and Emotion* 29 (6): 1126–1136.

Kim, ShinWoo, and Gregory L. Murphy. 2011. "Ideals and Category Typicality." *Journal of Experimental Psychology: Learning, Memory, and Cognition* 37 (5): 1092–1112.

Kimhy, David, Julia Vakhrusheva, Samira Khan, Rachel W. Chang, Marie C. Hansen, Jacob S. Ballon, Dolores Malaspina, and James J. Gross. 2014. "Emotional Granularity and Social Functioning in Individuals with Schizophrenia: An Experience Sampling Study." *Journal of Psychiatric Research* 53: 141–148.

Kircanski, K., M. D. Lieberman, and M. G. Craske. 2012. "Feelings into Words: Contributions of Language to Exposure Therapy." *Psychological Science* 23 (10): 1086–1091.

Kirsch, Irving. 2010. *The Emperor's New Drugs: Exploding the Antidepressant Myth.* New York: Basic Books.

Kitzbichler, Manfred G., Richard N. A. Henson, Marie L. Smith, Pradeep J. Nathan, and Edward T. Bullmore. 2011. "Cognitive Effort Drives Workspace Configuration of Human Brain Functional Networks." *Journal of Neuroscience* 31 (22): 8259–8270.

Klatzky, Roberta L., James W. Pellegrino, Brian P. McCloskey, and Sally Doherty. 1989. "Can You Squeeze a Tomato? The Role of Motor Representations in Semantic Sensibility Judgments." *Journal of Memory and Language* 28 (1): 56–77.

Kleckner, I. R., J. Zhang, A. Touroutoglou, L. Chanes, C. Xia, W. K. Simmons, B. C. Dickerson, and L. F. Barrett. Under review. "Evidence for a Large-Scale Brain System Supporting Interoception in Humans."

Klüver, Heinrich, and Paul C. Bucy. 1939. "Preliminary Analysis of Functions of the Temporal Lobes in Monkeys." *Archives of Neurology and Psychiatry* 42: 979–1000.

Kober, H., L. F. Barrett, J. Joseph, E. Bliss-Moreau, K. Lindquist, and T. D. Wager. 2008. "Functional Grouping and Cortical-Subcortical Interactions in Emotion: A Meta-Analysis of Neuroimaging Studies." *Neuroimage* 42 (2): 998–1031.

Koch, Kristin, Judith McLean, Ronen Segev, Michael A. Freed, Michael J. Berry, Vijay Balasubramanian, and Peter Sterling. 2006. "How Much the Eye Tells the Brain." *Current Biology* 16 (14): 1428–1434.

Kohut, Andrew. 2015. "Despite Lower Crime Rates, Support for Gun Rights Increases." *Pew Research Center,* April 17. http://www.pewresearch.org/fact-tank/2015/04/17/despite-lower-crime-rates-support-for-gun-rights-increases.

Kolodny, Andrew, David T. Courtwright, Catherine S. Hwang, Peter Kreiner, John L. Eadie, Thomas W. Clark, and G. Caleb Alexander. 2015. "The Prescription Opioid and Heroin Crisis: A Public Health Approach to an Epidemic of Addiction." *Annual Review of Public Health* 36: 559–574.

Koopman, Frieda A., Susanne P. Stoof, Rainer H. Straub, Marjolein A. van Maanen, Margriet J. Vervoordeldonk, and Paul P. Tak. 2011. "Restoring the Balance of the Autonomic Nervous System as an Innovative Approach to the Treatment of Rheumatoid Arthritis." *Molecular Medicine* 17 (9): 937–948.

Kopchia, Karen L., Harvey J. Altman, and Randall L. Commissaris. 1992. "Effects of Lesions of the Central Nucleus of the Amygdala on Anxiety-Like Behaviors in the Rat." *Pharmacology Biochemistry and Behavior* 43 (2): 453–461.

Kostović, I., and M. Judaš. 2015. "Embryonic and Fetal Development of the Human Cerebral Cortex." In *Brain Mapping, An Encyclopedic Reference, Volume 2: Anatomy and Physiology, Systems,* edited by Arthur W. Toga, 167–175. San Diego: Academic Press.

Kragel, Philip A., and Kevin S. LaBar. 2013. "Multivariate Pattern Classification Reveals Autonomic and Experiential Representations of Discrete Emotions." *Emotion* 13 (4): 681–690.

Kreibig, S. D. 2010. "Autonomic Nervous System Activity in Emotion: A Review." *Biological Psychology* 84 (3): 394–421.

Kring, A. M., and A. H. Gordon. 1998. "Sex Differences in Emotion: Expression, Experience, and Physiology." *Journal of Personality and Social Psychology* 74 (3): 686–703.

Krugman, Paul. 2014. "The Dismal Science: 'Seven Bad Ideas' by Jeff Madrick." *New York Times,* September 25. http://www.nytimes.com/2014/09/28/books/review/seven-bad-ideas-by-jeff-madrick.html.

Kuhl, Patricia K. 2007. "Is Speech Learning 'Gated' by the Social Brain?" *Developmental Science* 10 (1): 110–120.

———. 2014. "Early Language Learning and the Social Brain." *Cold Spring Harbor Symposia on Quantitative Biology* 79: 211–220.

Kuhl, Patricia, and Maritza Rivera-Gaxiola. 2008. "Neural Substrates of Language Acquisition." *Annual Review of Neuroscience* 31: 511–534.

Kuhn, Thomas S. 1966. *The Structure of Scientific Revolutions.* Chicago: University of Chicago Press.

Kundera, Milan. 1994. *The Book of Laughter and Forgetting.* New York: HarperCollins.

Kupfer, Alexander, Hendrik Müller, Marta M. Antoniazzi, Carlos Jared, Hartmut Greven, Ronald A. Nussbaum, and Mark Wilkinson. 2006. "Parental Investment by Skin Feeding in a Caecilian Amphibian." *Nature* 440 (7086): 926–929.

Kuppens, P., F. Tuerlinckx, J. A. Russell, and L. F. Barrett. 2013. "The Relationship Between Valence and Arousal in Subjective Experience." *Psychological Bulletin* 139: 917–940.

Kuppens, Peter, Iven Van Mechelen, Dirk J. M. Smits, Paul De Boeck, and Eva Ceulemans. 2007. "Individual Differences in Patterns of Appraisal and Anger Experience." *Cognition and Emotion* 21 (4): 689–713.

LaBar, Kevin S., J. Christopher Gatenby, John C. Gore, Joseph E. LeDoux, and Elizabeth A. Phelps. 1998. "Human Amygdala Activation During Conditioned Fear Acquisition and Extinction: A Mixed-Trial fMRI Study." *Neuron* 20 (5): 937–945.

Lakoff, George. 1990. *Women, Fire, and Dangerous Things: What Categories Reveal About the Mind.* Chicago: University of Chicago Press.

Laland, Kevin N., and Gillian R. Brown. 2011. *Sense and Nonsense: Evolutionary Perspectives on Human Behaviour.* Oxford: Oxford University Press.

Lane, Richard D., Geoffrey L. Ahern, Gary E. Schwartz, and Alfred W. Kaszniak. 1997. "Is Alexithymia the Emotional Equivalent of Blindsight?" *Biological Psychiatry* 42 (9): 834–844.

Lane, Richard D., and David A. S. Garfield. 2005. "Becoming Aware of Feelings: Integration of Cognitive-Developmental, Neuroscientific, and Psychoanalytic Perspectives." *Neuropsychoanalysis* 7 (1): 5–30.

Lane, Richard D., Lee Sechrest, Robert Riedel, Daniel E. Shapiro, and Alfred W. Kaszniak. 2000. "Pervasive Emotion Recognition Deficit Common to Alexithymia and the Repressive Coping Style." *Psychosomatic Medicine* 62 (4): 492–501.

Lang, Peter J., Mark K. Greenwald, Margaret M. Bradley, and Alfons O. Hamm. 1993. "Looking at Pictures: Affective, Facial, Visceral, and Behavioral Reactions." *Psychophysiology* 30 (3): 261–273.

Laukka, Petri, Hillary Anger Elfenbein, Nela Söder, Henrik Nordström, Jean Althoff, Wanda Chui, Frederick K. Iraki, Thomas Rockstuhl, and Nutankumar S. Thingujam. 2013. "Cross-Cultural Decoding of Positive and Negative Non-Linguistic Emotion Vocalizations." *Frontiers in Psychology* 4 (353): 185–192.

Lawrence, T. E. (1922) 2015. *Seven Pillars of Wisdom.* Toronto: Aegitas.

Lazarus, R. S. 1998. "From Psychological Stress to the Emotions: A History of Changing Outlooks." In *Personality: Critical Concepts in Psychology,* vol. 4, edited by Cary L. Cooper and Lawrence A. Pervin, 179–200. London: Routledge.

Lea, Stephen E. G. 2010. "Concept Learning in Nonprimate Mammals: In Search of Evidence." In *The Making of Human Concepts,* edited by Denis Mareschal, Paul Quinn, and Stephen E. G. Lea, 173–199. New York: Oxford University Press.

Lebois, Lauren A. M., Christine D. Wilson-Mendenhall, and Lawrence W. Barsalou. 2015. "Are Automatic Conceptual Cores the Gold Standard of Semantic Processing? The Context-Dependence of Spatial Meaning in Grounded Congruency Effects." *Cognitive Science* 39 (8): 1764–1801.

Lebrecht, S., M. Bar., L. F. Barrett, and M. J. Tarr. 2012. "Micro-Valences: Perceiving Affective Valence in Everyday Objects." *Frontiers in Perception Science* 3 (107): 1–5.

Lecours, S., G. Robert, and F. Desruisseaux. 2009. "Alexithymia and Verbal Elaboration of Affect in Adults Suffering from a Respiratory Disorder." *European Review of Applied Psychology–Revue européenne de psychologie appliquée* 59 (3): 187–195.

LeDoux, Joseph E. 2014. "Coming to Terms with Fear." *Proceedings of the National Academy of Sciences* 111 (8): 2871–2878.

———. 2015. *Anxious: Using the Brain to Understand and Treat Fear and Anxiety.* New York: Penguin.

Lee, Marion, Sanford Silverman, Hans Hansen, and Vikram Patel. 2011. "A Comprehensive Review of Opioid-Induced Hyperalgesia." *Pain Physician* 14: 145–161.

Leffel, Kristin, and Dana Suskind. 2013. "Parent-Directed Approaches to Enrich the Early Language Environments of Children Living in Poverty." *Seminars in Speech and Language* 34 (4): 267–278.

Leppänen, Jukka M., and Charles A. Nelson. 2009. "Tuning the Developing Brain to Social Signals of Emotions." *Nature Reviews Neuroscience* 10 (1): 37–47.

Levenson, Robert W. 2011. "Basic Emotion Questions." *Emotion Review* 3 (4): 379–386.

Levenson, Robert W., Paul Ekman, and Wallace V. Friesen. 1990. "Voluntary Facial Action Generates Emotion-Specific Autonomic Nervous System Activity." *Psychophysiology* 27 (4): 363–384.

Levenson, Robert W., Paul Ekman, Karl Heider, and Wallace V. Friesen. 1992. "Emotion and Autonomic Nervous System Activity in the Minangkabau of West Sumatra." *Journal of Personality and Social Psychology* 62 (6): 972–988.

Levy, Robert I. 1975. *Tahitians: Mind and Experience in the Society Islands.* Chicago: University of Chicago Press.

———. 2014. "The Emotions in Comparative Perspective." In *Approaches to Emotion,* edited by K. Scherer and P. Ekman, 397–412. Hillsdale, NJ: Erlbaum.

Lewontin, Richard. 1991. *Biology as Ideology: The Doctrine of DNA.* New York: HarperPerennial.

Li, Susan Shi Yuan, and Gavan P. McNally. 2014. "The Conditions That Promote Fear Learning: Prediction Error and Pavlovian Fear Conditioning." *Neurobiology of Learning and Memory* 108: 14–21.

Liberman, Alvin M., Franklin S. Cooper, Donald P. Shankweiler, and Michael Studdert-Kennedy. 1967. "Perception of the Speech Code." *Psychological Review* 74 (6): 431–461.

Lieberman, M. D., N. I. Eisenberger, M. J. Crockett, S. M. Tom, J. H. Pfeifer, and B. M. Way. 2007. "Putting Feelings into Words: Affect Labeling Disrupts Amygdala Activity in Response to Affective Stimuli." *Psychological Science* 18 (5): 421–428.

Lieberman, M. D., A. Hariri, J. M. Jarcho, N. I. Eisenberger, and S. Y. Bookheimer. 2005. "An fMRI Investigation of Race-Related Amygdala Activity in African-American and Caucasian-American Individuals." *Nature Neuroscience* 8 (6): 720–722.

Lin, Pei-Ying. 2013. "Unspeakableness: An Intervention of Language Evolution and Human Communication." http://uniquelang.peiyinglin.net/01untranslatable.html.

Lindquist, Kristen A., and Lisa Feldman Barrett. 2008. "Emotional Complexity." In *Handbook of Emotions*, 3rd edition, edited by Michael Lewis, Jeannette M. Haviland-Jones, and Lisa Feldman Barrett, 513–530. New York: Guilford Press.

———. 2012. "A Functional Architecture of the Human Brain: Emerging Insights from the Science of Emotion." *Trends in Cognitive Sciences* 16 (11): 533–540.

Lindquist, Kristen A., Lisa Feldman Barrett, Eliza Bliss-Moreau, and James A. Russell. 2006. "Language and the Perception of Emotion." *Emotion* 6 (1): 125–138.

Lindquist, Kristen A., Maria Gendron, Lisa Feldman Barrett, and Bradford C. Dickerson. 2014. "Emotion Perception, but Not Affect Perception, Is Impaired with Semantic Memory Loss." *Emotion* 14 (2): 375–387.

Lindquist, Kristen A., Ajay B. Satpute, Tor D. Wager, Jochen Weber, and Lisa Feldman Barrett. 2015. "The Brain Basis of Positive and Negative Affect: Evidence from a Meta-Analysis of the Human Neuroimaging Literature." *Cerebral Cortex* 26 (5): 1910–1922.

Lindquist, Kristen A., Tor D. Wager, Hedy Kober, Eliza Bliss-Moreau, and Lisa Feldman Barrett. 2012. "The Brain Basis of Emotion: A Meta-Analytic Review." *Behavioral and Brain Sciences* 35 (3): 121–143.

Llinás, Rodolfo Riascos. 2001. *I of the Vortex: From Neurons to Self.* Cambridge, MA: MIT Press.

Lloyd-Fox, Sarah, Borbála Széplaki-Köllőd, Jun Yin, and Gergely Csibra. 2015. "Are You Talking to Me? Neural Activations in 6-Month-Old Infants in Response to Being Addressed During Natural Interactions." *Cortex* 70: 35–48.

Lochmann, Timm, and Sophie Deneve. 2011. "Neural Processing as Causal Inference." *Current Opinion in Neurobiology* 21 (5): 774–781.

Loftus, Elizabeth F., and J. C. Palmer. 1974. "Reconstruction of Automobile Destruction: An Example of the Interaction Between Language and Memory." *Journal of Verbal Learning and Verbal Behavior* 13 (5): 585–589.

Lokuge, Sonali, Benicio N. Frey, Jane A. Foster, Claudio N. Soares, and Meir Steiner. 2011. "Commentary: Depression in Women: Windows of Vulnerability and New Insights into the Link Between Estrogen and Serotonin." *Journal of Clinical Psychiatry* 72 (11): 1563–1569.

Lorch, Marjorie Perlman. 2008. "The Merest Logomachy: The 1868 Norwich Discussion of Aphasia by Hughlings Jackson and Broca." *Brain* 131 (6): 1658–1670.

Louveau, Antoine, Igor Smirnov, Timothy J. Keyes, Jacob D. Eccles, Sherin J. Rouhani, J. David Peske, Noel C. Derecki, David Castle, James W. Mandell, and Kevin S. Lee. 2015. "Structural and Functional Features of Central Nervous System Lymphatic Vessels." *Nature* 523: 337–341.

Lujan, J. Luis, Ashutosh Chaturvedi, Ki Sueng Choi, Paul E. Holtzheimer, Robert E. Gross, Helen S. Mayberg, and Cameron C. McIntyre. 2013. "Tractography-Activation Models Applied to Subcallosal Cingulate Deep Brain Stimulation." *Brain Stimulation* 6 (5): 737–739.

Luminet, Olivier, Bernard Rimé, R. Michael Bagby, and Graeme Taylor. 2004. "A Multimodal Investigation of Emotional Responding in Alexithymia." *Cognition and Emotion* 18 (6): 741–766.

Lutz, Catherine. 1980. *Emotion Words and Emotional Development on Ifaluk Atoll.* Ph.D. diss., Harvard University, 003878556.

——. 1983. "Parental Goals, Ethnopsychology, and the Development of Emotional Meaning." *Ethos* 11 (4): 246–262.

Lynch, Mona, and Craig Haney. 2011. "Looking Across the Empathic Divide: Racialized Decision Making on the Capital Jury." *Michigan State Law Review* 2011: 573–607.

Ma, Lili, and Fei Xu. 2011. "Young Children's Use of Statistical Sampling Evidence to Infer the Subjectivity of Preferences." *Cognition* 120 (3): 403–411.

MacLean, P. D., and V. A. Kral. 1973. *A Triune Concept of the Brain and Behavior.* Toronto: University of Toronto Press.

Madrick, Jeff. 2014. *Seven Bad Ideas: How Mainstream Economists Have Damaged America and the World.* New York: Vintage.

Maihöfner, Christian, Clemens Forster, Frank Birklein, Bernhard Neundörfer, and Hermann O. Handwerker. 2005. "Brain Processing During Mechanical Hyperalgesia in Complex Regional Pain Syndrome: A Functional MRI Study." *Pain* 114 (1): 93–103.

Malik, Bilal R., and James J. L. Hodge. 2014. "Drosophila Adult Olfactory Shock Learning." *Journal of Visualized Experiments* (90): 1–5. doi:10.3791/50107.

Malt, Barbara, and Phillip Wolff. 2010. *Words and the Mind: How Words Capture Human Experience.* New York: Oxford University Press.

Marder, E., and A. L. Taylor. 2011. "Multiple Models to Capture the Variability in Biological Neurons and Networks." *Nature Neuroscience* 14: 133–138.

Marder, Eve. 2012. "Neuromodulation of Neuronal Circuits: Back to the Future." *Neuron* 76 (1): 1–11.

Mareschal, Denis, Mark H. Johnson, Sylvain Sirois, Michael Spratling, Michael S. C. Thomas, and Gert Westermann. 2007. *Neuroconstructivism-I: How the Brain Constructs Cognition.* New York: Oxford University Press.

Mareschal, Denis, Paul C. Quinn, and Stephen E. G. Lea. 2010. *The Making of Human Concepts.* New York: Oxford University Press.

Marmi, Josep, Jaume Bertranpetit, Jaume Terradas, Osamu Takenaka, and Xavier Domingo-Roura. 2004. "Radiation and Phylogeography in the Japanese Macaque, Macaca Fuscata." *Molecular Phylogenetics and Evolution* 30 (3): 676–685.

Martin, Alia, and Laurie R. Santos. 2014. "The Origins of Belief Representation: Monkeys Fail to Automatically Represent Others' Beliefs." *Cognition* 130 (3): 300–308.

Martin, René, Ellen E. I. Gordon, and Patricia Lounsbury. 1998. "Gender Disparities in the Attribution of Cardiac-Related Symptoms: Contribution of Common Sense Models of Illness." *Health Psychology* 17 (4): 346–357.

Martin, René, Catherine Lemos, Nan Rothrock, S. Beth Bellman, Daniel Russell, Toni Tripp-Reimer, Patricia Lounsbury, and Ellen Gordon. 2004. "Gender Disparities in Common Sense Models of Illness Among Myocardial Infarction Victims." *Health Psychology* 23 (4): 345–353.

Martins, Nicole. 2013. "Televised Relational and Physical Aggression and Children's Hostile Intent Attributions." *Journal of Experimental Child Psychology* 116 (4): 945–952.

Martins, Nicole, Marie-Louise Mares, Mona Malacane, and Alanna Peebles. In press.

"Liked Characters Get a Moral Pass: Young Viewers' Evaluations of Social and Physical Aggression in Tween Sitcoms." *Communication Research.*

Martins, Nicole, and Barbara J. Wilson. 2011. "Genre Differences in the Portrayal of Social Aggression in Programs Popular with Children." *Communication Research Reports* 28 (2): 130–140.

———. 2012a. "Mean on the Screen: Social Aggression in Programs Popular with Children." *Journal of Communication* 62 (6): 991–1009.

———. 2012b. "Social Aggression on Television and Its Relationship to Children's Aggression in the Classroom." *Human Communication Research* 38 (1): 48–71.

Massachusetts General Hospital Center for Law, Brain, and Behavior. 2013. "Memory in the Courtroom: Fixed, Fallible or Fleeting?" http://clbb.mgh.harvard.edu/memory-in-the -courtroom-fixed-fallible-or-fleeting.

Master, Sarah L., David M. Amodio, Annette L. Stanton, Cindy M. Yee, Clayton J. Hilmert, and Shelley E. Taylor. 2009. "Neurobiological Correlates of Coping Through Emotional Approach." *Brain, Behavior, and Immunity* 23 (1): 27–35.

Mathers, Colin, Doris Ma Fat, and Jan Ties Boerma. 2008. *The Global Burden of Disease: 2004 Update.* Geneva: World Health Organization.

Mathis, Diane, and Steven E. Shoelson. 2011. "Immunometabolism: An Emerging Frontier." *Nature Reviews Immunology* 11 (2): 81–83.

Matsumoto, David, Dacher Keltner, Michelle N. Shiota, Maureen O'Sullivan, and Mark Frank. 2008. "Facial Expressions of Emotion." In *Handbook of Emotions,* 3rd edition, edited by Michael Lewis, Jeannette M. Haviland-Jones, and Lisa Feldman Barrett, 211–234. New York: Guilford Press.

Matsumoto, David, Seung Hee Yoo, and Johnny Fontaine. 2008. "Mapping Expressive Differences Around the World: The Relationship Between Emotional Display Rules and Individualism Versus Collectivism." *Journal of Cross-Cultural Psychology* 39 (1): 55–74.

Matsuzawa, Tetsuro. 2010. "Cognitive Development in Chimpanzees: A Trade-Off Between Memory and Abstraction." In *The Making of Human Concepts,* edited by Denis Mareschal, Paul C. Quinn, and Stephen E. G. Lea, 227–244. New York: Oxford University Press.

Mayberg, Helen S. 2009. "Targeted Electrode-Based Modulation of Neural Circuits for Depression." *Journal of Clinical Investigation* 119 (4): 717–725.

Maye, Jessica, Janet F. Werker, and LouAnn Gerken. 2002. "Infant Sensitivity to Distributional Information Can Affect Phonetic Discrimination." *Cognition* 82 (3): B101–B111.

Mayr, Ernst. 1982. *The Growth of Biological Thought: Diversity, Evolution, and Inheritance.* Cambridge, MA: Harvard University Press.

———. 2007. *What Makes Biology Unique? Considerations on the Autonomy of a Scientific Discipline.* New York: Cambridge University Press.

McEwen, Bruce S., Nicole P. Bowles, Jason D. Gray, Matthew N. Hill, Richard G. Hunter, Ilia N. Karatsoreos, and Carla Nasca. 2015. "Mechanisms of Stress in the Brain." *Nature Neuroscience* 18 (10): 1353–1363.

McEwen, Bruce S., and Peter J. Gianaros. 2011. "Stress- and Allostasis-Induced Brain Plasticity." *Annual Review of Medicine* 62: 431–445.

McGlone, Francis, Johan Wessberg, and Håkan Olausson. 2014. "Discriminative and Affective Touch: Sensing and Feeling." *Neuron* 82 (4): 737–755.

McGrath, Callie L., Mary E. Kelley, Boadie W. Dunlop, Paul E. Holtzheimer III, W. Edward Craighead, and Helen S. Mayberg. 2014. "Pretreatment Brain States Identify

Likely Nonresponse to Standard Treatments for Depression." *Biological Psychiatry* 76 (7): 527–535.

McKelvey, Tara. 2015. "Boston in Shock over Tsarnaev Death Penalty." *BBC News*, May 16. http://www.bbc.com/news/world-us-canada-32762999.

McMenamin, Brenton W., Sandra J. E. Langeslag, Mihai Sirbu, Srikanth Padmala, and Luiz Pessoa. 2014. "Network Organization Unfolds over Time During Periods of Anxious Anticipation." *Journal of Neuroscience* 34 (34): 11261–11273.

McNally, Gavan P., Joshua P. Johansen, and Hugh T. Blair. 2011. "Placing Prediction into the Fear Circuit." *Trends in Neurosciences* 34 (6): 283–292.

Meganck, Reitske, Stijn Vanheule, Ruth Inslegers, and Mattias Desmet. 2009. "Alexithymia and Interpersonal Problems: A Study of Natural Language Use." *Personality and Individual Differences* 47 (8): 990–995.

Mena, Jesus D., Ryan A. Selleck, and Brian A. Baldo. 2013. "Mu-Opioid Stimulation in Rat Prefrontal Cortex Engages Hypothalamic Orexin/Hypocretin-Containing Neurons, and Reveals Dissociable Roles of Nucleus Accumbens and Hypothalamus in Cortically Driven Feeding." *Journal of Neuroscience* 33 (47): 18540–18552.

Mennin, Douglas S., Richard G. Heimberg, Cynthia L. Turk, and David M. Fresco. 2005. "Preliminary Evidence for an Emotion Dysregulation Model of Generalized Anxiety Disorder." *Behaviour Research and Therapy* 43 (10): 1281–1310.

Menon, V. 2011. "Large-Scale Brain Networks and Psychopathology: A Unifying Triple Network Model." *Trends in Cognitive Science* 15 (10): 483–506.

Mervis, Carolyn B., and Eleanor Rosch. 1981. "Categorization of Natural Objects." *Annual Review of Psychology* 32 (1): 89–115.

Merz, Emily C., Tricia A. Zucker, Susan H. Landry, Jeffrey M. Williams, Michael Assel, Heather B. Taylor, Christopher J. Lonigan, Beth M. Phillips, Jeanine Clancy-Menchetti, and Marcia A. Barnes. 2015. "Parenting Predictors of Cognitive Skills and Emotion Knowledge in Socioeconomically Disadvantaged Preschoolers." *Journal of Experimental Child Psychology* 132: 14–31.

Mesman, Judi, Harriet Oster, and Linda Camras. 2012. "Parental Sensitivity to Infant Distress: What Do Discrete Negative Emotions Have to Do with It?" *Attachment and Human Development* 14 (4): 337–348.

Mesquita, Batja, and Nico H. Frijda. 1992. "Cultural Variations in Emotions: A Review." *Psychological Bulletin* 112 (2): 179–204.

Mesulam, M.-Marcel. 2002. "The Human Frontal Lobes: Transcending the Default Mode Through Contingent Encoding." In *Principles of Frontal Lobe Function*, edited by Donald T. Stuss and Robert T. Knight, 8–30. New York: Oxford University Press.

Metti, Andrea L., Howard Aizenstein, Kristine Yaffe, Robert M. Boudreau, Anne Newman, Lenore Launer, Peter J. Gianaros, Oscar L. Lopez, Judith Saxton, and Diane G. Ives. 2015. "Trajectories of Peripheral Interleukin-6, Structure of the Hippocampus, and Cognitive Impairment over 14 Years in Older Adults." *Neurobiology of Aging* 36 (11): 3038–3044.

Miller, Andrew H., Ebrahim Haroon, Charles L. Raison, and Jennifer C. Felger. 2013. "Cytokine Targets in the Brain: Impact on Neurotransmitters and Neurocircuits." *Depression and Anxiety* 30 (4): 297–306.

Miller, Antonia Elise. 2010. "Inherent (Gender) Unreasonableness of the Concept of Reasonableness in the Context of Manslaughter Committed in the Heat of Passion." *William and Mary Journal of Women and the Law* 17: 249.

Miller, Gregory E., and Edith Chen. 2010. "Harsh Family Climate in Early Life Presages the

Emergence of a Proinflammatory Phenotype in Adolescence." *Psychological Science* 21 (6): 848–856.

Mitchell, Robert W., Nicholas S. Thompson, and H. Lyn Miles. 1997. *Anthropomorphism, Anecdotes, and Animals.* Albany, NY: SUNY Press.

Mobbs, Dean, Hakwan C. Lau, Owen D. Jones, and Christopher D. Frith. 2007. "Law, Responsibility, and the Brain." *PLOS Biology* 5 (4): e103. doi:10.1371/journal.pbio.0050103.

Montgomery, Ben. 2012. "Florida's 'Stand Your Ground' Law Was Born of 2004 Case, but Story Has Been Distorted." *Tampa Bay Times,* April 14. http://www.tampabay.com/news/publicsafety/floridas-stand-your-ground-law-was-born-of-2004-case-but-story-has-been/1225164.

Monyak, Suzanne. 2015. "Jury Awards $2.2M Verdict Against Food Storage Company in 'Defecator' DNA Case." *Daily Report,* June 22. http://www.dailyreportonline.com/id=1202730177957/Jury-Awards-22M-Verdict-Against-Food-Storage-Company-in-Defecator-DNA-Case.

Moon, Christine, Hugo Lagercrantz, and Patricia K. Kuhl. 2013. "Language Experienced in Utero Affects Vowel Perception After Birth: A Two-Country Study." *Acta paediatrica* 102 (2): 156–160.

Moore, Shelby A. D. 1994. "Battered Woman Syndrome: Selling the Shadow to Support the Substance." *Howard Law Journal* 38 (2): 297.

Morell, Virginia. 2013. *Animal Wise: How We Know Animals Think and Feel.* New York: Broadway Books.

Moriguchi, Y., A. Negreira, M. Weierich, R. Dautoff, B. C. Dickerson, C. I. Wright, and L. F. Barrett. 2011. "Differential Hemodynamic Response in Affective Circuitry with Aging: An fMRI Study of Novelty, Valence, and Arousal." *Journal of Cognitive Neuroscience* 23 (5): 1027–1041.

Moriguchi, Yoshiya, Alexandra Touroutoglou, Bradford C. Dickerson, and Lisa Feldman Barrett. 2013. "Sex Differences in the Neural Correlates of Affective Experience." *Social Cognitive and Affective Neuroscience* 9 (5): 591–600.

Morrison, Adele M. 2006. "Changing the Domestic Violence (Dis) Course: Moving from White Victim to Multi-Cultural Survivor." *UC Davis Law Review* 39: 1061–1120.

Murai, Chizuko, Daisuke Kosugi, Masaki Tomonaga, Masayuki Tanaka, Tetsuro Matsuzawa, and Shoji Itakura. 2005. "Can Chimpanzee Infants (Pan Troglodytes) Form Categorical Representations in the Same Manner as Human Infants (Homo Sapiens)?" *Developmental Science* 8 (3): 240–254.

Murphy, G. L. 2002. *The Big Book of Concepts.* Cambridge, MA: MIT Press.

Mysels, David J., and Maria A. Sullivan. 2010. "The Relationship Between Opioid and Sugar Intake: Review of Evidence and Clinical Applications." *Journal of Opioid Management* 6 (6): 445–452.

Naab, Pamela J., and James A. Russell. 2007. "Judgments of Emotion from Spontaneous Facial Expressions of New Guineans." *Emotion* 7 (4): 736–744.

Nadler, Janice, and Mary R. Rose. 2002. "Victim Impact Testimony and the Psychology of Punishment." *Cornell Law Review* 88: 419.

National Institute of Mental Health. 2015. "Research Domain Criteria (RDoC)." https://www.nimh.nih.gov/research-priorities/rdoc/.

National Institute of Neurological Disorders and Stroke. 2013. "Complex Regional Pain Syndrome Fact Sheet." http://www.ninds.nih.gov/disorders/reflex_sympathetic_dystrophy/detail_reflex_sympathetic_dystrophy.htm.

National Sleep Foundation. 2011. "Annual Sleep in America Poll Exploring Connections

with Communications Technology Use and Sleep." https://sleepfoundation.org/media-center/press-release/annual-sleep-america-poll-exploring-connections-communica
tions-technology-use.

Nauert, Rick. 2013. "70 Percent of Americans Take Prescription Drugs." *PsychCentral,* June 20. http://psychcentral.com/news/2013/06/20/70-percent-of-americans-take-prescrip
tion-drugs/56275.html.

Neisser, Ulric. 2014. *Cognitive Psychology, Classic Edition.* New York: Psychology Press.

Neuroskeptic. 2011. "Neurology vs Psychiatry." *Neuroskeptic Blog.* http://blogs.discover
magazine.com/neuroskeptic/2011/04/07/neurology-vs-psychiatry.

New Jersey Courts, State of New Jersey. 2012. "Identification: In-Court and Out-of-Court Identifications." http://www.judiciary.state.nj.us/criminal/charges/idinout.pdf.

Nielsen, Mark. 2009. "12-Month-Olds Produce Others' Intended but Unfulfilled Acts." *Infancy* 14 (3): 377–389.

Nisbett, Richard E., and Dov Cohen. 1996. *Culture of Honor: The Psychology of Violence in the South.* Boulder, CO: Westview Press.

Noble, Kimberly G., Suzanne M. Houston, Natalie H. Brito, Hauke Bartsch, Eric Kan, Joshua M. Kuperman, Natacha Akshoomoff, David G. Amaral, Cinnamon S. Bloss, and Ondrej Libiger. 2015. "Family Income, Parental Education and Brain Structure in Children and Adolescents." *Nature Neuroscience* 18 (5): 773–778.

Nobler, Mitchell S., Maria A. Oquendo, Lawrence S. Kegeles, Kevin M. Malone, Carl Campbell, Harold A. Sackeim, and J. John Mann. 2001. "Decreased Regional Brain Metabolism After ECT." *American Journal of Psychiatry* 158 (2): 305–308.

Nokia, Miriam S., Sanna Lensu, Juha P. Ahtiainen, Petra P. Johansson, Lauren G. Koch, Steven L. Britton, and Heikki Kainulainen. 2016. "Physical Exercise Increases Adult Hippocampal Neurogenesis in Male Rats Provided It Is Aerobic and Sustained." *Journal of Physiology* 594 (7): 1–19.

Norenzayan, Ara, and Steven J. Heine. 2005. "Psychological Universals: What Are They and How Can We Know?" *Psychological Bulletin* 131 (5): 763–784.

Nummenmaa, Lauri, Enrico Glerean, Riitta Hari, and Jari K. Hietanen. 2014. "Bodily Maps of Emotions." *Proceedings of the National Academy of Sciences* 111 (2): 646–651.

Obrist, Paul A. 1981. *Cardiovascular Psychophysiology: A Perspective.* New York: Plenum.

Obrist, Paul A., Roger A. Webb, James R. Sutterer, and James L. Howard. 1970. "The Cardiac-Somatic Relationship: Some Reformulations." *Psychophysiology* 6 (5): 569–587.

Ochsner, K. N., and J. J. Gross. 2005. "The Cognitive Control of Emotion." *Trends in Cognitive Science* 9 (5): 242–249.

Okamoto-Barth, Sanae, and Masaki Tomonaga. 2006. "Development of Joint Attention in Infant Chimpanzees." In *Cognitive Development in Chimpanzees,* edited by T. Matsuzawa, M. Tomanaga, and M. Tanaka, 155–171. Tokyo: Springer.

Olausson, Håkan, Johan Wessberg, Francis McGlone, and Åke Vallbo. 2010. "The Neurophysiology of Unmyelinated Tactile Afferents." *Neuroscience and Biobehavioral Reviews* 34 (2): 185–191.

Olfson, Mark, and Steven C. Marcus. 2009. "National Patterns in Antidepressant Medication Treatment." *Archives of General Psychiatry* 66 (8): 848–856.

Oosterwijk, Suzanne, Kristen A. Lindquist, Morenikeji Adebayo, and Lisa Feldman Barrett. 2015. "The Neural Representation of Typical and Atypical Experiences of Negative Images: Comparing Fear, Disgust and Morbid Fascination." *Social Cognitive and Affective Neuroscience* 11 (1): 11–22.

Opendak, Maya, and Elizabeth Gould. 2015. "Adult Neurogenesis: A Substrate for Experience-Dependent Change." *Trends in Cognitive Sciences* 19 (3): 151–161.

Ortony, Andrew, Gerald L. Clore, and Allan Collins. 1990. *The Cognitive Structure of Emotions*. New York: Cambridge University Press.

Osgood, Charles Egerton, George John Suci, and Percy H. Tannenbaum. 1957. *The Measurement of Meaning*. Urbana: University of Illinois Press.

Oster, Harriet. 2005. "The Repertoire of Infant Facial Expressions: An Ontogenetic Perspective." In *Emotional Development: Recent Research Advances*, edited by J. Nadel and D. Muir, 261–292. New York: Oxford University Press.

———. 2006. "Baby FACS: Facial Action Coding System for infants and Young Children." Unpublished monograph and coding manual. New York University.

Owren, Michael J., and Drew Rendall. 2001. "Sound on the Rebound: Bringing Form and Function Back to the Forefront in Understanding Nonhuman Primate Vocal Signaling." *Evolutionary Anthropology: Issues, News, and Reviews* 10 (2): 58–71.

Palumbo, R. V., M. E. Marraccini, L. L. Weyandt, O. Wilder-Smith, H. A. McGee, S. Liu, and M. S. Goodwin. In press. "Interpersonal Autonomic Physiology: A Systematic Review of the Literature." *Personality and Social Psychology Review*.

Panayiotou, Aalexia. 2004. "Bilingual Emotions: The Untranslatable Self." *Estudios de sociolingüística: Linguas, sociedades e culturas* 5 (1): 1–20.

Panksepp, J. 1998. *Affective Neuroscience: The Foundations of Human and Animal Emotions*. New York: Oxford University Press.

———. 2011. "The Basic Emotional Circuits of Mammalian Brains: Do Animals Have Affective Lives?" *Neuroscience and Biobehavioral Reviews* 35 (9): 1791–1804.

Panksepp, Jaak, and Jules B. Panksepp. 2013. "Toward a Cross-Species Understanding of Empathy." *Trends in Neurosciences* 36 (8): 489–496.

Parise, Eugenio, and Gergely Csibra. 2012. "Electrophysiological Evidence for the Understanding of Maternal Speech by 9-Month-Old Infants." *Psychological Science* 23 (7): 728–733.

Park, Hae-Jeong, and Karl Friston. 2013. "Structural and Functional Brain Networks: From Connections to Cognition." *Science* 342 (6158): 1238411.

Park, Seong-Hyun, and Richard H. Mattson. 2009. "Ornamental Indoor Plants in Hospital Rooms Enhanced Health Outcomes of Patients Recovering from Surgery." *Journal of Alternative and Complementary Medicine* 15 (9): 975–980.

Parker, George Howard. 1919. *The Elementary Nervous System*. Philadelphia: J. B. Lippincott.

Parr, Lisa A., Bridget M. Waller, Sarah J. Vick, and Kim A. Bard. 2007. "Classifying Chimpanzee Facial Expressions Using Muscle Action." *Emotion* 7 (1): 172–181.

Passingham, Richard. 2009. "How Good Is the Macaque Monkey Model of the Human Brain?" *Current Opinion in Neurobiology* 19 (1): 6–11.

Paulus, Martin P., and Murray B. Stein. 2010. "Interoception in Anxiety and Depression." *Brain Structure and Function* 214 (5–6): 451–463.

Pavlenko, Aneta. 2009. "Conceptual Representation in the Bilingual Lexicon and Second Language Vocabulary Learning." In *The Bilingual Mental Lexicon: Interdisciplinary Approaches*, edited by Aneta Pavlenko, 125–160. Bristol, UK: Multilingual Matters.

———. 2014. *The Bilingual Mind: And What It Tells Us About Language and Thought*. Cambridge: Cambridge University Press.

Peelen, M. V., A. P. Atkinson, and P. Vuilleumier. 2010. "Supramodal Representations of Perceived Emotions in the Human Brain." *Journal of Neuroscience* 30 (30): 10127–10134.

Percy, Elise J., Joseph L. Hoffmann, and Steven J. Sherman. 2010. "Sticky Metaphors and the Persistence of the Traditional Voluntary Manslaughter Doctrine." *University of Michigan Journal of Law Reform* 44: 383.

Perfors, Amy, Joshua B. Tenenbaum, Thomas L. Griffiths, and Fei Xu. 2011. "A Tutorial Introduction to Bayesian Models of Cognitive Development." *Cognition* 120 (3): 302–321.

Perissinotto, Carla M., Irena Stijacic Cenzer, and Kenneth E. Covinsky. 2012. "Loneliness in Older Persons: A Predictor of Functional Decline and Death." *Archives of Internal Medicine* 172 (14): 1078–1084.

Pessoa, L., E. Thompson, and A. Noe. 1998. "Finding Out About Filling-In: A Guide to Perceptual Completion for Visual Science and the Philosophy of Perception." *Behavioral and Brain Sciences* 21 (6): 723–802.

Pillsbury, Samuel H. 1989. "Emotional Justice: Moralizing the Passions of Criminal Punishment." *Cornell Law Review* 74: 655–710.

Pimsleur. 2014. "Words We Wish Existed in English." *Pimsleur Approach.* https://www.pimsleurapproach.com/words-we-wish-existed-in-english/.

Pinker, Steven. 1997. *How the Mind Works.* New York: Norton.

———. 2002. *The Blank Slate: The Modern Denial of Human Nature.* New York: Penguin.

Pinto, A., D. Di Raimondo, A. Tuttolomondo, C. Buttà, G. Milio, and G. Licata. 2012. "Effects of Physical Exercise on Inflammatory Markers of Atherosclerosis." *Current Pharmaceutical Design* 18 (28): 4326–4349.

Pisotta, Iolanda, and Marco Molinari. 2014. "Cerebellar Contribution to Feedforward Control of Locomotion." *Frontiers in Human Neuroscience* 8: 1–5.

Planck, Max. 1931. *The Universe in the Light of Modern Physics.* London: Allen and Unwin.

Ploghaus, Alexander, Charvy Narain, Christian F. Beckmann, Stuart Clare, Susanna Bantick, Richard Wise, Paul M. Matthews, J. Nicholas P. Rawlins, and Irene Tracey. 2001. "Exacerbation of Pain by Anxiety Is Associated with Activity in a Hippocampal Network." *Journal of Neuroscience* 21 (24): 9896–9903.

Pollack, Irwin, and James M. Pickett. 1964. "Intelligibility of Excerpts from Fluent Speech: Auditory vs. Structural Context." *Journal of Verbal Learning and Verbal Behavior* 3 (1): 79–84.

Pond, Richard S., Jr., Todd B. Kashdan, C. Nathan DeWall, Antonina Savostyanova, Nathaniel M. Lambert, and Frank D. Fincham. 2012. "Emotion Differentiation Moderates Aggressive Tendencies in Angry People: A Daily Diary Analysis." *Emotion* 12 (2): 326–337.

Posner, M. I., C. R. Snyder, and B. J. Davidson. 1980. "Attention and the Detection of Signals." *Journal of Experimental Psychology* 109 (2): 160–174.

Posner, Michael I., and Steven W. Keele. 1968. "On the Genesis of Abstract Ideas." *Journal of Experimental Psychology* 77 (July): 353–363.

Power, Jonathan D., Alexander L. Cohen, Steven M. Nelson, Gagan S. Wig, Kelly Anne Barnes, Jessica A. Church, Alecia C. Vogel, Timothy O. Laumann, Fran M. Miezin, and Bradley L. Schlaggar. 2011. "Functional Network Organization of the Human Brain." *Neuron* 72 (4): 665–678.

Pratt, Maayan, Magi Singer, Yaniv Kanat-Maymon, and Ruth Feldman. 2015. "Infant Negative Reactivity Defines the Effects of Parent-Child Synchrony on Physiological and Behavioral Regulation of Social Stress." *Development and Psychopathology* 27 (4, part 1): 1191–1204.

Prebble, S. C., D. R. Addis, and L. J. Tippett. 2012. "Autobiographical Memory and Sense of Self." *Psychological Bulletin* 139 (4): 815–840.

Press, Clare, and Richard Cook. 2015. "Beyond Action-Specific Simulation: Domain-General Motor Contributions to Perception." *Trends in Cognitive Sciences* 19 (4): 176–178.

Pribram, Karl H. 1958. "Comparative Neurology and the Evolution of Behavior." In *Behavior and Evolution,* edited by Anne Roe and George Gaylord Simpson, 140–164. New Haven, CT: Yale University Press.

Quaranta, A., M. Siniscalchi, and G. Vallortigara. 2007. "Asymmetric Tail-Wagging Responses by Dogs to Different Emotive Stimuli." *Current Biology* 17 (6): R199–R201.

Quattrocki, E., and Karl Friston. 2014. "Autism, Oxytocin and Interoception." *Neuroscience and Biobehavioral Reviews* 47: 410–430.

Quoidbach, Jordi, June Gruber, Moïra Mikolajczak, Alexsandr Kogan, Ilios Kotsou, and Michael I. Norton. 2014. "Emodiversity and the Emotional Ecosystem." *Journal of Experimental Psychology: General* 143 (6): 2057–2066.

Raichle, M. E. 2010. "Two Views of Brain Function." *Trends in Cognitive Science* 14 (4): 180–190.

Ramon y Cajal, Santiago. 1909–1911. *Histology of the Nervous System of Man and Vertebrates.* Translated by Neeley Swanson and Larry W. Swanson. New York: Oxford University Press.

Ranganathan, Rajiv, and Les G. Carlton. 2007. "Perception-Action Coupling and Anticipatory Performance in Baseball Batting." *Journal of Motor Behavior* 39 (5): 369–380.

Range, Friederike, Ulrike Aust, Michael Steurer, and Ludwig Huber. 2008. "Visual Categorization of Natural Stimuli by Domestic Dogs." *Animal Cognition* 11 (2): 339–347.

Raz, G., T. Touroutoglou, C. Wilson-Mendenhall, G. Gilam, T. Lin, T. Gonen, Y. Jacob, S. Atzil, R. Admon, M. Bleich-Cohen, A. Maron-Katz, T. Hendler, and L. F. Barrett. 2016. "Functional Connectivity Dynamics During Film Viewing Reveal Common Networks for Different Emotional Experiences." *Cognitive, Affective, and Behavioral Neuroscience* 16 (4): 709–723.

Redelmeier, Donald A., and Simon D. Baxter. 2009. "Rainy Weather and Medical School Admission Interviews." *Canadian Medical Association Journal* 181 (12): 933.

Repacholi, Betty M., and Alison Gopnik. 1997. "Early Reasoning About Desires: Evidence from 14- and 18-Month-Olds." *Developmental Psychology* 33 (1): 12–21.

Repetti, Rena L., Shelley E. Taylor, and Teresa E. Seeman. 2002. "Risky Families: Family Social Environments and the Mental and Physical Health of Offspring." *Psychological Bulletin* 128 (2): 330–366.

Reynolds, Gretchen. 2015. "How Walking in Nature Changes the Brain." *New York Times,* July 22. http://well.blogs.nytimes.com/2015/07/22/how-nature-changes-the-brain/.

Reynolds, S. M., and K. C. Berridge. 2008. "Emotional Environments Retune the Valence of Appetitive Versus Fearful Functions in Nucleus Accumbens." *Nature Neuroscience* 11 (4): 423–425.

Richerson, Peter J., and Robert Boyd. 2008. *Not by Genes Alone: How Culture Transformed Human Evolution.* Chicago: University of Chicago Press.

Rieke, Fred. 1999. *Spikes: Exploring the Neural Code.* Cambridge, MA: MIT Press.

Rigotti, Mattia, Omri Barak, Melissa R. Warden, Xiao-Jing Wang, Nathaniel D. Daw, Earl K. Miller, and Stefano Fusi. 2013. "The Importance of Mixed Selectivity in Complex Cognitive Tasks." *Nature* 497 (7451): 585–590.

Rimmele, Ulrike, Lila Davachi, Radoslav Petrov, Sonya Dougal, and Elizabeth A. Phelps. 2011. "Emotion Enhances the Subjective Feeling of Remembering, Despite Lower Accuracy for Contextual Details." *Emotion* 11 (3): 553–562.

Riva-Posse, Patricio, Ki Sueng Choi, Paul E. Holtzheimer, Cameron C. McIntyre, Robert E. Gross, Ashutosh Chaturvedi, Andrea L. Crowell, Steven J. Garlow, Justin K. Rajendra, and Helen S. Mayberg. 2014. "Defining Critical White Matter Pathways Mediating Successful Subcallosal Cingulate Deep Brain Stimulation for Treatment-Resistant Depression." *Biological Psychiatry* 76 (12): 963–969.

Roberson, Debi, Jules Davidoff, Ian R. L. Davies, and Laura R. Shapiro. 2005. "Color Categories: Evidence for the Cultural Relativity Hypothesis." *Cognitive Psychology* 50 (4): 378–411.

Rosch, Eleanor. 1978. "Principles of Categorization." In *Cognition and Categorization*, edited by Eleanor Rosch and Barbara B. Lloyd, 2–48. Hillsdale, NJ: Erlbaum.

Roseman, I. J. 1991. "Appraisal Determinants of Discrete Emotions." *Cognition and Emotion* 5 (3): 161–200.

———. 2011. "Emotional Behaviors, Emotivational Goals, Emotion Strategies: Multiple Levels of Organization Integrate Variable and Consistent Responses." *Emotion Review* 3: 1–10.

Rossi, Alexandre Pongrácz, and César Ades. 2008. "A Dog at the Keyboard: Using Arbitrary Signs to Communicate Requests." *Animal Cognition* 11 (2): 329–338.

Rottenberg, Jonathan. 2014. *The Depths: The Evolutionary Origins of the Depression Epidemic*. New York: Basic Books.

Rowe, Meredith L., and Susan Goldin-Meadow. 2009. "Differences in Early Gesture Explain SES Disparities in Child Vocabulary Size at School Entry." *Science* 323 (5916): 951–953.

Roy, M., D. Shohamy, N. Daw, M. Jepma, G. E. Wimmer, and T. D. Wager. 2014. "Representation of Aversive Prediction Errors in the Human Periaqueductal Gray." *Nature Neuroscience* 17 (11): 1607–1612.

Roy, Mathieu, Mathieu Piché, Jen-I Chen, Isabelle Peretz, and Pierre Rainville. 2009. "Cerebral and Spinal Modulation of Pain by Emotions." *Proceedings of the National Academy of Sciences* 106 (49): 20900–20905.

Russell, J. A. 1991a. "Culture and the Categorization of Emotions." *Psychological Bulletin* 110 (3): 426–450.

———. 1991b. "In Defense of a Prototype Approach to Emotion Concepts." *Journal of Personality and Social Psychology* 60 (1): 37–47.

———. 1994. "Is There Universal Recognition of Emotion from Facial Expressions? A Review of the Cross-Cultural Studies." *Psychological Bulletin* 115 (1): 102–141.

———. 2003. "Core Affect and the Psychological Construction of Emotion." *Psychological Review* 110 (1): 145–172.

Russell, J. A., and L. F. Barrett. 1999. "Core Affect, Prototypical Emotional Episodes, and Other Things Called Emotion: Dissecting the Elephant." *Journal of Personality and Social Psychology* 76 (5): 805–819.

Rychlowska, Magdalena, Yuri Miyamoto, David Matsumoto, Ursula Hess, Eva Gilboa-Schechtman, Shanmukh Kamble, Hamdi Muluk, Takahiko Masuda, and Paula Marie Niedenthal. 2015. "Heterogeneity of Long-History Migration Explains Cultural Differences in Reports of Emotional Expressivity and the Functions of Smiles." *Proceedings of the National Academy of Sciences* 112 (19): E2429–E2436.

Sabra, Abdelhamid I. 1989. *The Optics of Ibn al-Haytham, Books I–III: On Direct Vision*. Vol. 1. London: Warburg Institute, University of London.

Safina, Carl. 2015. *Beyond Words: What Animals Think and Feel*. New York: Macmillan.

Salerno, Jessica M., and Bette L. Bottoms. 2009. "Emotional Evidence and Jurors' Judgments: The Promise of Neuroscience for Informing Psychology and Law." *Behavioral Sciences and the Law* 27 (2): 273–296.

Salminen, Jouko K., Simo Saarijärvi, Erkki Äärelä, Tuula Toikka, and Jussi Kauhanen. 1999. "Prevalence of Alexithymia and Its Association with Sociodemographic Variables in the General Population of Finland." *Journal of Psychosomatic Research* 46 (1): 75–82.

Salter, Michael W., and Simon Beggs. 2014. "Sublime Microglia: Expanding Roles for the Guardians of the CNS." *Cell* 158 (1): 15–24.

Sanchez, Raf, and Peter Foster. 2015. "'You Rape Our Women and Are Taking over Our Country,' Charleston Church Gunman Told Black Victims." *Telegraph,* June 18. http://www.telegraph.co.uk/news/worldnews/northamerica/usa/11684957/You-rape-our-women-and-are-taking-over-our-country-Charleston-church-gunman-told-black-victims.html.

Sauter, Disa A., Frank Eisner, Paul Ekman, and Sophie K. Scott. 2010. "Cross-Cultural Recognition of Basic Emotions Through Nonverbal Emotional Vocalizations." *Proceedings of the National Academy of Sciences* 107 (6): 2408–2412.

———. 2015. "Emotional Vocalizations Are Recognized Across Cultures Regardless of the Valence of Distractors." *Psychological Science* 26 (3): 354–356.

Sbarra, David A., and Cindy Hazan. 2008. "Coregulation, Dysregulation, Self-Regulation: An Integrative Analysis and Empirical Agenda for Understanding Adult Attachment, Separation, Loss, and Recovery." *Personality and Social Psychology Review* 12 (2): 141–167.

Scalia, Antonin, and Bryan A. Garner. 2008. *Making Your Case: The Art of Persuading Judges.* St. Paul, MN: Thomson/West.

Schacter, D. L., D. R. Addis, D. Hassabis, V. C. Martin, R. N. Spreng, and K. K. Szpunar. 2012. "The Future of Memory: Remembering, Imagining, and the Brain." *Neuron* 76 (4): 677–694.

Schacter, Daniel L. 1996. *Searching for Memory: The Brain, the Mind, and the Past.* New York: Basic Books.

Schacter, Daniel L., and Elizabeth F. Loftus. 2013. "Memory and Law: What Can Cognitive Neuroscience Contribute?" *Nature Neuroscience* 16 (2): 119–123.

Schachter, Stanley, and Jerome Singer. 1962. "Cognitive, Social, and Physiological Determinants of Emotional State." *Psychological Review* 69 (5): 379–399.

Schatz, Howard, and Beverly J. Ornstein. 2006. *In Character: Actors Acting.* Boston: Bulfinch Press.

Schilling, Elizabeth A., Robert H. Aseltine, and Susan Gore. 2008. "The Impact of Cumulative Childhood Adversity on Young Adult Mental Health: Measures, Models, and Interpretations." *Social Science and Medicine* 66 (5): 1140–1151.

Schnall, Simone, Kent D. Harber, Jeanine K. Stefanucci, and Dennis R. Proffitt. 2008. "Social Support and the Perception of Geographical Slant." *Journal of Experimental Social Psychology* 44 (5): 1246–1255.

Scholz, Joachim, and Clifford J. Woolf. 2007. "The Neuropathic Pain Triad: Neurons, Immune Cells and Glia." *Nature Neuroscience* 10 (11): 1361–1368.

Schumann, Karina, Jamil Zaki, and Carol S. Dweck. 2014. "Addressing the Empathy Deficit: Beliefs About the Malleability of Empathy Predict Effortful Responses When Empathy Is Challenging." *Journal of Personality and Social Psychology* 107 (3): 475–493.

Schuster, Mary Lay, and Amy Propen. 2010. "Degrees of Emotion: Judicial Responses to Victim Impact Statements." *Law, Culture and the Humanities* 6 (1): 75–104.

Schwarz, Norbert, and Gerald L. Clore. 1983. "Mood, Misattribution, and Judgments of Well-Being: Informative and Directive Functions of Affective States." *Journal of Personality and Social Psychology* 45 (3): 513–523.

Schyns, P. G., R. L. Goldstone, and J. P. Thibaut. 1998. "The Development of Features in Object Concepts." *Behavioral and Brain Sciences* 21 (1): 1–17, 17–54.

Searle, John R. 1995. *The Construction of Social Reality*. New York: Simon and Schuster.

Selby, Edward A., Stephen A. Wonderlich., Ross D. Crosby, Scott G. Engel, Emily Panza, James E. Mitchell, Scott J. Crow, Carol B. Peterson, and Daniel Le Grange. 2013. "Nothing Tastes as Good as Thin Feels: Low Positive Emotion Differentiation and Weight-Loss Activities in Anorexia Nervosa." *Clinical Psychological Science* 2 (4): 514–531.

Seminowicz, D. A., H. S. Mayberg, A. R. McIntosh, K. Goldapple, S. Kennedy, Z. Segal, and S. Rafi-Tari. 2004. "Limbic-Frontal Circuitry in Major Depression: A Path Modeling Metanalysis." *Neuroimage* 22 (1): 409–418.

Seo, M.-G., B. Goldfarb, and L. F. Barrett. 2010. "Affect and the Framing Effect Within Individuals Across Time: Risk Taking in a Dynamic Investment Game." *Academy of Management Journal* 53: 411–431.

Seruga, Bostjan, Haibo Zhang, Lori J. Bernstein, and Ian F. Tannock. 2008. "Cytokines and Their Relationship to the Symptoms and Outcome of Cancer." *Nature Reviews Cancer* 8 (11): 887–899.

Settle, Ray H., Barbara A. Sommerville, James McCormick, and Donald M. Broom. 1994. "Human Scent Matching Using Specially Trained Dogs." *Animal Behaviour* 48 (6): 1443–1448.

Shadmehr, Reza, Maurice A. Smith, and John W. Krakauer. 2010. "Error Correction, Sensory Prediction, and Adaptation in Motor Control." *Annual Review of Neuroscience* 33: 89–108.

Sharrock, Justine. 2013. "How Facebook, A Pixar Artist, and Charles Darwin Are Reinventing the Emoticon." *Buzzfeed*, February 8. http://www.buzzfeed.com/justineshar rock/how-facebook-a-pixar-artist-and-charles-darwin-are-reinventi?utm_term= .iglrx82Ky#.hxRb0da4w.

Shenhav, Amitai, Matthew M. Botvinick, and Jonathan D. Cohen. 2013. "The Expected Value of Control: An Integrative Theory of Anterior Cingulate Cortex Function." *Neuron* 79 (2): 217–240.

Shepard, Roger N., and Lynn A. Cooper. 1992. "Representation of Colors in the Blind, Color-Blind, and Normally Sighted." *Psychological Science* 3 (2): 97–104.

Sheridan, Margaret A., and Katie A. McLaughlin. 2014. "Dimensions of Early Experience and Neural Development: Deprivation and Threat." *Trends in Cognitive Sciences* 18 (11): 580–585.

Siegel, E. H., M. K. Sands, P. Condon, Y. Chang, J. Dy, K. S. Quigley, and L. F. Barrett. Under review. "Emotion Fingerprints or Emotion Populations? A Meta-Analytic Investigation of Autonomic Features of Emotion Categories."

Silva, B. A., C. Mattucci, P. Krzywkowski, E. Murana, A. Illarionova, V. Grinevich, N. S. Canteras, D. Ragozzino, and C. T. Gross. 2013. "Independent Hypothalamic Circuits for Social and Predator Fear." *Nature Neuroscience* 16 (12): 1731–1733.

Simon, Herbert A. 1962. "The Architecture of Complexity." *Proceedings of the American Philosophical Society* 106 (6): 467–482.

Simon, Jonathan. 2007. *Governing Through Crime: How the War on Crime Transformed American Democracy and Created a Culture of Fear*. New York: Oxford University Press.

Sinha, Pawan, Margaret M. Kjelgaard, Tapan K. Gandhi, Kleovoulos Tsourides, Annie L. Cardinaux, Dimitrios Pantazis, Sidney P. Diamond, and Richard M. Held. 2014. "Autism as a Disorder of Prediction." *Proceedings of the National Academy of Sciences* 111 (42): 15220–15225.

Siniscalchi, Marcello, Rita Lusito, Giorgio Vallortigara, and Angelo Quaranta. 2013. "Seeing Left- or Right-Asymmetric Tail Wagging Produces Different Emotional Responses in Dogs." *Current Biology* 23 (22): 2279–2282.

Skerry, Amy E., and Rebecca Saxe. 2015. "Neural Representations of Emotion Are Organized Around Abstract Event Features." *Current Biology* 25 (15): 1945–1954.

Slavich, George M., and Steven W. Cole. 2013. "The Emerging Field of Human Social Genomics." *Clinical Psychological Science* 1 (3): 331–348.

Slavich, George M., and Michael R. Irwin. 2014. "From Stress to Inflammation and Major Depressive Disorder: A Social Signal Transduction Theory of Depression." *Psychological Bulletin* 140 (3): 774.

Sloan, Erica K., John P. Capitanio, Ross P. Tarara, Sally P. Mendoza, William A. Mason, and Steve W. Cole. 2007. "Social Stress Enhances Sympathetic Innervation of Primate Lymph Nodes: Mechanisms and Implications for Viral Pathogenesis." *Journal of Neuroscience* 27 (33): 8857–8865.

Sloutsky, Vladimir M., and Anna V. Fisher. 2012. "Linguistic Labels: Conceptual Markers or Object Features?" *Journal of Experimental Child Psychology* 111 (1): 65–86.

Smith, Dylan M., George Loewenstein, Aleksandra Jankovic, and Peter A. Ubel. 2009. "Happily Hopeless: Adaptation to a Permanent, but Not to a Temporary, Disability." *Health Psychology* 28 (6): 787–791.

Smith, Edward E., and Douglas L. Medin. 1981. *Categories and Concepts.* Cambridge, MA: Harvard University Press.

So Bad So Good. 2012. "25 Handy Words that Simply Don't Exist in English." April 29. http://sobadsogood.com/2012/04/29/25-words-that-simply-dont-exist-in-english/.

Somerville, Leah H., and Paul J. Whalen. 2006. "Prior Experience as a Stimulus Category Confound: An Example Using Facial Expressions of Emotion." *Social Cognitive and Affective Neuroscience* 1 (3): 271–274.

Soni, Mira, Valerie H. Curran, and Sunjeev K. Kamboj. 2013. "Identification of a Narrow Post-Ovulatory Window of Vulnerability to Distressing Involuntary Memories in Healthy Women." *Neurobiology of Learning and Memory* 104: 32–38.

Soskin, David P., Clair Cassiello, Oren Isacoff, and Maurizio Fava. 2012. "The Inflammatory Hypothesis of Depression." *Focus* 10 (4): 413–421.

Sousa, Cláudia, and Tetsuro Matsuzawa. 2006. "Token Use by Chimpanzees (Pan Troglodytes): Choice, Metatool, and Cost." In *Cognitive Development in Chimpanzees,* edited by T. Matsuzawa, M. Tomanaga, and M. Tanaka, 411–438. Tokyo: Springer.

Southgate, Victoria, and Gergely Csibra. 2009. "Inferring the Outcome of an Ongoing Novel Action at 13 Months." *Developmental Psychology* 45 (6): 1794–1798.

Spiegel, Alix. 2012. "What Vietnam Taught Us About Breaking Bad Habits." *National Public Radio,* January 2. http://www.npr.org/sections/health-shots/2012/01/02/144431794/what-vietnam-taught-us-about-breaking-bad-habits.

Sporns, Olaf. 2011. *Networks of the Brain.* Cambridge, MA: MIT Press.

Spunt, R. P., E. B. Falk, and M. D. Lieberman. 2010. "Dissociable Neural Systems Support Retrieval of How and Why Action Knowledge." *Psychological Science* 21 (11): 1593–1598.

Spunt, R. P., and M. D. Lieberman. 2012. "An Integrative Model of the Neural Systems

Supporting the Comprehension of Observed Emotional Behavior." *Neuroimage* 59 (3): 3050–3059.

Spyridaki, Eirini C., Panagiotis Simos, Pavlina D. Avgoustinaki, Eirini Dermitzaki, Maria Venihaki, Achilles N. Bardos, and Andrew N. Margioris. 2014. "The Association Between Obesity and Fluid Intelligence Impairment Is Mediated by Chronic Low-Grade Inflammation." *British Journal of Nutrition* 112 (10): 1724–1734.

Srinivasan, Ramprakash, Julie D. Golomb, and Aleix M. Martinez. In press. "A Neural Basis of Facial Action Recognition in Humans." *Journal of Neuroscience.*

Stanton, Annette L., Sharon Danoff-Burg, Christine L. Cameron, Michelle Bishop, Charlotte A. Collins, Sarah B. Kirk, Lisa A. Sworowski, and Robert Twillman. 2000. "Emotionally Expressive Coping Predicts Psychological and Physical Adjustment to Breast Cancer." *Journal of Consulting and Clinical Psychology* 68 (5): 875.

Stanton, Annette L., Sharon Danoff-Burg, and Melissa E. Huggins. 2002. "The First Year After Breast Cancer Diagnosis: Hope and Coping Strategies as Predictors of Adjustment." *Psycho-Oncology* 11 (2): 93–102.

Steiner, Adam P., and A. David Redish. 2014. "Behavioral and Neurophysiological Correlates of Regret in Rat Decision-Making on a Neuroeconomic Task." *Nature Neuroscience* 17 (7): 995–1002.

Stellar, Jennifer E., Neha John-Henderson, Craig L. Anderson, Amie M. Gordon, Galen D. McNeil, and Dacher Keltner. 2015. "Positive Affect and Markers of Inflammation: Discrete Positive Emotions Predict Lower Levels of Inflammatory Cytokines." *Emotion* 15 (2): 129–133.

Stephens, C. L., I. C. Christie, and B. H. Friedman. 2010. "Autonomic Specificity of Basic Emotions: Evidence from Pattern Classification and Cluster Analysis." *Biological Psychology* 84 (3): 463–473.

Sterling, Peter. 2012. "Allostasis: A Model of Predictive Regulation." *Physiology and Behavior* 106 (1): 5–15.

Sterling, Peter, and Simon Laughlin. 2015. *Principles of Neural Design.* Cambridge, MA: MIT Press.

Stevenson, Seth. 2015. "Tsarnaev's Smirk." *Slate.com,* April 21. http://www.slate.com/ar ticles/news_and_politics/dispatches/2015/04/tsarnaev_trial_sentencing_phase_pros ecutor_makes_case_that_dzhokhar_tsarnaev.html.

Stolk, Arjen, Lennart Verhagen, and Ivan Toni. 2016. "Conceptual Alignment: How Brains Achieve Mutual Understanding." *Trends in Cognitive Sciences* 20 (3): 180–191.

Striedter, Georg F. 2006. "Précis of Principles of Brain Evolution." *Behavioral and Brain Sciences* 29 (1): 1–12.

Styron, William. 2010. *Darkness Visible: A Memoir of Madness.* New York: Open Road Media.

Sullivan, Michael J. L., Mary E. Lynch, and A. J. Clark. 2005. "Dimensions of Catastrophic Thinking Associated with Pain Experience and Disability in Patients with Neuropathic Pain Conditions." *Pain* 113 (3): 310–315.

Susskind, Joshua M., Daniel H. Lee, Andrée Cusi, Roman Feiman, Wojtek Grabski, and Adam K. Anderson. 2008. "Expressing Fear Enhances Sensory Acquisition." *Nature Neuroscience* 11 (7): 843–850.

Suvak, M. K., and L. F. Barrett. 2011. "Considering PTSD from the Perspective of Brain Processes: A Psychological Construction Analysis." *Journal of Traumatic Stress* 24: 3–24.

Suvak, M. K., B. T. Litz, D. M. Sloan, M. C. Zanarini, L. F. Barrett, and S. G. Hofmann. 2011.

"Emotional Granularity and Borderline Personality Disorder." *Journal of Abnormal Psychology* 120 (2): 414–426.

Swanson, Larry W. 2012. *Brain Architecture: Understanding the Basic Plan.* New York: Oxford University Press.

Tabibnia, Golnaz, Matthew D. Lieberman, and Michelle G. Craske. 2008. "The Lasting Effect of Words on Feelings: Words May Facilitate Exposure Effects to Threatening Images." *Emotion* 8 (3): 307–317.

Tagkopoulos, Ilias, Yir-Chung Liu, and Saeed Tavazoie. 2008. "Predictive Behavior Within Microbial Genetic Networks." *Science* 320 (5881): 1313–1317.

Tamir, Maya. 2009. "What Do People Want to Feel and Why? Pleasure and Utility in Emotion Regulation." *Current Directions in Psychological Science* 18 (2): 101–105.

Tanaka, Masayuki. 2011. "Spontaneous Categorization of Natural Objects in Chimpanzees." In *Cognitive Development in Chimpanzees,* edited by T. Matsuzawa, M. Tomanaga, and M. Tanaka, 340–367. Tokyo: Springer.

Tang, Yi-Yuan, Britta K. Hölzel, and Michael I. Posner. 2015. "The Neuroscience of Mindfulness Meditation." *Nature Reviews Neuroscience* 16 (4): 213–225.

Tassinary, Louis G., and John T. Cacioppo. 1992. "Unobservable Facial Actions and Emotion." *Psychological Science* 3 (1): 28–33.

Tassinary, Louis G., John T. Cacioppo, and Eric J. Vanman. 2007. "The Skeletomotor System: Surface Electromyography." In *Handbook of Psychophysiology,* 3rd edition, edited by John T. Cacioppo and Louis G. Tassinary, 267–300. New York: Cambridge University Press.

Taumoepeau, Mele, and Ted Ruffman. 2006. "Mother and Infant Talk About Mental States Relates to Desire Language and Emotion Understanding." *Child Development* 77 (2): 465–481.

———. 2008. "Stepping Stones to Others' Minds: Maternal Talk Relates to Child Mental State Language and Emotion Understanding at 15, 24, and 33 Months." *Child Development* 79 (2): 284–302.

TedMed. 2015. "Great Challenges." http://www.tedmed.com/greatchallenges.

Teicher, Martin H., Susan L. Andersen, Ann Polcari, Carl M. Anderson, and Carryl P. Navalta. 2002. "Developmental Neurobiology of Childhood Stress and Trauma." *Psychiatric Clinics* 25 (2): 397–426.

Teicher, Martin H., Susan L. Andersen, Ann Polcari, Carl M. Anderson, Carryl P. Navalta, and Dennis M. Kim. 2003. "The Neurobiological Consequences of Early Stress and Childhood Maltreatment." *Neuroscience and Biobehavioral Reviews* 27 (1): 33–44.

Teicher, Martin H., and Jacqueline A. Samson. 2016. "Annual Research Review: Enduring Neurobiological Effects of Childhood Abuse and Neglect." *Journal of Child Psychology and Psychiatry* 57 (3): 241–266.

Teicher, Martin H., Jacqueline A. Samson, Ann Polcari, and Cynthia E. McGreenery. 2006. "Sticks, Stones, and Hurtful Words: Relative Effects of Various Forms of Childhood Maltreatment." *American Journal of Psychiatry* 163: 993–1000.

Tejero-Fernández, Victor, Miguel Membrilla-Mesa, Noelia Galiano-Castillo, and Manuel Arroyo-Morales. 2015. "Immunological Effects of Massage After Exercise: A Systematic Review." *Physical Therapy in Sport* 16 (2): 187–192.

Tenenbaum, Joshua B., Charles Kemp, Thomas L. Griffiths, and Noah D. Goodman. 2011. "How to Grow a Mind: Statistics, Structure, and Abstraction." *Science* 331 (6022): 1279–1285.

Tiedens, Larissa Z. 2001. "Anger and Advancement Versus Sadness and Subjugation: The

Effect of Negative Emotion Expressions on Social Status Conferral." *Journal of Personality and Social Psychology* 80 (1): 86–94.

Tomasello, Michael. 2014. *A Natural History of Human Thinking*. Cambridge, MA: Harvard University Press.

Tomkins, Silvan S., and Robert McCarter. 1964. "What and Where Are the Primary Affects? Some Evidence for a Theory." *Perceptual and Motor Skills* 18 (1): 119–158.

Tononi, Giulio, and Gerald M. Edelman. 1998. "Consciousness and Complexity." *Science* 282 (5395): 1846–1851.

Touroutoglou, A., E. Bliss-Moreau, J. Zhang, D. Mantini, W. Vanduffel, B. Dickerson, and L. F. Barrett. 2016. "A Ventral Salience Network in the Macaque Brain." *Neuroimage* 132: 190–197.

Touroutoglou, A., K. A. Lindquist, B. C. Dickerson, and L. F. Barrett. 2015. "Intrinsic Connectivity in the Human Brain Does Not Reveal Networks for 'Basic' Emotions." *Social Cognitive and Affective Neuroscience* 10 (9): 1257–1265.

Tovote, Philip, Jonathan Paul Fadok, and Andreas Lüthi. 2015. "Neuronal Circuits for Fear and Anxiety." *Nature Reviews Neuroscience* 16 (6): 317–331.

Tracey, Irene. 2010. "Getting the Pain You Expect: Mechanisms of Placebo, Nocebo and Reappraisal Effects in Humans." *Nature Medicine* 16 (11): 1277–1283.

Tracy, Jessica L., and Daniel Randles. 2011. "Four Models of Basic Emotions: A Review of Ekman and Cordaro, Izard, Levenson, and Panksepp and Watt." *Emotion Review* 3 (4): 397–405.

Tranel, Daniel, Greg Gullickson, Margaret Koch, and Ralph Adolphs. 2006. "Altered Experience of Emotion Following Bilateral Amygdala Damage." *Cognitive Neuropsychiatry* 11 (3): 219–232.

Traub, Richard J., Dong-Yuan Cao, Jane Karpowicz, Sangeeta Pandya, Yaping Ji, Susan G. Dorsey, and Dean Dessem. 2014. "A Clinically Relevant Animal Model of Temporomandibular Disorder and Irritable Bowel Syndrome Comorbidity." *Journal of Pain* 15 (9): 956–966.

Triandis, Harry Charalambos. 1994. *Culture and Social Behavior*. New York: McGraw-Hill.

Trivedi, Bijal P. 2004. "'Hot Tub Monkeys' Offer Eye on Nonhuman 'Culture.'" *National Geographic News*, February 6. http://news.nationalgeographic.com/news/2004/02/0206 _040206_tvmacaques.html.

Trumble, Angus. 2004. *A Brief History of the Smile*. New York: Basic Books.

Tsai, Jeanne L. 2007. "Ideal Affect: Cultural Causes and Behavioral Consequences." *Perspectives on Psychological Science* 2 (3): 242–259.

Tsuda, Makoto, Simon Beggs, Michael W. Salter, and Kazuhide Inoue. 2013. "Microglia and Intractable Chronic Pain." *Glia* 61 (1): 55–61.

Tucker, Mike, and Rob Ellis. 2001. "The Potentiation of Grasp Types During Visual Object Categorization." *Visual Cognition* 8 (6): 769–800.

———. 2004. "Action Priming by Briefly Presented Objects." *Acta psychologica* 116 (2): 185–203.

Turati, Chiara. 2004. "Why Faces Are Not Special to Newborns: An Alternative Account of the Face Preference." *Current Directions in Psychological Science* 13 (1): 5–8.

Turcsán, Borbála, Flóra Szánthó, Ádám Miklósi, and Enikő Kubinyi. 2015. "Fetching What the Owner Prefers? Dogs Recognize Disgust and Happiness in Human Behaviour." *Animal Cognition* 18 (1): 83–94.

Turkheimer, Eric, Erik Pettersson, and Erin E. Horn. 2014. "A Phenotypic Null Hypothesis for the Genetics of Personality." *Annual Review of Psychology* 65: 515–540.

U.S. Census Bureau. 2015. "Families and Living Arrangements." http://www.census.gov/hhes/families.

Vallacher, Robin R., and Daniel M. Wegner. 1987. "What Do People Think They're Doing? Action Identification and Human Behavior." *Psychological Review* 94 (1): 3–15.

Van de Cruys, Sander, Kris Evers, Ruth Van der Hallen, Lien Van Eylen, Bart Boets, Lee de-Wit, and Johan Wagemans. 2014. "Precise Minds in Uncertain Worlds: Predictive Coding in Autism." *Psychological Review* 121 (4): 649–675.

Van den Heuvel, Martijn P., and Olaf Sporns. 2011. "Rich-Club Organization of the Human Connectome." *Journal of Neuroscience* 31 (44): 15775–15786.

———. 2013. "An Anatomical Substrate for Integration Among Functional Networks in Human Cortex." *Journal of Neuroscience* 33 (36): 14489–14500.

Van der Laan, L. N., D. T. de Ridder, M. A. Viergever, and P. A. Smeets. 2011. "The First Taste Is Always with the Eyes: A Meta-Analysis on the Neural Correlates of Processing Visual Food Cues." *Neuroimage* 55 (1): 296–303.

Van Essen, David C., and Donna Dierker. 2007. "On Navigating the Human Cerebral Cortex: Response to 'In Praise of Tedious Anatomy'." *Neuroimage* 37 (4): 1050–1054.

Vauclair, Jacques, and Joël Fagot. 1996. "Categorization of Alphanumeric Characters by Guinea Baboons: Within—and Between—Class Stimulus." *Cahiers de psychologie cognitive* 15 (5): 449–462.

Vernon, Michael L., Shir Atzil, Paula Pietromonaco, and Lisa Feldman Barrett. 2016. "Love Is a Drug: Parallel Neural Mechanisms in Love and Drug Addiction." Unpublished manuscript, University of Massachusetts, Amherst.

Verosupertramp85. 2012. "Lost in Translation." January 13. http://verosupertram.wordpress.com/2012/01/13/lost-in-translation.

Voorspoels, Wouter, Wolf Vanpaemel, and Gert Storms. 2011. "A Formal Ideal-Based Account of Typicality." *Psychonomic Bulletin and Review* 18 (5): 1006–1014.

Vouloumanos, Athena, Kristine H. Onishi, and Amanda Pogue. 2012. "Twelve-Month-Old Infants Recognize That Speech Can Communicate Unobservable Intentions." *Proceedings of the National Academy of Sciences* 109 (32): 12933–12937.

Vouloumanos, Athena, and Sandra R. Waxman. 2014. "Listen Up! Speech Is for Thinking During Infancy." *Trends in Cognitive Sciences* 18 (12): 642–646.

Wager, T. D., J. Kang, T. D. Johnson, T. E. Nichols, A. B. Satpute, and L. F. Barrett. 2015. "A Bayesian Model of Category-Specific Emotional Brain Responses." *PLOS Computational Biology* 11 (4): e1004066.

Wager, Tor D., and Lauren Y. Atlas. 2015. "The Neuroscience of Placebo Effects: Connecting Context, Learning and Health." *Nature Reviews Neuroscience* 16 (7): 403–418.

Wager, Tor D., Lauren Y. Atlas, Martin A. Lindquist, Mathieu Roy, Choong-Wan Woo, and Ethan Kross. 2013. "An fMRI-Based Neurologic Signature of Physical Pain." *New England Journal of Medicine* 368 (15): 1388–1397.

Walker, A. K., A. Kavelaars, C. J. Heijnen, and R. Dantzer. 2014. "Neuroinflammation and Comorbidity of Pain and Depression." *Pharmacological Reviews* 66 (1): 80–101.

Walker, Suellen M., Linda S. Franck, Maria Fitzgerald, Jonathan Myles, Janet Stocks, and Neil Marlow. 2009. "Long-Term Impact of Neonatal Intensive Care and Surgery on Somatosensory Perception in Children Born Extremely Preterm." *Pain* 141 (1): 79–87.

Walløe, Solveig, Bente Pakkenberg, and Katrine Fabricius. 2014. "Stereological Estimation of Total Cell Numbers in the Human Cerebral and Cerebellar Cortex." *Frontiers in Human Neuroscience* 8: 508.

Wang, Jing, Ronald J. Iannotti, and Tonja R. Nansel. 2009. "School Bullying Among Adolescents in the United States: Physical, Verbal, Relational, and Cyber." *Journal of Adolescent Health* 45 (4): 368–375.

Waters, Sara F., Tessa V. West, and Wendy Berry Mendes. 2014. "Stress Contagion Physiological Covariation Between Mothers and Infants." *Psychological Science* 25 (4): 934–942.

Waxman, Sandra R., and Susan A. Gelman. 2010. "Different Kinds of Concepts and Different Kinds of Words: What Words Do for Human Cognition." In *The Making of Human Concepts*, edited by Denis Mareschal, Paul C. Quinn, and Stephen E. G. Lea, 101–130. New York: Oxford University Press.

Waxman, Sandra R., and Dana B. Markow. 1995. "Words as Invitations to Form Categories: Evidence from 12- to 13-Month-Old Infants." *Cognitive Psychology* 29 (3): 257–302.

Wegner, Daniel M., and Kurt Gray. 2016. *The Mind Club: Who Thinks, What Feels, and Why It Matters.* New York: Viking.

Wei, Qiang, Hugh M. Fentress, Mary T. Hoversten, Limei Zhang, Elaine K. Hebda-Bauer, Stanley J. Watson, Audrey F. Seasholtz, and Huda Akil. 2012. "Early-Life Forebrain Glucocorticoid Receptor Overexpression Increases Anxiety Behavior and Cocaine Sensitization." *Biological Psychiatry* 71 (3): 224–231.

Weierich, M. R., C. I. Wright, A. Negreira, B. C. Dickerson, and L. F. Barrett. 2010. "Novelty as a Dimension in the Affective Brain." *Neuroimage* 49 (3): 2871–2878.

Weisleder, Adriana, and Anne Fernald. 2013. "Talking to Children Matters: Early Language Experience Strengthens Processing and Builds Vocabulary." *Psychological Science* 24 (11): 2143–2152.

Westermann, Gert, Denis Mareschal, Mark H. Johnson, Sylvain Sirois, Michael W. Spratling, and Michael S. C. Thomas. 2007. "Neuroconstructivism." *Developmental Science* 10 (1): 75–83.

Whitacre, James, and Axel Bender. 2010. "Degeneracy: A Design Principle for Achieving Robustness and Evolvability." *Journal of Theoretical Biology* 263 (1): 143–153.

Whitacre, James M., Philipp Rohlfshagen, Axel Bender, and Xin Yao. 2012. "Evolutionary Mechanics: New Engineering Principles for the Emergence of Flexibility in a Dynamic and Uncertain World." *Natural Computing* 11 (3): 431–448.

Widen, Sherri C. In press. "The Development of Children's Concepts of Emotion." In *Handbook of Emotions,* 4th edition, edited by Lisa Feldman Barrett, Michael Lewis, and Jeannette M. Haviland-Jones, 307–318. New York: Guilford Press.

Widen, Sherri C., Anita M. Christy, Kristen Hewett, and James A. Russell. 2011. "Do Proposed Facial Expressions of Contempt, Shame, Embarrassment, and Compassion Communicate the Predicted Emotion?" *Cognition and Emotion* 25 (5): 898–906.

Widen, Sherri C., and James A. Russell. 2013. "Children's Recognition of Disgust in Others." *Psychological Bulletin* 139 (2): 271–299.

Wiech, Katja, Chia-shu Lin, Kay H. Brodersen, Ulrike Bingel, Markus Ploner, and Irene Tracey. 2010. "Anterior Insula Integrates Information About Salience into Perceptual Decisions About Pain." *Journal of Neuroscience* 30 (48): 16324–16331.

Wiech, Katja, and Irene Tracey. 2009. "The Influence of Negative Emotions on Pain: Behavioral Effects and Neural Mechanisms." *Neuroimage* 47 (3): 987–994.

Wierzbicka, Anna. 1986. "Human Emotions: Universal or Culture-Specific?" *American Anthropologist* 88 (3): 584–594.

———. 1999. *Emotions Across Languages and Cultures: Diversity and Universals.* Cambridge: Cambridge University Press.

Wikan, Unni. 1990. *Managing Turbulent Hearts: A Balinese Formula for Living.* Chicago: University of Chicago Press.

Williams, David M., Shira Dunsiger, Ernestine G. Jennings, and Bess H. Marcus. 2012. "Does Affective Valence During and Immediately Following a 10-Min Walk Predict Concurrent and Future Physical Activity?" *Annals of Behavioral Medicine* 44 (1): 43–51.

Williams, J. Bradley, Diana Pang, Bertha Delgado, Masha Kocherginsky, Maria Tretiakova, Thomas Krausz, Deng Pan, Jane He, Martha K. McClintock, and Suzanne D. Conzen. 2009. "A Model of Gene-Environment Interaction Reveals Altered Mammary Gland Gene Expression and Increased Tumor Growth Following Social Isolation." *Cancer Prevention Research* 2 (10): 850–861.

Wilson, Craig J., Caleb E. Finch, and Harvey J. Cohen. 2002. "Cytokines and Cognition—The Case for a Head-to-Toe Inflammatory Paradigm." *Journal of the American Geriatrics Society* 50 (12): 2041–2056.

Wilson, Timothy D., Dieynaba G. Ndiaye, Cheryl Hahn, and Daniel T. Gilbert. 2013. "Still a Thrill: Meaning Making and the Pleasures of Uncertainty." In *The Psychology of Meaning,* edited by Keith D. Markman and Travis Proulx, 421–443. Washington, DC: American Psychological Association.

Wilson-Mendenhall, Christine D., Lisa Feldman Barrett, and Lawrence W. Barsalou. 2013. "Situating Emotional Experience." *Frontiers in Human Neuroscience* 7: 1–16.

———. 2015. "Variety in Emotional Life: Within-Category Typicality of Emotional Experiences Is Associated with Neural Activity in Large-Scale Brain Networks." *Social Cognitive and Affective Neuroscience* 10 (1): 62–71.

Wilson-Mendenhall, Christine D., Lisa Feldman Barrett, W. Kyle Simmons, and Lawrence W. Barsalou. 2011. "Grounding Emotion in Situated Conceptualization." *Neuropsychologia* 49: 1105–1127.

Winkielman, P., K. C. Berridge, and J. L. Wilbarger. 2005. "Unconscious Affective Reactions to Masked Happy Versus Angry Faces Influence Consumption Behavior and Judgments of Value." *Personality and Social Psychology Bulletin* 31 (1): 121–135.

Wistrich, Andrew J., Jeffrey J. Rachlinski, and Chris Guthrie. 2015. "Heart versus Head: Do Judges Follow the Law or Follow Their Feelings." *Texas Law Review* 93: 855–923.

Wittgenstein, Ludwig. 1953. *Philosophical Investigations.* London: Blackwell.

Wolpe, Noham, and James B. Rowe. 2015. "Beyond the 'Urge to Move': Objective Measures for the Study of Agency in the Post-Libet Era." In *Sense of Agency: Examining Awareness of the Acting Self,* edited by Nicole David, James W. Moore, and Sukhvinder Obhi, 213–235. Lausanne, Switzerland: Frontiers Media.

Woo, Choong-Wan, Mathieu Roy, Jason T. Buhle, and Tor D. Wager. 2015. "Distinct Brain Systems Mediate the Effects of Nociceptive Input and Self-Regulation on Pain." *PLOS Biology* 13 (1): e1002036. doi:10.1371/journal.pbio.1002036.

Wood, Wendy, and Dennis Rünger. 2016. "Psychology of Habit." *Annual Review of Psychology* 67: 289–314.

Wu, L. L., and L. W. Barsalou. 2009. "Perceptual Simulation in Conceptual Combination: Evidence from Property Generation." *Acta psychologica (amst)* 132 (2): 173–189.

Xu, Fei. 2002. "The Role of Language in Acquiring Object Kind Concepts in Infancy." *Cognition* 85 (3): 223–250.

Xu, Fei, Melissa Cote, and Allison Baker. 2005. "Labeling Guides Object Individuation in 12-Month-Old Infants." *Psychological Science* 16 (5): 372–377.

Xu, Fei, and Tamar Kushnir. 2013. "Infants Are Rational Constructivist Learners." *Current Directions in Psychological Science* 22 (1): 28–32.

Yang, Yang Claire, Courtney Boen, Karen Gerken, Ting Li, Kristen Schorpp, and Kathleen Mullan Harris. 2016. "Social Relationships and Physiological Determinants of Longevity Across the Human Life Span." *Proceedings of the National Academy of Sciences* 113 (3): 578–583.

Yeager, Mark P., Patricia A. Pioli, and Paul M. Guyre. 2011. "Cortisol Exerts Bi-Phasic Regulation of Inflammation in Humans." *Dose Response* 9 (3): 332–347.

Yeo, B. T., et al. 2011. "The Organization of the Human Cerebral Cortex Estimated by Intrinsic Functional Connectivity." *Journal of Neurophysiology* 106 (3): 1125–1165.

Yeo, B. T. Thomas, Fenna M. Krienen, Simon B. Eickhoff, Siti N. Yaakub, Peter T. Fox, Randy L. Buckner, Christopher L. Asplund, and Michael W. L. Chee. 2014. "Functional Specialization and Flexibility in Human Association Cortex." *Cerebral Cortex* 25 (10): 3654–3672.

Yeomans, Martin R., Lucy Chambers, Heston Blumenthal, and Anthony Blake. 2008. "The Role of Expectancy in Sensory and Hedonic Evaluation: The Case of Smoked Salmon Ice-Cream." *Food Quality and Preference* 19 (6): 565–573.

Yik, Michelle S. M., Zhaolan Meng, and James A. Russell. 1998. "Brief Report: Adults' Freely Produced Emotion Labels for Babies' Spontaneous Facial Expressions." *Cognition and Emotion* 12 (5): 723–730.

Yin, Jun, and Gergely Csibra. 2015. "Concept-Based Word Learning in Human Infants." *Psychological Science* 26 (8): 1316–1324.

Yoshikubo, Shin'ichi. 1985. "Species Discrimination and Concept Formation by Rhesus Monkeys (Macaca Mulatta)." *Primates* 26 (3): 285–299.

Younger, Jarred, Arthur Aron, Sara Parke, Neil Chatterjee, and Sean Mackey. 2010. "Viewing Pictures of a Romantic Partner Reduces Experimental Pain: Involvement of Neural Reward Systems." *PLOS One* 5 (10): e13309. doi:10.1093/cercor/bhv001.

Zachar, Peter. 2014. *A Metaphysics of Psychopathology.* Cambridge, MA: MIT Press.

Zachar, Peter, and Kenneth S. Kendler. 2007. "Psychiatric Disorders: A Conceptual Taxonomy." *American Journal of Psychiatry* 164: 557–565.

Zaki, J., N. Bolger, and K. Ochsner. 2008. "It Takes Two: The Interpersonal Nature of Empathic Accuracy." *Psychological Science* 19 (4): 399–404.

Zavadski, Katie. 2015. "Everything Known About Charleston Church Shooting Suspect Dylann Roof." *Daily Beast,* June 20. http://www.thedailybeast.com/articles/2015/06/18/everything-known-about-charleston-church-shooting-suspect-dylann-roof.html.

Zhang, F., H. Fung, T. Sims, and J. L. Tsai. 2013. "The Role of Future Time Perspective in Age Differences in Ideal Affect." 66th Annual Scientific Meeting of the Gerontological Society of America, New Orleans, November 20–24.

Zhuo, Min. 2016. "Neural Mechanisms Underlying Anxiety–Chronic Pain Interactions." *Trends in Neurosciences* 39 (3): 136–145.

Zilles, Karl, Hartmut Mohlberg, Katrin Amunts, Nicola Palomero-Gallagher, and Sebastian Bludau. 2015. "Cytoarchitecture and Maps of the Human Cerebral Cortex." In *Brain Mapping: An Encyclopedic Reference,* volume 2, edited by Arthur W. Toga, 115–136. Cambridge, MA: Academic Press.

Notes

This book has extended endnotes on the web at how-emotions-are-made.com, providing additional scientific details, commentary, and stories about the construction of emotion and related topics.

Many of the printed endnotes include a web link to heam.info (example: heam.info/malloy). These links are shortcuts to the appropriate pages at how-emotions-are-made.com.

Introduction:
The Two-Thousand-Year-Old Assumption

1. *"sacrificed their lives protecting students"*: See the video and transcript at heam.info/malloy.
2. *by chance or by custom*: Tracy and Randles 2011; Ekman and Cordaro 2011; Roseman 2011.
3. *newspaper articles that discuss emotion*: From a study by my lab; see heam.info/magazines. *emoticons inspired by Darwin's writings*: Sharrock 2013. See also heam.info/facebook-1.
4. *through "emotion analytics"*: See references at heam.info/analytics-1. *"team chemistry" from facial expressions*: ESPN 2014. See also heam.info/bucks. *training on the classical view*: Until recently, the FBI National Academy offered a training course based on Paul Ekman's research.
5. *a product of human agreement*: Searle 1995.
6. *cost taxpayers $900 million*: Government Accountability Office 2013. SPOT's reincarnation, called HIDE (Hostile Intent Detection and Evaluation), may be consistent with newer evidence; see heam.info/spot-1.
7. *men . . . with fatal consequences*: This differential treatment persists even when physicians are told that women are at high risk of a heart attack (Martin et al. 1998; Martin et al. 2004).
8. *and hundreds of coalition forces*: Triandis 1994, 29.

1. The Search for Emotion's "Fingerprints"

1. *they feel anxious:* Higgins 1987.
2. *I described as* emotional granularity: The discovery of emotional granularity inspired a new domain of emotion research; see heam.info/granularity-1.
3. *part of universal human nature:* This book has had tremendous influence in psychology; see heam.info/darwin-1.
4. *small muscles on each side:* Tassinary et al. 2007.
5. *sadness, and happiness:* Ekman et al. 1969; Izard 1971; Tomkins and McCarter 1964.
6. *that best matches the face:* E.g., Ekman et al. 1969; Izard 1971. *face best matches the story:* E.g., Ekman and Friesen 1971. This is called the "Dashiell" method, after its inventor, the psychologist John Dashiell (1927).
7. *(language) to posed faces:* Ekman and Friesen 1971; Ekman et al. 1987. *expected emotion words and stories:* Ekman et al. 1969; Ekman and Friesen 1971. For an overview of the research program with the Fore of New Guinea, see Russell 1994. *such as Japan and Korea:* Russell 1994; Elfenbein and Ambady 2002.
8. *diagnostic fingerprints of emotion:* "The strongest evidence for distinguishing one emotion from another comes from research on facial expressions. There is robust, consistent evidence of a universal facial expression for anger, fear, enjoyment, sadness and disgust" (Ekman 1992, 175–176).
9. *how much, and how often:* Tassinary and Cacioppo 1992. *each muscle during each emotion:* Calculations control for random movements, or movements during a nonemotional comparison period.
10. *pleasant versus unpleasant feeling:* Cacioppo et al. 2000.
11. *facial movements as they occur:* Ekman and Friesen 1984. FACS was adapted from a method first developed by Swedish anatomist Carl-Herman Hjortsjö in 1969; see heam.info/FACS. *consistently match the posed photos:* Matsumoto, Keltner, et al. 2008. There are hundreds of published studies on emotional expressions, but this research was able to report only twenty-five studies where spontaneous facial movements were measured. Only half of those using FACS coding found that these movements matched the expected configurations, whereas all of those using a more relaxed version of FACS found a match. All found evidence supporting the claim that people make spontaneous facial movements during emotion matching the expected facial expressions. See heam.info/FACS.
12. *learn rules of social appropriateness:* The classical view calls them "display rules" (Matsumoto, Yoo, et al. 2008). *the two situations was indistinguishable:* Camras et al. 2007. The FACS method in this study was specially designed for babies (Oster 2006). For more on infant emotions, see heam.info/infants-2. *seeing facial movements at all:* Babies show cultural differences as well; see heam.info/camras-1.
13. *to offending smells and tastes:* Their facial movements have also been linked to nonemotional factors such as gaze direction, head position, and respiration (Oster 2005). *from the basic emotion method:* See heam.info/newborns-1. Nor do infants have distinctive cries for each emotion; see heam.info/newborns-2.
14. *disgust rather than anger:* Aviezer et al. 2008.
15. *asked actors to portray them:* Silvan S. Tomkins and Robert McCarter (1964) created the photos by drawing on earlier photos taken by the French neurologist Guillaume-

Benjamin-Amand Duchenne, who was cited in Darwin ([1872] 2005); see also Widen and Russell 2013.

16. *emotion experts—accomplished actors:* This work was conducted by my former graduate student and now postdoctoral fellow Maria Gendron. *to match written scenarios:* Schatz and Ornstein 2006.

17. *her brow is slightly knitted:* Sadly, Ms. Leo's publicist declined my request to reproduce this instructive photograph.

18. *to improve your peripheral vision:* Susskind et al. 2008.

19. *instruments of social communication:* Fridlund 1991; Fernández-Dols and Ruiz-Belda 1995. *the same each time:* Barrett 2011b; Barrett et al. 2011. *has a diagnostic facial expression:* For evidence on whether non-human primates are similar to humans in their expressions, see heam.info/primates-1. For evidence on whether people blind since birth make facial expressions, see heam.info/blind-2.

20. *the journal* Science *in 1983:* Ekman et al. 1983. *in the autonomic nervous system:* The autonomic nervous system controls the body's internal organs, such as the heart, the lungs, etc. It is part of the peripheral nervous system (in contrast to the brain and spinal cord, which are considered the central nervous system). *(a measure of sweat):* Also known as an electrodermal response or a galvanic skin response; see heam.info/galvanic-1.

21. *to move particular facial muscles:* A second task was used as well; see heam.info/recall-1.

22. *can be evoked this way:* Facial muscles may move during emotion perception; see heam.info/faces-2.

23. *surprise, and disgust:* Some of these results were unsurprising and others a mystery; see heam.info/body-1.

24. *target emotions from these instructions:* Levenson et al. 1990, Study 4. *when these studies were conducted:* Barsalou et al. 2003. See heam.info/simulation-1. *the Minangkabau of West Sumatra:* Levenson et al. 1992. These experiments not only established reliability but also improved specificity, supporting the classical view. *than the Western subjects did:* It is not clear that the African subjects shared the same Westernized emotion concepts; see heam.info/sumatra-1.

25. *and other bodily functions:* See heam.info/body-4. *[no] bodily changes that distinguished emotions:* Distinctions were of affect only; see heam.info/body-2. *exactly the same film clips:* Kragel and LaBar 2013; Stephens et al. 2010.

26. *22,000 test subjects:* This work was conducted by my former graduate student Erika Siegel as her Ph.D. dissertation. Siegel et al., under review. *emotion fingerprints in the body:* For details on these meta-analyses, see heam.info/meta-analysis-1.

27. *No, they don't:* Some versions of the classical view are designed to explain this variation; e.g., classical appraisal theories (chapter 8) propose that a person has to evaluate the situation in a particular way to trigger anger. See heam.info/appraisal-1. *body for its own sake:* The sympathetic and parasympathetic nervous systems are together called the autonomic nervous system. They evolved to support your body's movement (e.g., so you don't faint when you stand up). It is well known that sympathetic activity is mobilized for the metabolic demands associated with actual movement behavior (cardio-somatic coupling; Obrist et al. 1970) or expected conditions (e.g., supra-metabolic activity; Obrist 1981). See also heam.info/threat-1. *an angry person's physiological response:* Kassam and Mendes 2013; Harmon-Jones and Peterson 2009.

28. category involves different bodily responses: Test subjects reported feeling a given emotion (e.g., sadness) at a time when the experimenter expected it but were measured as having a variety of bodily responses. *not uniformity, is the norm:* See heam. info/variation-1.

29. thinking, *which was proposed by Darwin:* Darwin (1859) 2003. *in abstract, statistical terms:* Mayr 2007. *able to identify one:* The average size of the American family in 2015 was 3.14 people (U.S. Census Bureau 2015).

30. *they'd avoided before the surgery:* Klüver and Bucy (1939) called this "psychic blindness"; see heam.info/kluver-1.

31. *no strong feelings of fear:* Adolphs and Tranel 2000; Tranel et al. 2006; Feinstein et al. 2011. *difficulty identifying them as fearful:* Adolphs et al. 1994.

32. *learn to fear new objects:* Bechara et al. 1995.

33. *and hear fear in voices:* Adolphs and Tranel 1999; Atkinson et al. 2007. SM also had difficulty seeing fear in scenes only when they contained faces; see Adolphs and Tranel 2003. SM's difficulties have other explanations not related to fear; see heam.info/SM -1. *even without her amygdalae:* SM could perceive fear in faces under some circumstances; see heam.info/SM-2.

34. *very different profiles regarding fear:* Becker et al. 2012. *compensating for her missing amygdalae:* Ibid. See also heam.info/twins-1.

35. *results have been similarly variable:* In general, studying emotion via brain lesions is problematic; see heam.info/lesions-1.

36. *can produce the same outcome:* Edelman and Gally 2001. Degeneracy even applies to an individual experience of emotion; see heam.info/degeneracy-1.

37. *anterior insula and early visual cortex:* Whenever scientists speak of an "increase" in brain activity, it always means an increase relative to some control. For brevity, I do not write "relative to some control" throughout the text. Also, a phrase like "increased brain activity" is a simplification. Scientifically speaking, brain imaging (specifically functional magnetic resonance imaging, or fMRI) measures changes in magnetic fields, which come from changes in blood flow, which are themselves linked to changes in neural activity. I will continue to speak of increases and decreases in "activity" as a convenient shorthand. See heam.info/fMRI. *linked to visual cortex:* Moriguchi et al. 2013. *an example of degeneracy:* More details about the study are at heam. info/degeneracy-2.

38. *serve more than one purpose:* Barrett and Satpute 2013. The philosopher Mike Anderson calls them multi-use, meaning multipurpose (Anderson 2014). *to many different mental states:* One to many also exists at the level of individual brain regions, e.g., Yeo et al. 2014.

39. *related to firing neurons:* fMRI is much like an MRI that you might receive at a doctor's office, with a few tweaks. See heam.info/fMRI.

40. *viewed faces with neutral expressions:* Breiter et al. 1996.

41. *triggering "fear" stimulus:* Fischer et al. 2003.

42. *not seen it before:* This effect was first observed by Dubois et al. (1999); see heam.info/ novelty. *them in brain-imaging experiments:* Somerville and Whalen 2006. *the brain locus of fear:* An early experiment on fear followed a similar trajectory; see heam.info/ amygdala-1.

43. *fingerprints once and for all:* This work was completed as a Ph.D. dissertation by Kristen A. Lindquist, a former graduate student in my lab (Lindquist et al. 2012).

44. fingerprint for any single emotion: For more details on our meta-analysis, see heam
    .info/meta-analysis-2. *(a brain network):* Touroutoglou et al. 2015. *stimulate individ-*
    ual neurons with electricity: Guillory and Bujarski 2014. *such as monkeys and rats:*
    Barrett, Lindquist, Bliss-Moreau, et al. 2007. See heam.info/stimulation-1.
45. *pretty tricky to pull off:* Levenson 2011.
46. *to the environment or context:* This variability is not infinite, of course, but constrained
    by the patterns that are possible in the body and available in one's culture. For ev-
    idence that emotions have no vocal signatures and hormone signatures, see heam
    .info/vocal-1. Two papers from my lab illustrate different patterns of brain activity
    within an emotion category: Wilson-Mendenhall et al. 2011, and Wilson-Mendenhall
    et al. 2015.
47. *mathematical average for the norm:* Clark-Polner, Johnson, et al., in press. Pattern
    classification is misapplied in the search for emotion fingerprints; see heam.info/pat-
    tern-1.
48. *brain-imaging studies of emotion:* Wager et al. 2015.

2. Emotions Are Constructed

1. *We will call it* simulation: Barsalou 1999; Barsalou 2008b. As is typical in science,
   different psychologists have called this mental feat by different names, depending on
   their research interests. Examples are "perceptual inference" and "perceptual completion" (Pessoa et al. 1998), "embodied cognition," and "grounded cognition."
2. *with a hint of sweetness:* Sensory neurons also fire during motion, and motor neurons
   during sensation; e.g., Press and Cook 2015; Graziano 2016. *using sensory and motor*
   neurons: Barsalou 1999.
3. *gagged from the simulated smell:* Simulation explains how ancient Greeks saw gods
   and monsters in the stars; see heam.info/simulation-2.
4. *not reactions to it:* For a review, see Chanes and Barrett 2016.
5. *what a "Bee" is:* Barsalou 2003, 2008a.
6. *because they're useful or desirable:* For a similar analogy, see Boghossian 2006.
7. *(assuming you enjoy salmon):* Yeomans et al. 2008.
8. *the defendant cannot be trusted:* Danziger et al. 2011.
9. *an emotion on the spot:* My experience in the coffee shop was typically Jamesian; see
   heam.info/coffee. *the* theory of constructed emotion: In my academic papers, I called
   it the "Conceptual Act Theory of Emotion." Thank goodness for editors.
10. *in a bunch of blobs:* Scientists call this "affective misattribution"; see heam.info/
    affect-9. *using the same manufacturing process:* Some cultures lack emotion concepts
    and instead experience physical illness; you'll learn this in chapter 7.
11. *and, of course, emotion:* For references on construction, see heam.info/construction-1.
12. *of changes in their environment:* Freddolino and Tavazoie 2012; Tagkopoulos et al.
    2008.
13. *and act in the world:* For the various incarnations of social construction, see Hacking
    1999. *depending on your social role:* Harré 1986.
14. *states like cognitions and emotions:* See more on these philosophers at heam.info/
    construction-2. *"such processes variously combined":* James 1884, 188. *on the context*
    surrounding them: Schachter and Singer 1962. The famous Schachter and Singer ex-
    periment is described at heam.info/arousal-1. *the mechanism and its product:* William
    James and Wilhelm Wundt, founding fathers of psychology, were skeptical of emotion

organs; see heam.info/james-wundt. *emotions and how they work*: For other examples of new psychological construction theories, see chapters in Barrett and Russell 2015; LeDoux 2014, 2015. *distinct from cognitions and perceptions*: The roots of construction stretch back further into mental philosophy; see heam.info/construction-3.

15. *treated you in a certain way*: The gross wiring of the brain comes from ancient Hox genes that are conserved in all vertebrate animals, even fish, but human activity influences the microwiring of a brain that incorporates experiences for later use (Donoghue and Purnell 2005). *your future experiences and perceptions*: Mareschal et al. 2007; Karmiloff-Smith 2009; Westermann et al. 2007.

16. *population of unique individuals*: James wrote, "There is no limit to the number of possible different emotions which may exist, and why the emotions of different individuals may vary indefinitely, both as to their constitution and as to objects which call them forth" (1894, 454).

17. *time spent chilling the dough*: See heam.info/chocolate-1 for some revealing examples.

18. *sugar, and salt*: Barrett 2009.

19. *participate at any given time*: Marder and Taylor 2011.

20. *in for a difficult time*: Imagine trying to reverse-engineer a croissant by tasting one; see heam.info/croissant. Problems with reverse engineering are a clue that you are dealing with emergence (Barrett 2011a), i.e., that a system has properties beyond the sum of its components. See also heam.info/emergence-1.

21. *and expert bakers know this*: This is called the "norm of reaction" in genetics; see heam.info/holism-1.

22. *greater robustness for survival*: Whitacre and Bender 2010; Whitacre et al. 2012. *computational power of the brain*: Rigotti et al. 2013; Balasubramanian 2015. *a flexible mind without fingerprints*: Degeneracy is a prerequisite for natural selection; see heam.info/degeneracy-3.

23. *time of day at which they are eaten*: Cupcakes and muffins are both snacks, however. And banana bread, which is a breakfast food and a dessert, is virtually identical to a banana muffin or cupcake except for the shape.

24. *your body metabolizes it differently*: Crum et al. 2011.

25. *perceptions exist within the perceiver*: By contrast, it is possible to measure how "accurately" a person detects a facial muscle movement because these movements can be measured electrically as you saw in chapter 1. See also Srinivasan et al., in press.

3. The Myth of Universal Emotions

1. *U.S. Open tennis finals*: For a similar example, see Barrett, Lindquist, and Gendron 2007. Also see Aviezer et al. 2012. For more details, see heam.info/aviezer-1.

2. *make meaning from the image*: A similar phenomenon occurs with the McGurk Effect, in which when someone speaks to you, what you see (mouth movements) influences what you hear (the sounds you perceive); see heam.info/mcgurk.

3. *fearful, and so on*: You even need knowledge of a person to recognize him in different photos; see heam.info/faces-4.

4. *world can recognize from birth*: E.g., Izard 1994.

5. *of the time on average*: In the basic emotion method, choosing the expected emotion word is called "accuracy," which is a misnomer; see heam.info/bem-1. *about 72 percent of the time*: Russell 1994, table 2. See heam.info/bem-2.

6. *the results were even lower*: See, e.g., Widen et al. 2011.

7.   *certain emotions and not others:* See heam.info/priming-1. This process is called prim-
     ing: It's like when someone says, "Try not to think of a white bear"; see heam.info/
     wegner-1. *"screaming in terror":* For a fascinating example of simulation, see Gosselin
     and Schyns 2003. *the posed faces they see:* This study was conducted by my former
     graduate student Maria Gendron for her master's thesis (Gendron et al. 2012). *use the
     basic emotion method:* You can experience this priming yourself by listening to music
     backward; see heam.info/stairway.

8.   *impair emotion perception even more:* This study was conducted by my former gradu-
     ate student Kristen Lindquist for her undergraduate honors thesis (Lindquist et al.
     2006).

9.   *lasts less than one second:* You can temporarily deactivate your own emotion concepts
     in the same way; see heam.info/satiate-1. *yes/no decisions were incorrect:* Test subjects
     literally see faces differently depending on which concepts are called to mind by the
     emotion words provided in an experiment; see heam.info/gendron-1.

10. *that was meaningful to them:* Lindquist et al. 2014. All test subjects sorted the faces
     by the feeling being depicted, and all were confident that the people in the same pile
     felt exactly the same way. Patients were also asked to sort the photos by actor to make
     sure they could understand and carry out our instructions. *sadness pile, and so on:* In
     other experiments, the patients produced random piles; see heam.info/dementia-1.
     pleasant versus unpleasant feeling: We studied three patients in this sample; see heam.
     info/dementia-2.

11. *who exhibit low emotional granularity:* Widen, in press; see heam.info/widen-1. *that
     infants picked up on:* Caron et al. 1985. This phenomenon is called "toothiness"; see
     heam.info/teeth-1.

12. *posed stereotypes are supposedly displaying:* Subjects do even worse when viewing real,
     spontaneous facial movements during emotional experiences, rather than the posed
     photos of the basic emotion method. Agreement is quite abysmal (Crivelli et al. 2015;
     Naab and Russell 2007; Yik et al. 1998).

13. *cognitive psychologist Debi Roberson:* Roberson et al. 2005. Roberson has shown that
     people do not perceive colors in a universal way; for more on whether color categories
     are universal, see heam.info/color-1. *in Opuwo, northern Namibia:* See heam.info/
     himba-1.

14. *didn't look like Himba tribespeople:* Lacking any Himba in Massachusetts, we had to
     construct this photo set carefully; see heam.info/himba-2. *mixtures of the remaining
     faces:* Gendron et al. 2014b. *inferring mental states or feelings:* Vallacher and Weg-
     ner 1987. *to give evidence of universality:* In an additional experiment, we provided
     emotion words to guide the sorting task. The resulting piles looked a bit more like
     the results we would get with the basic emotion method but not dramatically so. See
     Gendron et al. 2014b.

15. *of photos of posed faces:* Sauter et al. 2010. Sauter's procedure is described at heam.
     info/sauter-1. *that emotion perception was universal:* Several others have replicated
     Sauter et al.'s findings (Laukka et al. 2013; Cordaro et al. 2016). *rather than "happy":*
     Gendron et al. 2014a. See heam.info/himba-3 for more details.

16. *in the story was feeling:* "Each participant was asked, after each story, how the target
     person was feeling, in order to ensure that the participant had understood the story
     correctly" (Sauter et al. 2015, 355). Sauter et al. have called this extra step a "ma-
     nipulation check"; see heam.info/himba-4. "in their own words": Sauter et al. 2015,

355 (emphasis added). *the corresponding English emotion concepts:* Gendron et al. 2014a.

17. *the better match for sadness:* Himba participants had to "explain the intended emotion in their own words, before they proceeded to the experimental trials for that story" (Sauter et al. 2015, 355). That is, all the trials were delivered one right after the other, in what scientists call a "block" of trials; see heam.info/himba-4.

18. *invention of the Middle Ages:* Trumble 2004, 89. *became more accessible and affordable:* Jones 2014. *"heavily freighted with significance":* Beard 2014, 75. See also heam .info/smile-1. *happiness is simply not universal:* Smiles mean different things in different cultures (Rychlowska et al. 2015); see heam.info/smile-2.

19. *"that conforms to Ekman":* Fischer 2013.

20. *used the basic emotion method:* People worldwide can perceive pleasant versus unpleasant feeling in experiments that don't use the basic emotion method; see heam. info/valence-2.

21. *Trobriand Islands in New Guinea:* Crivelli et al. 2016.

22. *provided strong evidence for universality:* For a summary, see Russell 1994; Gendron et al. 2014b. *still claiming it as fact:* To read about the key condition, see Norenzayan and Heine 2005.

23. *"something the Fore didn't do":* Ekman 2007, 7. *a set of facial movements:* Kudos to the social psychologist Robert Zajonc, who pointed out the embedded assumptions in the word "expression." *of certain Japanese emotion concepts:* For examples, see heam.info/ japanese-1. *emotions as transactions between people:* Lutz 1980; Lutz 1983.

24. *catalogued many of the concerns:* Russell 1994.

25. *hypothetical substance called luminiferous ether:* Firestein 2012, 22.

26. *in the face, body, and voice:* The project began with one intrepid young psychologist, David Cordaro; see heam.info/cordaro.

4. The Origin of Feeling

1. *and displeasure feel qualitatively different:* Pleasure and displeasure are like a sixth sense; see heam.info/pleasure-1. *waking moment of your life:* Every human language that has been studied has words for "feels good" and "feels bad" (Wierzbicka 1999). Words in a variety of human languages also connote good and bad (Osgood et al. 1957). Findings like these have led psychologists like J. A. Russell to claim that valence and arousal properties are universal (Russell 1991a). See heam.info/ pleasure-2.

2. *and your immune system:* Your body is a confusing array of "systems"; see heam.info/ systems-1.

3. *the brain operated similarly:* The roots of this analogy run deep; see heam.info/stimu-lus-1.

4. *awaiting a jump-start:* Walløe et al. 2014; see heam.info/neurons-1. *continue from birth until death:* E.g., Llinás 2001; Raichle 2010; Swanson 2012.

5. *called intrinsic networks:* Yeo et al. 2011. Some of these networks are in your brain at birth, and others develop in the first few years of your life, as you interact with the physical and social environments (e.g., Gao et al. 2009; Gao, Alcauter, et al. 2014; Gao, Elton, et al. 2014). *producing the same basic function:* Marder and Taylor 2011; Marder 2012. It is best to think about function at the network level rather than the

module/hub level. See heam.info/network-1. *discoveries of the past decade:* See heam. info/intrinsic-1.

6. *called simulation in chapter 2:* Intrinsic activity is also called default mode activity and the resting state; see heam.info/resting-1.

7. *a dark, silent box:* This observation is somewhat different than that offered by Fred Rieke (1999) and others, that the brain itself is a black box that only has access to its own states. *smells, and other sensations:* Bar 2007.

8. *your brain makes* predictions: Clark 2013; Hohwy 2013; Friston 2010; Bar 2009; Lochmann and Deneve 2011.

9. *brain's primary mode of operation:* Memory works similarly; see heam.info/memory-1.

10. *skull but* explain *it:* Clark 2013; Hohwy 2013; Deneve and Jardri 2016. *would be a visual prediction:* If you can taste the apple (is it tart or sweet?), then neurons in taste cortex have changed their firing patterns as a gustatory prediction. If you hear the crunch of biting into an apple and feel the juice dribbling down your chin, then neural firing in auditory cortex and somatosensory cortex has also changed as auditory and somatosensory predictions.

11. *intent about moving your body:* Wolpe and Rowe 2015. *"illusion of free will":* Entertaining books on the illusion of free will are at heam.info/free-1.

12. *connection in every waking moment:* Koch et al. 2006. Sensory input that reaches your brain from the outside world is incomplete; see heam.info/vision-1. *interconnections than it could maintain:* Sterling and Laughlin 2015; Balasubramanian 2015.

13. *Only a small fraction:* In the bottom figure, the arrows are not meant to imply that predictions are carried from a single neuron to V1. More details on this example are at heam.info/vision-2.

14. *the signal represents intrinsic activity:* Raichle 2010. This intrinsic activity is metabolically expensive; see heam.info/expensive-1.

15. *baseball in a typical game:* On a regulation-sized baseball diamond, you have about 688 milliseconds to move into position, unless you are a professional baseball player, in which case you have more like 400 milliseconds. See heam.info/baseball-1.

16. *Prediction makes the game possible:* Ranganathan and Carlton 2007. This is also true for basketball; see Aglioti et al. 2008. *using your past experience:* Locating objects in space and preparing to act on them more heavily involves the dorsal part of your visual system; it transmits prediction error from the world a bit faster than the ventral part of the visual system, which is more important for conscious seeing (Barrett and Bar 2009). See heam.info/dorsal-1. *and you catch it:* Your brain initiates your catch well before you consciously see the ball in the predicted location. You become aware of your intention to move your arm at about the same time as you become aware of seeing the ball in its current location, however, so it seems as if you see the ball and then move your arm to catch it. See heam.info/ventral-1.

17. *sensory input of mashed carrots:* Another example might be inattentional blindness: see heam.info/blind-1.

18. *influencing and constraining each other:* Chanes and Barrett 2016. There is evidence in studies with rats that taste works by prediction, but currently there are no experiments in humans; my examples in chapter 2 of my daughter's gross foods birthday party and the salmon ice cream experiment demonstrate both olfactory (smell) and gustatory (taste) predictions in action.

19. *prediction and sensory input:* Carhart-Harris et al. 2016; Barrett and Simmons 2015; Chanes and Barrett 2016. See heam.info/LSD.

20. *and changes your blood pressure:* Along with your autonomic nervous system, your brain commands two other systems within the body that make physical movements possible. Your endocrine system regulates your metabolism, ions (like sodium), etc., through hormones, and your immune system protects your body against disease. See heam.info/interoception-7. *remember, is called interoception:* Interoception was originally defined by Sir Charles Scott Sherrington; for a readable and comprehensive update, see Craig 2015; heam.info/interoception-1.

21. *that is noisy and ambiguous:* Interoceptive information is noisy and ambiguous; see heam.info/interoception-2. *of movements inside your body:* Barrett and Simmons, 2015.

22. *heart pounding in your chest:* Even an inflamed organ might not produce a sensation; see heam.info/interoception-3. Self-reports of bodily sensations rarely correspond to actual sensitivity; see heam.info/interoception-6. *experience these sensations with precision:* See heam.info/interoception-2.

23. *be an instance of emotion:* Scientists still don't understand why intense interoceptive sensations are sometimes experienced as physical symptoms and other times as emotions.

24. *hearing, and other senses:* Kleckner et al., under review. The interoceptive network is made up of two overlapping networks that go by many other names, depending on the interests of the scientists who named them; see heam.info/interoception-12. *your body in the world:* Interoception is actually a whole-brain process anchored in this network; see heam.info/interoception-9.

25. *and the default mode network:* Many studies seem to show that the default mode and salience networks work in opposition: a brain can be in an internal mode, with the default mode network "activated" and the salience network "deactivated" (meaning one is sending more signal than during the rest period and the other is showing less), or the brain can be in an external mode with the opposite pattern. This opposition is an analysis artifact. The two networks can work together or in opposition. For a detailed list of cortical and subcortical regions in the interoceptive network, see heam.info/regions-1.

26. *called your* primary interoceptive cortex: For more on the primary interoceptive cortex, see heam.info/interoception-10.

27. *simulated in the usual way:* Barrett and Simmons 2015. Every other intrinsic network in the brain overlaps with the interoceptive network in at least one of its regions (van den Heuvel and Sporns 2013). So the interoceptive network doesn't create all of its predictions by itself; see heam.info/interoception-11.

28. *a budget for your body:* Scientists call this budget-balancing act "allostasis" (Sterling 2012). See heam.info/allostasis-1.

29. *region within the interoceptive network:* These regions are called "limbic" and include the amygdala; the nucleus accumbens and the rest of the ventral striatum; the anterior, mid, and posterior cingulate cortices; the ventromedial prefrontal cortex (part of the orbitofrontal cortex); the anterior insula; and more.

30. *times when you are stressed:* For more on cortisol, see heam.info/cortisol-1.

31. *of everyone else around us:* If we had also measured Erika's endocrine and immune responses, we would have found them elevated. For example, body-budgeting circuitry instructs the autonomic nervous system to regulate your immune response to avoid joint inflammation as you move. See Koopman et al. 2011.

32. *and other objects and scenes:* The stimuli are from the International Affective Pic-

ture System (Lang et al. 1993). *blood vessels dilate:* See heam.info/galvanic-1. *controlling these inner-body movements:* Weierich et al. 2010; Moriguchi et al. 2011. See also heam.info/fMRI.

33. *anything else relevant to you:* My lab has demonstrated this in collaboration with the cognitive scientist Larry Barsalou and Christy Wilson-Mendenhall (Larry's former Ph.D. student, who completed a postdoctoral fellowship in my lab). We asked test subjects to imagine some scenarios we provided while we observed their brain activity using fMRI scanning (Wilson-Mendenhall et al. 2011). See heam.info/scenarios. *simulation strongly drives their feelings:* Killingsworth and Gilbert 2010.

34. *leading to tangible benefits:* Palumbo et al., in press. Synchrony can also incur costs if one person is stressed; see Waters et al. 2014; Pratt et al. 2015. *less bothered by pain:* Scientists have seen this in experiments using electric shocks (Coan et al. 2006; Younger et al. 2010). For reviews, see Eisenberger 2012; Eisenberger and Cole 2012. *than if you are alone:* Schnall et al. 2008. *supportive person in your life:* John-Henderson, Stellar, et al. 2015. See chapter 10 and heam.info/children-2 for additional discussion. *helping to regulate your budget:* Sbarra and Hazan 2008; Hofer 1984, 2006.

35. *that you experience every day:* This is what some people call "mood."

36. *simpler feeling with two features:* Scholars and scientists have confused affect and emotion for centuries. See heam.info/affect-1. In the science of emotion, the term "affect" is sometimes used to mean anything emotional. In this book, we limit the term to a specific meaning: a change in your internal environment that you experience as feelings of valence and arousal. This modern conception of affect was developed by Wilhelm Wundt; see heam.info/wundt-1. *which scientists call* valence: Barrett and Bliss-Moreau 2009a; Russell 2003. The word "valence" has other meanings in science; see heam.info/valence-1.

37. *basic features of human experience:* Eastern and Western philosophy describe valence and arousal as basic to human experience; see heam.info/affect-2. *with fully formed emotions:* Infants experience affect, even as there is no consistent evidence that they experience emotions (Mesman et al. 2012); see heam.info/affect-3.

38. *from birth until death:* Barrett and Bliss-Moreau 2009a; Quattrocki and Friston 2014. See heam.info/affect-4.

39. *the great mysteries of science:* The structure of the cortex provides some hints to the mystery of affect; see heam.info/cortex-2. *to regulate your body budget:* People believe interoception is "for" feeling because feelings are important to people, and scientists, as people, create causal hypotheses to explain what is important to them. See heam.info/teleology. *if so, how desperately:* Unpleasant affect might be the brain's signal for an unbalanced body budget; see heam.info/budget-1.

40. *to search for explanations:* For example, arousal is a cue to learn (i.e., process prediction error; Johansen and Fields 2004; Fields and Margolis 2015; McNally et al. 2011). With learning comes better prediction and categorization, and therefore a specific action plan. *are collectively your* affective niche: A similar concept is "ecological niche," which is all the aspects of a creature's physical environment that are relevant to its survival.

41. *from the origin representing intensity:* A circumplex represents relationships through the geometry of a circle (Barrett and Russell 1999); see heam.info/circumplex.

42. *"unpleasant, low arousal":* Hundreds of studies over the past thirty years have demonstrated that feelings can be characterized as points within this affective circumplex (Russell and Barrett 1999; Barrett and Bliss-Moreau 2009a). Some people feel changes

of valence and arousal together, whereas for others the two properties are independent (Kuppens et al. 2013). *cultures like China and Japan:* Tsai 2007; Zhang et al. 2013.

43. *that your judgment was correct:* Philosophers call this "world-focused" affect; see heam.info/affect-8. *a 2011 study of judges:* Danziger et al. 2011. In laboratory experiments, when test subjects use strong affect to make harsh sentencing decisions, we see increased activity in a visceromotor region of the interoceptive network (Buckholtz et al. 2008).

44. *based on gut feelings:* Huntsinger et al. 2014. People use affect as information about whatever is in one's focus of attention; see heam.info/realism-3. *explicitly asked about the weather:* Schwarz and Clore 1983. *negatively when it is rainy:* Interview candidates receive lower ratings on rainy days; see Redelmeier and Baxter 2009; heam.info/realism-4. *maybe it's just lunchtime:* People have invented the concept "hangry" to cover this experience. *anything to do with you:* Even simple actions like taking a drink become moments of affective realism (Winkielman et al. 2005). See heam.info/realism-5.

45. *see the person's face differently:* Anderson et al. 2012. Affect takes as its object whatever is in mind at the time; see heam.info/realism-1.

46. *"bring it on themselves":* Affective realism lets us sidestep responsibility; see heam.info/realism-2.

47. *when they are performing well:* Shenhav et al. 2013; Inzlicht et al. 2015.

48. *camera to be a gun:* Reuters journalist Namir Noor-Eldeen, driver Saeed Chmagh, and several others were killed; see heam.info/gunner-1.

49. *as a weapon:* Fachner et al. 2015, 27–30.

50. *in preparation to run:* Your arteries contain special cells called baroreceptors; see heam.info/budget-2. *predict those sensations as well:* See heam.info/interoception-8.

51. *your interoceptive predictions are not:* Barrett and Simmons 2015. *the predicted need is over:* See heam.info/cortex-1.

52. *sluggish to correct their predictions:* Sometimes your body-budgeting regions can act quickly to change their predictions, like when your life is on the line. If you're driving on the highway and another driver cuts you off, those body-budgeting regions let you correct your trajectory plenty fast.

53. *simulations in your interoceptive network:* Barrett and Simmons 2015. My lab has evidence that affect is largely prediction; see heam.info/affect-5.

54. *ideas are not just speculation:* Is it possible to peer into a person's brain and see exactly how interoceptive predictions are transformed into affect during brain imaging? The answer, I'm afraid, is not yet. But a meta-analysis conducted by members of my lab examining over four hundred brain-imaging studies found that body-budgeting regions in the interoceptive network, which issue interoceptive predictions, consistently increase in activity when people report strong changes in their affective feelings (Lindquist et al. 2015). *from treatment-resistant depression:* Holtzheimer et al. 2012; Lujan et al. 2013. *in the patient's interoceptive network:* Specifically, the bundles of axons that connect body-budgeting regions within the interoceptive network; see heam.info/mayberg-1. *in synchrony with the stimulation:* Choi et al. 2015. *new treatments for mental illness:* The neurons stimulated by Mayberg are not specific for affect, however; see heam.info/affect-6.

55. *destroyed by a rare illness:* Feinstein et al. 2010. See heam.info/HSE. *difficulty with smell and taste:* These last items are no surprise because limbic tissue regulates these

bodily functions. *producing the same outcome:* Since Roger has working autonomic nervous, endocrine, and immune systems, and much of the subcortical circuitry involved in interoception is still intact (like regions of his brainstem and hypothalamus), he still has sensory inputs coming to his interoceptive cortex from the body, which can be used to compute prediction error. See heam.info/roger. *based primarily on uncorrected predictions:* These patients still have interoceptive perceptions; see heam.info/PAF.

56. *predictors in your entire brain:* van den Heuvel and Sporns 2011, 2013. *are wired to listen:* Chanes and Barrett 2016. According to the noted neuroanatomist Helen Barbas, body-budgeting regions (also called "limbic" regions) are the most powerful feedback system in the brain, based on the pattern of their connections to other cortical regions. Another name for "feedback" is "prediction." See Barbas and Rempel-Clower 1997, and heam.info/cortex-1.

57. *through affect-colored glasses:* Seo et al. 2010. Neuroeconomics seeks to understand how the brain estimates the value of different choices to allow decision-making. Value and affect are related concepts. See heam.info/neuroeconomics.

58. *for wisdom:* Damasio 1994. *the fabric of every decision:* Certainly other philosophers, such as David Hume, have held that view; see heam.info/affect-7.

59. *continued to guide economic practice:* Notably, the last century of seesawing among crisis, increased regulation, complaints, decreased regulation, followed by another crisis. See also heam.info/econ-1.

60. *up to the Great Recession:* Madrick 2014. *people are rational decision makers:* Krugman 2014. Another condition is that people are assumed to have all the price and product information they need, a situation that rarely occurs in practice; Marshall Sonenshine, professor of finance and economics at Columbia University, personal communication, May 10–July 31, 2013. *lurking beneath the surface:* Other economic disasters may have been precipitated by the anatomy of the human cortex. See heam.info/crises.

61. *uniquely human cortex:* The "neocortex" is not really new to the mammalian brain; see heam.info/triune-1. *successful misconceptions in human biology:* MacLean and Kral 1973. See heam.info/triune-2. *in his bestseller* Emotional Intelligence: Goleman 2006. He continues to rely on a version of the triune brain in his newer books. *brain evolution knows:* Evolutionary biologist Georg Striedter, editor of the scholarly journal *Brain, Behavior and Evolution* and author of *Principles of Brain Evolution* (2005), writes, "Many 'classic' notions about how vertebrate brains evolved (e.g., by adding neocortex to an ancestral 'smell-brain') continue to hold sway among many non-specialists, even though they have long been disproved" (2006, 2). *"present in all vertebrates":* See more quotes from Finlay at heam.info/finlay-1. *keep themselves efficient and nimble:* Striedter 2006; Finlay and Uchiyama 2015. For more on brain evolution, see heam.info/evolution-1.

5. Concepts, Goals, and Words

1. *to perceive bands of color:* This process is called "categorical perception"; see heam.info/rainbow-1.

2. *sound into syllables and words:* In an unfamiliar spoken language, you might not even discern word boundaries; see heam.info/speech-1.

3. *context within the same speaker:* Many thanks to Larry Barsalou for this description; Barsalou 1992, chapter 9. *when presented in isolation:* Pollack and Pickett 1964. *to communicate with others:* Foulke and Sticht 1969; Liberman et al. 1967.

4. *like complex objects and scenes:* Grill-Spector and Weiner 2014.

5. *incapable of learning:* Jorge Luis Borges's story "Funes the Memorious" dramatizes this condition; see heam.info/funes.

6. *really out there:* William James used the phrase "blooming, buzzing confusion" to describe the world as a newborn infant would perceive it.

7. *means to be human:* There is an animated and important debate about whether a few concepts are inborn, such as number and cause. This debate is not central to our discussion here, because it doesn't change the theory of constructed emotion nor any interpretations of experiments. I do flag the debate where relevant, however.

8. *events by categorizing using concepts:* The philosopher Immanuel Kant wrote that we perceive the world in terms of concepts; see heam.info/kant-2.

9. *describing necessary and sufficient features:* Smith and Medin 1981; Murphy 2002.

10. *from antiquity until the 1970s:* Murphy 2002. *an ostrich a representative bird:* Philosopher Ludwig Wittgenstein also pointed out that most concepts cannot be defined by necessary and sufficient features and instead preferred to use family resemblances (Wittgenstein 1953; also see Murphy 2002; Lakoff 1990). *view of concepts finally collapsed:* Murphy 2002.

11. *known as the* prototype: Rosch 1978; Mervis and Rosch 1981; Posner and Keele 1968. *majority of the category's features:* Also known as family resemblance; see heam.info/prototype-1.

12. *of a given emotion category:* J. A. Russell, for example, has a prototype view of emotion concepts (Russell 1991b); see heam.info/russell-1.

13. *rarely found in real life:* In my research, I call this state of affairs "the emotion paradox" (Barrett 2006b); see heam.info/paradox-1.

14. *need them, on the spot:* Your brain is engaged in conceptual combination, discussed later in this chapter and at heam.info/combination-1. *that best fits the situation:* Your brain is using something like pattern classification; see heam.info/pattern-2.

15. *finding similarities in the variations:* Posner and Keele 1968. *constructed in the same manner:* Some scientists still believe that each emotion concept is a fixed prototype in the brain, however; see heam.info/prototype-2.

16. *with four wheels nailed on:* Barsalou 1985; Voorspoels et al. 2011; but see Kim and Murphy 2011. For a discussion, see Murphy 2002.

17. *in a particular situation:* Your brain combines bits and pieces of past experience to create a concept that is the best fit to the sensory cues of the current situation; this allows you to achieve your goal in this situation. Barsalou (1985) demonstrated that concepts are constructed dynamically and flexibly; see heam.info/goals-1.

18. *your goal in the moment:* These ideas are similar, although not identical, to those found in Edelman 1987; see heam.info/edelman-1.

19. *a process called* statistical learning: Xu and Kushnir 2013; Tenenbaum et al. 2011. See more on statistical learning at heam.info/stats-1.

20. *I won't enter that debate:* This is the nativism/empiricism debate; see heam.info/concepts-1.

21. *interest in listening to speech:* Vouloumanos and Waxman 2014. *and even in utero:* Moon et al. 2013. *a few minutes of exposure:* See Maye et al. 2002, in Kuhl 2007.

Whether the patterning of certain sound concepts (phonemes) is learned from experience or triggered by experience (i.e., is innate) is a matter of great debate. For an excellent treatment of the nativist view, see Berent 2013. For a discussion of the empiricist view, how concepts can be learned from similarity, see Goldstone 1994. See also heam.info/concepts-5. *heard spoken by live humans:* The neural connections that are not used are likely pruned away. For more on the world's tuning of language, see Kuhl and Rivera-Gaxiola 2008.

22. *association between color and sound:* Gweon et al. 2010.

23. *color was in the majority:* Denison and Xu 2010. Infants are sensitive to probabilities as young as six months old (Denison et al. 2013), and can use probabilities to make predictions and decisions (Denison and Xu 2014).

24. *environment but anticipate them:* Freddolino and Tavazoie 2012. of the people around them: Keil and Newman 2010; Gelman 2009. The information that resides in the minds of others is the similarities created by their conceptual systems.

25. *in the world does too:* Repacholi and Gopnik 1997. *interesting, colorful Slinky toys:* Ma and Xu 2011.

26. *randomly versus with intent:* Details on this experiment are at heam.info/ball-1. *will occur several minutes later:* Southgate and Csibra 2009; Vouloumanos et al. 2012. Infants as young as eight months old can infer goals (Hamlin et al. 2009; Nielsen 2009; Brandone and Wellman 2009).

27. *and strong eye contact:* Vouloumanos and Waxman 2014; Vouloumanos et al. 2012; Keil and Newman 2010; Lloyd-Fox et al. 2015; Golinkoff et al. 2015.

28. *regularity that speeds concept learning:* Sloutsky and Fisher 2012. *infant to form a concept:* Waxman and Gelman 2010; Waxman and Markow 1995.

29. *the effect never materialized:* Other sounds don't work either; see heam.info/sounds-1.

30. *"equivalence that is mental":* Waxman and Gelman 2010.

31. *"wug" or "dak":* Xu et al. 2005. *to represent things as equivalent:* See heam.info/goals-2. *physical similarity without a word:* Yin and Csibra 2015. See experimental results at heam.info/goals-3.

32. *and so are experientially blind:* Turati 2004. See also heam.info/faces-1.

33. *understanding of facial expressions:* E.g., Denham 1998; Izard 1994; Leppänen and Nelson 2009.

34. *has put in your path:* Clore and Ortony 2008; Ceulemans et al. 2012; Roseman 2011.

35. *in all its sensory detail:* Schyns et al. 1998.

36. *to construct perceptions of anger:* This may be when children begin to learn that emotions cause actions; see heam.info/knowledge-1.

37. *or wishing to appear powerful:* For more on goals related to anger, see also heam.info/anger-1.

38. *until around age three:* The psychologists James A. Russell and Sherri C. Widen have a long program of research on children's emotion concepts; for a review, see Widen, in press; also see heam.info/russell-2. *to four months of age:* For the details on infant affect concepts, see heam.info/infants-1.

39. *and "hoot" faces:* Parr et al. 2007. *concepts for the face categories:* Fugate et al. 2010.

40. *"give me a hug":* Harris et al., in press.

41. *suffocation, and constriction:* Panayiotou 2004.

42. *no equivalent in English:* Pavlenko 2014. *ones from your primary language:* Pavlenko 2009. See also heam.info/language-1. *situation, known as* zlit'sia: Ibid., chapter 6.

43. *the English concept for guilt:* Victor Danilchenko, a Ukrainian immigrant and computer scientist who worked with my husband, tells me that native Russian speakers in the United States sometimes use English idioms while speaking in Russian. A favorite example is "to run out of sugar," literally translated as bursting forth at a run from a pile of sugar.

44. *special power called* conceptual combination: Wu and Barsalou 2009. See also heam .info/combination-1.

45. *those that describe the situation:* This is another point where the theory of constructed emotion diverges from the classical view, which would say that a person is "feeling several emotions at the same time," as if those emotions were objectively distinguishable, rather than constructing a completely new emotional experience.

46. *potent capability of the brain:* See heam.info/combination-1.

47. *to six objects in mind:* Feigenson and Halberda 2008.

48. *10 percent of the world's population:* Salminen et al. 1999. The word "alexithymia" comes from the roots "a" (lack), "lexis" (word), and "thymos" (mood). See Lindquist and Barrett 2008 for a review, and heam.info/alexithymia-1. *to experience them as emotional:* Lane et al. 1997; Lane and Garfield 2005. *emotion in others as well:* Lane et al. 2000. See heam.info/alexithymia-1. *have a restricted emotion vocabulary:* Lecours et al. 2009; Meganck et al. 2009. See heam.info/alexithymia-1. *have difficulty remembering emotion words:* Luminet et al. 2004.

49. *touches, and interoceptive sensations:* Frost et al. 2015.

50. *forgetting your best friend's birthday:* See heam.info/shepard-1.

51. *to use as predictions:* Using Bayesian rules of probability (Perfors et al. 2011). See also heam.info/bayes-1.

52. *architect of the whole experience:* Nonetheless, people actively construct the temporal order of events; see heam.info/causality-1.

6. How the Brain Makes Emotions

1. *predictions of "Anger" simultaneously:* A common goal for "Anger" in Western cultures is to defend oneself from threat or harm (Clore and Ortony 2008; Ceulemans et al. 2012).

2. *the promotion in your place:* See heam.info/anger-1.

3. *is exquisitely bright but diffuse:* Gopnik 2009. See also heam.info/gopnik-1. *other things in the dark:* Posner et al. 1980.

4. *predictions that span the senses:* Different senses play "supporting roles" for one another; see heam.info/multi-2.

5. *regardless of the sensory differences:* Many papers use faces as the textbook example for explaining concept formation, because the visual system has been well-studied and so is better understood than most other sensory systems, and because humans are experts at seeing faces in sensory inputs. For a well-written, accessible example using faces, see Hawkins and Blakeslee 2004; see also heam.info/muller-1.

6. *identical, groupings of neurons:* See more on these distributed response patterns of neurons at heam.info/concepts-2.

7. *of neurons on each occasion:* As you've read many times now, neurons are multipurpose; this is true even when it comes to concepts. Neurons alter their firing rate to participate in many different assemblies, so that a single neuron contributes to numerous

instances of the same concept, as well as different concepts. Multipurpose does not mean all-purpose, of course. Different instances of the same concept need not share the same neurons, and instances of different concepts need not be located in different groupings of neurons; different instances must be separable, not separate. See Grill-Spector and Weiner 2014, and heam.info/multi-1.

8. *in any concrete way:* See heam.info/multi-1.

9. *"a prediction" of happiness:* Also, when your brain "learns an instance of a concept," that is equivalent to saying your brain receives and processes sensory input, that is, prediction error, making the new instance more similar to some previous instances and less similar to others.

10. *sensory input before it arrives:* Chanes and Barrett 2016. If things "settle" and "predict" too quickly, then the prediction will not seem calibrated to context. This is probably a hallmark of psychopathology.

11. *such as the Korean word* jeong: Lin 2013.

12. *future, such as imagination:* Also called prospection (e.g., Schacter et al. 2012; Buckner 2012; Mesulam 2002). *experiences of the present moment:* Clark 2013; Friston 2010; Bar 2009; Bruner 1990; Barsalou 2009. See chapter 4, figure 4-3. As I explained in chapter 5, it is metabolically inefficient to compute perceptions and plan actions from scratch in every moment of your life. We have evolved an efficient nervous system that saves costs by minimizing redundancy (which is wasteful, metabolically speaking). The brain exploits the fact that certain patterns of sensations and events tend to recur with some regularity. It learns (i.e., changes neural firing rates, and eventually grows new neurons or connections) only what is novel and relevant to the body budget; this is why the brain predicts (i.e., reconstructs, infers, or guesses) those regularities, where possible, rather than squandering resources to detect them again and again. See heam.info/present-1. *"the remembered present":* Edelman 1990.

13. *are more probabilistic than that:* Since thousands of predictions are launched, many could be active at the same time, but the one that fits the incoming input best will become your experience and either confirm or correct your action. This is possibly one reason why a feeling of anger in the exact same situation might feel slightly different than it does on another occasion. The other predictions in the population might be different. Exact identity might require more precision — at the level of every single neuron — than the brain is capable of achieving (because of noise and context).

14. *known as your* control network: Scientists have identified three overlapping, intrinsic networks for this purpose (e.g., Power et al. 2011); see heam.info/control-4.

15. *number? — in each moment:* There are other selection mechanisms in the brain; see heam.info/selection-1.

16. *shape your perception and action:* I briefly discuss Edelman's Theory of Neural Darwinism at heam.info/edelman-1.

17. *perception, and launch an action:* In psychology, we have many names for describing this "tinkering," such as keeping a goal in mind, focusing attention, weeding out distractions, selecting the best action, and so on, and we refer to them as different processes, such as working memory, selective attention, and so on. See heam.info/control-5.

18. *I scream or not?:* See heam.info/selection-1.

19. *but refrain from punching him:* Gross and Barrett 2011; Ochsner and Gross 2005. See heam.info/regulation-1.

20. *the interoceptive and control networks:* This efficient structure is a small world architecture with rich club hubs; see heam.info/hubs-1.

21. *be a prerequisite for consciousness:* Chanes and Barrett 2016. See also heam.info/meg-1. *all associated with hub damage:* Particularly the anterior insula and anterior cingulate cortex (Menon 2011; Crossley et al. 2014).

22. *go beyond the information given:* Cognitive psychologist Jerome S. Bruner coined the term "acts of meaning" (Bruner 1990). See also heam.info/bruner-1.

7. Emotions as Social Reality

1. *can make those changes meaningful:* Some people believe that these vibrations are the essence of the sound because a sound cannot be heard without them. But this explanation misses the point. Vibrations are not sufficient for a sound to occur. Sounds do not have simple, single causes; see heam.info/sound-1.

2. *made meaningful by a brain:* All three cone types must work together to perceive a single category of color, like red; see heam.info/cones-1.

3. *learn from conversations and books:* Shepard and Cooper 1992. See heam.info/shepard-1. *up the continuous spectrum differently:* Roberson et al. 2005. See heam.info/color-1.

4. perceiver-independent *categories:* Philosophers call them "ontologically objective"; see heam.info/perceiver-1.

5. *your two-year-old child:* Even biologists' criteria for flowers and weeds are subjective; see heam.info/flower-1. *by the external world:* Einstein and Infeld 1938, 33. Or see Max Planck's more cynical take from *The Universe in the Light of Modern Physics* (1931, 58–59): "We have no right to assume that any physical laws exist, or if they have existed up to now, that they will continue to exist in a similar manner in the future."

6. *you sample the sensory world:* Susskind et al. 2008.

7. *be discarded for scientific endeavors:* The sixteenth-century philosopher Francis Bacon, for instance, warned about using common-sense language in science, reifying the referent of the word that doesn't warrant it. So did William James. Many scientists and philosophers since then have warned about the evils of "folk psychology." Common-sense concepts or words might not be the best flashlights to illuminate the search for underlying mechanisms. *I took this latter view:* Barrett 2006a.

8. *classic example of social reality:* Searle 1995. Ernst Cassirer anticipated the idea of social reality; see heam.info/reality-3.

9. *the core of social reality:* A concept is a population of instances that might be physically different but are treated as similar for some purpose; in social reality, that purpose is the set of functions that people impose that transcend the physical nature of the instances themselves (i.e., people treat the instances as mentally similar, despite the instances' physical differences).

10. *knowledge is called* collective intentionality: See more on collective intentionality at heam.info/collective-1.

11. *categorization as a cooperative act:* I created my lab through cooperative categorization. I gathered all the people working with me, gave us a name (so we identified ourselves as a group with a common goal), and poof: instant lab. T-shirts and mousepads with the lab logo didn't hurt either.

12. *of communication and social influence:* Tomasello 2014.

13. *the two concepts are named:* For more on how concept learning can occur without a word, see heam.info/concepts-3.

14. *the functions of an emotion:* The linguist George Lakoff calls emotion an essentially contested concept, because people in American culture agree that emotions exist, but they don't necessarily agree on the definition, and scientists are unable to settle the matter. It seems to me that contested concepts are casualities in a battle over social reality: whose concepts are going to win and define what exists?

15. *man acts as he does:* Tomasello 2014.

16. *the norms of our culture:* The emoter and the perceiver are not categorizing the same psychological moment; see heam.info/concepts-4.

17. *asking in the first place:* This is an example of a "category error." According to the philosopher Gilbert Ryle, a category error is an ontological error where things belonging to one category are mistaken for belonging to another. Here, social reality is mistaken for physical reality.

18. *comrades-in-arms:* Bourke 2000; Jamison 2005; Lawrence (1922) 2015. *needed for their military duties:* The psychologist Maya Tamir would refer to this as an example of instrumental emotion regulation. People construct unpleasant emotions because they are useful in a given context (Tamir 2009).

19. *and to create civilizations:* Boyd et al. 2011.

20. *on the African savanna:* Even universality does not necessarily imply innateness — think Coca-Cola. *a matter of cultural evolution:* Case in point: the Hadza of Tanzania, who have lived continuously on the African savanna for at least 150,000 years since the Pleistocene Epoch, do not recognize posed facial configurations of fear, based on my lab's visit in 2016. For excellent treatments of the relation between culture and evolution, see Laland and Brown 2011; Richerson and Boyd 2008; and Jablonka et al. 2014. See also heam.info/culture-1.

21. *people of the Kalahari Desert:* See more about the !Kung people, as well as languages that seem to lack a distinct word for "fear," at heam.info/kung-1.

22. *driving force behind human culture:* Social reality is embedded in the definition of culture. Zoologists Kevin N. Laland and Gillian R. Brown call culture "a cohesive set of mental representations, a collections of ideas, beliefs, and values that are transmitted among individuals and acquired through social learning" (Laland and Brown 2011, 9). The geneticist Eva Jablonka's definition adds behaviors and products (Jablonka et al. 2014). *be more fit to reproduce:* Boyd et al. 2011. This paper argues that biology and culture are not battling for control of human behavior (just as cognition and emotion are not waging a battle). This war is all in our minds — it is a social reality created by minds that are as much a consequence of culture as of genes. Robert Boyd and his colleagues write, "Culture is as much a part of human biology as our particular pelvis" (2011, 10924). The capacity to create emotion concepts, share them with others, and use them to construct social reality is a function of our biological makeup.

23. *word for rainbow,* радуга: To produce the word радуга with a non-Russian keyboard, visit translate.google.com and translate the word "rainbow" to Russian, then copy and paste.

24. *green are to an American:* Other cultural examples include the Himba, who categorize some shades of Western "green" and "blue" as a single color, and the Berinmo of Papua New Guinea, who have only five color categories.

25. *that don't exist in English:* Good summaries can be found in Russell 1991a; Mesquita

and Frijda 1992; and Pavlenko 2014. *calling it* "Forelsket": So Bad So Good 2012. *certain feeling of close friendship*: Verosupertramp85 2012. "Tocka" *is a spiritual anguish*: Ibid. *a strong, spiritual longing*: Wikipedia, s.v. "Saudade," last modified April 1, 2016, http://en.wikipedia.org/wiki/Saudade. *called* "Pena Ajena": So Bad So Good 2012.

26. *something that is unbearably adorable*: Garber 2013; So Bad So Good 2012.
27. *before the event takes place*: "Better Than English" 2016.
28. *looking worse after a haircut*: Pimsleur 2014.
29. *depending on context*: Lutz 1980; Russell 1991b. *"the desire for revenge"*: Kundera 1994. *required to be grateful anyway*: So Bad So Good 2012.
30. *no concept of "Anger"*: Briggs 1970. *no concept of "Sadness"*: Levy 1975; Levy 2014.
31. *individual, in the body*: Nummenmaa et al. 2014. Various scholars throughout history have also located emotion in the body; see heam.info/body-3. *require two or more people*: Pavlenko 2014. *Westerners lump together as emotional*: Ibid.
32. *"to basic psychological realities"*: Wierzbicka 1986, 584. *invention of the seventeenth century*: Danziger 1997.
33. *spatial relations, and causality*: Mapping words to conceptual representations is neither simple nor universal; see heam.info/concepts-13. *language to language is astonishing*: Malt and Wolff 2010, 7.
34. *had never smiled so much*: Victor Danilchenko, my husband's colleague, tells me that in his native Ukraine, habitual smiling is not the norm, and the term "American smile" means a fake and insincere smile. *prefer high arousal, pleasant states*: Tsai 2007.
35. *shame, and respect*: De Leersnyder et al. 2011.
36. *report more physical illness*: Consedine et al. 2014.

8. A New View of Human Nature

1. *lower salaries in the future*: The human brain develops until late adolescence, but the most sensitive time begins during the first trimester and continues throughout the first several years of life, particularly for brain regions important for body-budgeting, control, and learning (Hill et al. 2010). These brain regions are thinner (fewer connections between neurons, or even fewer neurons) in infants and young children raised in poverty. Importantly, their brains do not start off smaller but grow more slowly over the first three years of life (Hanson et al. 2013); the growth occurs particularly in the connections between neurons (Kostović and Judaš 2015), so reduced connectivity will limit conceptual development and speed of processing, which is strongly related to IQ. Social reality thus becomes physical reality; see heam.info/children-1.
2. *as in control or not*: The experience of being in control is often a function of affect and belief and is largely unrelated to the actual amount of control you have (Job et al. 2013; Inzlicht et al. 2015; Job et al. 2015; Barrett et al. 2004). See heam.info/control-7.
3. *for five months after training*: Halperin et al. 2013. The method of recategorization in these studies was called "reappraisal," which is defined as changing the meaning of a situation.
4. *and only partly predictable world*: Sporns 2011.
5. *from an early animal ancestor*: Darwin (1872) 2005.
6. *to refer to both possibilities*: Philosophers debate over the definition of essences; see heam.info/essences-1.
7. *subcortical regions of your brain*: See also Panksepp 1998; Pinker 2002, 220; Tracy and

Randles 2011. *is a set of genes:* Pinker 1997. Each emotion supposedly issues from a specialized "organ of computation" designed to solve a specific problem for your hominin ancestors on the African savanna, so your genes have a better chance of replicating themselves into the next generation. Much has been written on the idea of mental organs and evolution; see heam.info/organs-1. *a metaphorical "program":* Cosmides and Tooby 2000; Ekman and Cordaro 2011. Pinker does not go in for emotion programs as essences and takes a more nuanced approach. In *How the Mind Works,* he writes, "The problem with the emotions is not that they are untamed forces or vestiges of our animal past; it is that they were designed to propagate copies of the genes that built them rather than to promote happiness, wisdom, or moral values" (1997, 370). So, even though we are supposed to walk around with stone-age minds created by stone-age brains, emotions are not "burned so deeply into the brain that organisms are condemned to feel as their remote ancestors did" (371). *and events in the world:* Scientists debate which emotions should be considered basic; see heam.info/basic-1. *whether to trigger an emotion:* E.g., Frijda 1988; Roseman 1991.

8.　*biology into a modern science:* Darwin (1859) 2003. *"paralyzing grip of essentialism":* Mayr 1982, 87. See also heam.info/darwin-2.

9.　*with their own essences:* The types were strictly ordered and catalogued by how they looked to the naked eye, an arrangement known as a typology; see heam.info/typol ogy. *rich, dense, lustrous gold:* American Kennel Club 2016.

10.　*"survival of the fittest":* The scholar Herbert Spencer coined this term in 1864 after reading Darwin's *Origin. no essence at their core:* A species is a goal-based concept, where the goal is successful reproduction. There are different properties or mechanisms that can be used to anchor this concept; see Mayr 2007, chapter 10. Using the species concept to classify individuals as belonging to the same reproductive community makes those individuals a conceptual category. *to Darwin's theory of evolution: Origin* actually contained five conceptual innovations; see heam.info/origin-1.

11.　*greatest achievement by writing* Expression: What was the reason for Darwin's hypocrisy? See heam.info/darwin-3.

12.　*parts together to make sounds:* Darwin (1872) 2005, 188. *imbalance could cause frizzy hair:* This is a great example of the representativeness error; see heam.info/frizzy.

13.　*"fear of a bear":* James 1894, 206.

14.　*other popular books on emotion:* Damasio 1994. *like little bits of wisdom:* Damasio and Carvalho 2013. Damasio has further outlined his somatic marker hypothesis in his three bestselling books. See also heam.info/damasio-1. *are transformed into conscious feelings:* Damasio and Carvalho 2013.

15.　*logically impossible to prove false:* Hope can be dangerous in science; see heam.info/essentialism-1.

16.　*the psychological origin of essentialism:* The developmental psychologist Fei Xu, whom we met in chapter 5, refers to words as "essence placeholders" (Xu 2002). *"shall be the name":* James (1890) 2007, 195. *reflect firm boundaries in nature:* Philosophers use the term "natural kinds" to describe categories with essences. These categories have firm boundaries in nature. For example, if you assume that an emotion category is a natural kind, then its fingerprint is the set of necessary and sufficient features that describe all instances; it defines the kind of emotion by analogy. The emotion's underlying cause defines the category by homology (Barrett 2006a).

17.　*original red "blicket":* Gopnik and Sobel 2000. *extend concepts by ignoring variation:*

Early in life, infants have many concepts and therefore perform induction. E.g., Bergelson and Swingley 2012; Parise and Csibra 2012.

18. *compressing them into efficient summaries:* For more detail, see heam.info/finlay-2.

19. *"of his lowly origin":* Darwin (1871) 2004, 689. *pinnacle of the animal kingdom:* Aristotle, Darwin, and others weigh in at heam.info/beast-1.

20. *the world completely outside you:* Different branches of the classical view frame this boundary differently; see heam.info/boundary-1.

21. *"a common progenitor":* Darwin (1872) 2005, 11.

22. *a dozen times in* Expression: Ibid., 19 (twice), 25, 27 (twice), 30 (twice), 32, 39, 44 (three times), 46, 187 (twice). *his broader arguments about evolution:* A claim that infuriated many of his contemporaries; see heam.info/darwin-4.

23. *wrote extensively on Darwin's ideas:* Floyd Allport is not often discussed in modern psychology, but his brother, Gordon Allport, is a towering figure in social psychology who wrote important scientific works on personality and prejudice, and who trained some of the most influential psychologists of the twentieth century. *"which the latter develops":* Allport 1924, 215.

24. *"rest and . . . TV":* Gardner 1975.

25. *evidence that he was wrong:* Finger 2001. *a perfectly healthy Broca's area:* This matches other evidence available at the time; see heam.info/broca-1. *a healthy dose of essentialism:* Lorch 2008. See also heam.info/broca-2. *essentialist views of the mind:* The full story of Broca's area is at heam.info/broca-3.

26. *was sculpted by evolution:* See more on The Descent of Man at heam.info/darwin-5. *our "lowly origin":* Darwin (1871) 2004, 89, 689.

27. *that regulate mankind's animalistic emotions:* The term "limbic" originated in the murky world of seventeenth-century anatomy; see heam.info/limbic-1. *rationality as our crowning glory:* Darwin's ideas came from Plato and Aristotle; see heam.info/darwin-6.

28. *alone consider it a system:* Criticisms of the limbic system concept are at heam.info/limbic-2.

29. *charioteer wrangling two winged horses:* Plato called his model the tripartite soul; see heam.info/plato-1. *human constructions dependent on concepts:* Both views are in practice today (Dreyfus and Thompson 2007).

30. *world through judgment and inference:* Sabra 1989, cited in Hohwy 2013, 5. *imagination, and intelligence:* See more on these Christian theologians at heam.info/medieval-1. *"functions of the latter":* James (1890) 2007, 28. *an essentialist sort of evolution:* See heam.info/war-1.

31. *as blobs in the brain:* "I describe mental life by the metaphor of two agents, called System 1 and System 2, which respectively produce fast and slow thinking. I speak of the features of intuitive and deliberate thought as if there were traits and dispositions of two characters in your mind. In the picture that emerges from recent research, the intuitive System 1 is more influential than your experience tells you, and it is the secret author of many of the choices and judgments you make" (Kahneman 2011, 13). Like most ideas in psychology, System 1 and System 2 are metaphors or concepts of social reality that people use, in agreement, to refer to phenomena, not to processes or brain systems. System 1 refers to times when predictions are less corrected by prediction error. System 2 refers to times when predictions are more corrected by prediction error.

32. *a construction theory of memory:* Schacter 1996.

33. *shaped this way and that:* Pinker 2002. *to the environment we are:* E.g., Charney 2012; see heam.info/genes-1.
34. *"are learned or innate":* Pinker 2002, 40–41. *devil is in the details:* See heam.info/evolution-3.
35. *the "four F's":* These behaviors were mentioned as a group in 1958 by psychologist Karl H. Pribram, though he referred to the fourth "F" as "sex" (Pribram 1958).
36. *to function like a computer:* Neisser 2014; Fodor 1983; Chomsky 1980; Pinker 1997.
37. *emotion research was allegedly dead:* Duffy 1934, 1941. *colleagues had never heard of:* A short list of papers is at heam.info/chorus-1. *and speculating about constructionist ideas:* Gendron and Barrett 2009.
38. *"to reject science itself":* Kuhn 1966, 79. *how emotions are made:* Details on the face-reading initiatives of Microsoft, Apple, and the rest are at heam.info/faces-3.
39. *They are following an ideology:* Lewontin 1991.
40. *change who you become tomorrow:* We're not talking about radical transformations here but small, incremental changes.

9. Mastering Your Emotions

1. *if you try hard enough:* A popular theory of emotion regulation used in self-help books comes from psychologist James J. Gross. For a recent example, see Gross 2015. See also heam.info/gross-1.
2. *refined sugar and bad fats:* Kiecolt-Glaser 2010. *regularly sleep-deprived:* National Sleep Foundation 2011. *depression and other mental illnesses:* Cassoff et al. 2012; Banks and Dinges 2007; Harvey et al. 2011; Goldstein and Walker 2014. *toxic for your body budget:* Some evidence that people have unrealistic goals, from Rottenberg 2014: In 2006, over 25% of high school students said that earning a lot of money was extremely important to them, up from 16% in 1976 (Bachman et al. 2006); 31% said they had a goal to be famous one day (Halpern 2008); and the number of people having aesthetic procedures rose 20% in just 2015 alone and 500% between 1997 and 2007 (American Society for Aesthetic Plastic Surgery 2016). *time disrupts your sleeping patterns:* Chang, Aeschbach, et al. 2015. See heam.info/sleep-1.
3. *some form of distress:* TedMed 2015. A recent study by the Mayo Clinic confirmed this high number, reporting that 26% of Americans are taking prescription opioids or antidepressants (Nauert 2013). And between 80% and 90% surveyed believe that people take drugs to relieve stress (American Psychological Association 2012). There was a 200% increase in the use of opioids stronger than morphine over a ten-year period (from 2002 to 2012), and a majority of people taking prescription opioids (80%) are taking a morphine equivalent or something stronger; this was almost 7% of the adult population of the United States in 2012 (Center for Disease Control and Prevention 2015).
4. *work out vigorously and regularly:* Numerous studies show that exercise benefits health in many different ways (Gleeson et al. 2011; Denham et al. 2016; Erickson et al. 2011), particularly jogging, at least if you are a rat (Nokia et al. 2016). *and get plenty of sleep:* Goldstein and Walker 2014.
5. *way of your interoceptive network:* Olausson et al. 2010; McGlone et al. 2014. *might otherwise experience as unpleasant:* E.g., Tejero-Fernández et al. 2015.
6. *the slow-paced breathing:* Deep, slow breathing helps perk up your parasympathetic nervous system, which in turn has a calming effect. It's an easy way to control the

activation of your body-budgeting regions voluntarily. Quick, short breaths have the opposite effect. *harmful inflammation in your body:* Kiecolt-Glaser et al. 2014; Kiecolt-Glaser et al. 2010. *depression, and other illnesses:* Pinto et al. 2012; Ford 2002; Josefsson et al. 2014.

7. *psychiatric patients recover more quickly:* Park and Mattson 2009; Beukeboom et al. 2012. Also the toxic effects of uncontrollable noise, lack of green spaces, inconsistent temperature, crowding, lack of fresh vegetables, and other ills of poverty are well known, as we will discuss in chapter 10.

8. *also beneficial to the budget:* Crying, when it slows your breathing, will tweak your parasympathetic nervous system, which helps calm you; see heam.info/crying-1.

9. *you reap the benefits:* Dunn et al. 2011. See also Dunn and Norton 2013.

10. *Knitting works, apparently:* Clave-Brule et al. 2009.

11. *"as purely cognitive abilities":* Goleman 1998, 34. *then you're emotionally intelligent:* E.g., in Bourassa-Perron 2011.

12. *low to high emotional granularity:* For a review, see Barrett and Bliss-Moreau 2009a.

13. *fewer days hospitalized for illness:* Quoidbach et al. 2014, Study 2, with ten thousand test subjects.

14. *your experiences in new ways:* See heam.info/emotions-1.

15. *study about fear of spiders:* Kircanski et al. 2012. "Emotion labeling" or "affect labeling," as it is called, is associated with reduced activity in the interoceptive network's body-budgeting regions and greater activity in a control network region (Lieberman et al. 2007; Lieberman et al. 2005).

16. *flexible when regulating their emotions:* Barrett et al. 2001. This paper showed for the first time that intense negative affect, if categorized as emotional experience, is linked to improved emotion regulation. For a review, see Kashdan et al. 2015. Also see heam .info/negative-1. *to drink excessively when stressed:* They consumed about 40% less alcohol than their lower-granularity peers (Kashdan et al. 2010). *someone who has hurt them:* Twenty to fifty percent less likely (Pond et al. 2012). *correct action in social situations:* Kimhy et al. 2014.

17. *major depressive disorder:* Demiralp et al. 2012. *social anxiety disorder:* Kashdan and Farmer 2014. *eating disorders:* Selby et al. 2013. *autism spectrum disorders:* Erbas et al. 2013. *borderline personality disorder:* Suvak et al. 2011; Dixon-Gordon et al. 2014. *more anxiety and depressed feelings:* Mennin et al. 2005, Study 1; Erbas et al. 2014, Studies 2 and 3. *distinguishing positive from negative emotions:* Kimhy et al. 2014.

18. *new moments to cultivate positivity:* E.g., Emmons and McCullough 2003; Froh et al. 2008.

19. *anger before a big game:* Ford and Tamir 2012.

20. *and their movements and sounds:* Gottman et al. 1996; Katz et al. 2012. *well-developed conceptual system for emotion:* E.g., Taumoepeau and Ruffman 2006, 2008. For review, see Harris et al., in press.

21. *themselves for building emotion concepts:* Ensor and Hughes 2008.

22. *poised for greater academic success:* For a review, see Merz et al. 2015. *social behavior and academic performance:* Brackett et al. 2012. See also heam.info/yale-1. *better instructional support for students:* Hagelskamp et al. 2013.

23. *better vocabulary and reading comprehension:* Hart and Risley 1995. The details of these studies are at heam.info/words-1. *lag in the social world:* Fernald et al. 2013. *improves the children's school performance:* Merz et al. 2015; Weisleder and Fernald 2013; Leffel and Suskind 2013; Rowe and Goldin-Meadow 2009; Hirsh-Pasek et al. 2015.

24. *resources to deal with it:* Hart and Risley 2003.
25. *impacts the child's nervous system:* Infants also learn to perceive affect in a voice earlier than in faces; see heam.info/affect-10.
26. *a walk in a park:* Reynolds 2015; Bratman et al. 2015.
27. *10 percent of users avoid relapse:* Spiegel 2012. See also Wood and Rünger 2016.
28. *rather than for the nutrients:* Mysels and Sullivan 2010.
29. *tangible benefits to your life:* This topic is known as stress reappraisal (Jamieson, Mendes, et al. 2013). *that the body is coping:* Jamieson et al. 2010; Jamieson et al. 2012; Jamieson, Nock, et al. 2013. *generally make people feel crappy:* Crum et al. 2013. *so they perform better:* John-Henderson, Rheinschmidt, et al. 2015. *course grade through effective recategorization:* Jamieson et al. 2016. *struggle to make ends meet:* Only 27% of students in remedial math ever earn a bachelor's degree; for details see heam.info/math-1.
30. *the health benefits of continuing:* Cabanac and Leblanc 1983; Ekkekakis et al. 2013; Williams et al. 2012. Also thanks to Ian Kleckner for the Marines example.
31. *intensity of the pain does:* Sullivan et al. 2005. *and crave them less:* Garland et al. 2014. *symptoms with long-term use:* Chen 2014.
32. *the essence of you:* For more on Western psychology's take on the self, see heam .info/self-1.
33. *would call prolonged unpleasant affect:* Buddhism refers to self-affirming possessions, compliments, etc., as "mental poisons." Not only do they cause you to suffer (e.g., feeling like an imposter), but also you feel the urge to harm anything that might invalidate you or threaten to unmask your fictional self. For an example of a fictional self, see heam.info/self-2. *It is an enduring affliction:* It's also a good idea to give up the fiction that people remain the same; see heam.info/self-3.
34. *It depends on other people:* I am not saying that your "Self" is a mere reflection of how others see you or treat you. That is symbolic interactionism, proposed by the philosopher George Herbert Mead and sociologist C. H. Cooley. Still, do you ever find yourself acting and feeling very differently when you are in a new context where no one knows who you are (like when traveling on an airplane)? *be a self by yourself:* This is a signature phrase of social psychologist Hazel Markus. *Wilson out of a volleyball:* The volleyball had the name "Wilson" stamped on it because it was made by the Wilson Sporting Goods Company.
35. *"Stinging Insects," and "Fear":* The self is a concept, but not in the way that social psychologists mean it; see heam.info/self-4.
36. *that we have multiple selves:* After the pioneering research of psychologist Hazel Markus; see heam.info/markus-1. *goal shifts based on context:* Could it be that the population of instances which are "your self" are held together by a word — perhaps your name? See heam.info/self-5.
37. *relation to the same body:* Lebrecht et al. 2012. *of your sense of self:* Other scientists and philosophers have had similar intuitions (Damasio 1999; Craig 2015).
38. *lose your sense of self:* Prebble et al. 2012.
39. *"Wealth" become unnecessary:* Deconstructing the self means putting aside mental poisons to reveal the true nature of experience, i.e., the dharmas in the traditional Abhidarma Buddhist account.
40. *an antacid tablet in water:* Heartbreak from being dumped is a little trickier, because forming an attachment with someone means that you two are co-regulating each oth-

er's body budgets, so separation and loss actually involve some recalibration of your body budget to account for this.

41. *between these regions are stronger:* Tang et al. 2015; Creswell et al., in press. For a summary of the brain-related influences on three types of meditation practice, see heam .info/meditation-1. *[not all] have been well-controlled:* How meditation helps one deconstruct the self and be mindful is an open question; see heam.info/ meditation-2.

42. *something vastly greater than yourself:* Keltner and Haidt 2003. Awe in atheists is similar to faith in those who are believers (Caldwell-Harris et al. 2011).

43. *song comforting while falling asleep:* Only male crickets chirp, and they have different songs for different purposes, but mostly they are singing to attract females. So engage in a little mental inference and think of these sounds as rapturous love songs of nature.

44. *(nobody has proved cause and effect):* Stellar et al. 2015.

45. *a moment of affective realism:* Rimmele et al. 2011.

46. *predict and categorize* in synchrony: Gendron and Barrett, in press; Stolk et al. 2016.

47. *other's chests rising and falling:* For indirect supporting evidence, see Giuliano et al. 2015. *to prepare them for hypnosis:* Some scientists refer to this phenomenon as affective synchrony or affective contagion.

48. *bees, ants, and cockroaches:* Broly and Deneubourg 2015.

49. *to be a good sender:* Zaki et al. 2008.

10. Emotion and Illness

1. *25–40 percent get sick:* Cohen and Williamson 1991.

2. *from a noseful of germs:* Cohen et al. 2003.

3. *inflammation flares up:* Yeager et al. 2011. See more on inflammation at heam.info/ imflammation-1.

4. *to feel seriously like crap:* In a laboratory, when test subjects are injected with the typhoid vaccine, which causes a temporary increase in their proinflammatory cytokines, this was associated with increased activity in the interoceptive network, along with reports of feeling fatigued and very unpleasant (Eisenberger et al. 2010; Harrison, Brydon, Walker, Gray, Steptoe, and Critchley 2009; Harrison, Brydon, Walker, Gray, Steptoe, Dolan, et al. 2009). *cytokines that make inflammation worse:* Mathis and Shoelson 2011. *even get sick more often:* Yang et al. 2016; Cohen et al. 1997; Holt-Lunstad et al. 2010.

5. *the body into the brain:* Proinflammatory cytokines cross the blood-brain barrier (Dantzer et al. 2000; Wilson et al. 2002; Miller et al. 2013). *cells that secrete these cytokines:* Louveau et al. 2015. *particularly within your interoceptive network:* Soskin et al. 2012; Ganzel et al. 2010; McEwen and Gianaros 2011; McEwen et al. 2015. See heam .info/inflammation-2. *pay attention and remember things:* Karlsson et al. 2010. *lowering performance on IQ tests:* There is a vicious cycle: lower IQ, often associated with childhood adversity and poverty, predicts higher levels of inflammation in midlife (Calvin et al. 2011). See also Metti et al. 2015.

6. *flush with cortisol and cytokines:* See more on the relationship between cytokines and cortisol levels at heam.info/cortisol-2. *and chronic inflammation sets in:* Dantzer et al. 2014; Miller et al. 2013. This situation actually sensitizes you to interoceptive and nociceptive input (Walker et al. 2014).

7. *really, truly in trouble:* Dowlati et al. 2010; Slavich and Cole 2013; Slavich and Irwin 2014; Seruga et al. 2008.

8. *acts like fertilizer for disease:* Irwin and Cole 2011; Slavich and Cole 2013. See more on stress, genes, and cytokines at heam.info/cytokines-1. See also heam.info/glial-1. *Death from cancer comes sooner:* Stress-related increases in β-adrenergic sympathetic nervous system (SNS) activity encourage proinflammatory gene expression and discourage anti-viral immune gene expression as cells replicate (Irwin and Cole 2011). These transcriptional effects have been observed in breast tissue, lymph nodes, and the brain (Williams et al. 2009; Sloan et al. 2007; Drnevich et al. 2012). In this way, an acute physiological state can influence cellular makeup for days, weeks, months, or even years (Slavich and Cole 2013), enhancing vulnerability to cancer. Stress-related SNS activity also directly influences the micro-environment of tumor cells, enhancing metastasis, augmenting tumor cell potency, and increasing mortality (Antoni et al. 2006; Cole and Sood 2012).

9. *distinguished it from all others:* Zachar and Kendler 2007; Zachar 2014.

10. *all associated with hub damage:* Menon 2011; Crossley et al. 2014; Goodkind et al. 2015.

11. *poverty, abuse, or loneliness:* For a discussion of childhood adversity and earlier mortality in adulthood, see Danese and McEwen 2012. For loneliness-related death, see Perissinotto et al. 2012. For the link between poverty and brain development, see Hanson et al. 2013, and for the link between childhood poverty and premature adult mortality (independent of family history, ethnicity, cigarette smoking, and other risk factors), see Hertzman and Boyce 2010. Also see Adler et al. 1994.

12. *stress and emotion are independent:* For a rare counterexample, see Lazarus 1998.

13. *circuitry that regulates the budget:* Ganzel et al. 2010; McEwen and Gianaros 2011; McEwen et al. 2015.

14. *accurately regulate your body budget:* E.g., Danese and McEwen 2012; Sheridan and McLaughlin 2014; Schilling et al. 2008; Ansell et al. 2012; Hart and Rubia 2012; Teicher and Samson 2016; Felitti et al. 1998. For more on how childhood adversity wires the brain, see heam.info/adversity-1. *trajectory toward chronic disease:* Miller and Chen 2010. *childhood abuse or neglect:* Teicher et al. 2002; Teicher et al. 2003; Teicher et al. 2006; Teicher and Samson 2016. *the target of a bully:* Teicher et al. 2002; Teicher et al. 2003; Teicher et al. 2006. *psychiatric and physical diseases:* Copeland et al. 2014. *cancer, and other diseases:* Repetti et al. 2002. For more on the bad effects of stress, see heam.info/stress-3.

15. *during recovery from prostate cancer:* Hoyt et al. 2013. *or after a stressful event:* Master et al. 2009. *affect that they didn't label:* Hoyt et al. 2013. *for cancer-related symptoms:* Stanton et al. 2000; Stanton et al. 2002. *that lead to poor health:* Labeling reduced sympathetic nervous system reactivity to negative images for up to a week (Tabibnia et al. 2008).

16. *brain predicts damage is imminent:* International Association for the Study of Pain 2012. The IASP now defines pain as an emotional experience and writes that "pain is always subjective. Each individual learns the application of the word through experiences related to injury in early life." Translation: pain is a population of perceptions that vary, one from the next, and the concept needed to construct these perceptions is learned early in life. Sounds like the theory of constructed emotion, doesn't it?

17. *sensations and made them meaningful:* For an example of body-budgeting regions processing nociceptive prediction errors, see Roy et al. 2014.

18. *process nociception change their activity:* E.g., Wiech et al. 2010. For a review, see Tracey 2010; Wager and Atlas 2015. *treatment like a sugar pill:* Büchel et al. 2014; Tracey 2010; Wager and Atlas 2015. *and other opiate drugs:* Opioids are not the only neurotransmitters responsible for the placebo effect. Also involved is cholecystokinin (CCK), which acts on endogenous cannabinoid receptors in your brain, the same as marijuana. CCK tunes up nociception, whereas opioids tune it down (Wager and Atlas 2015). *"your internal medicine cabinet":* Benedetti et al. 2006; Benedetti 2014; Tracey 2010; Wager and Atlas 2015. See also heam.info/opioids-1. Many people believe that dopamine is the neurochemical that is linked to positivity and reward; for more on that, see heam.info/dopamine-1.

19. *interoceptive and control networks:* For another example of how these same brain networks configure to make meaning of nociceptive input during the construction of pain experiences, see Woo et al. 2015. For more on similarities between the construction of pain and emotion, see heam.info/pain-1. *is a form of interoception:* The prominent neuroanatomist A. D. (Bud) Craig, who knows more about this circuitry than just about anyone else, argues that nociception is a form of interoception (Craig 2015). See heam.info/craig-1.

20. *you'd report feeling more pain:* E.g., Wiech and Tracey 2009; Roy et al. 2009; Bushnell et al. 2013; Ellingsen et al. 2013. *like a volume control:* For a partial outline of some of the circuitry, see Wager and Atlas 2015. *status reports to your brain:* See more on nociceptive pathways at heam.info/pain-2. *you could develop a stomachache:* E.g., Traub et al. 2014.

21. *and chronic back pain:* Chronic pain can be neuropathic, inflammatory, or idiopathic; see heam.info/pain-3. *$635 billion each year:* American Academy of Pain Medicine 2012. *more than half the time:* Apkarian et al. 2013 estimates that 50 million Americans are either partially or totally disabled by pain. *today's great medical mysteries:* One part of the mystery: opioid drugs taken to relieve pain actually have a hand in transforming acute pain into chronic pain; see Lee et al. 2011 for a comprehensive review of opioid-induced hyperalgesia. See also heam.info/opioids-2.

22. *with its roots in inflammation:* Borsook 2012; Scholz and Woolf 2007; Tsuda et al. 2013. The International Association for the Study of Pain defines chronic pain (which they call "neuropathic pain") as "pain caused by a lesion or disease of the somatosensory system" (IASP 2012). Aberrant predictions count as "a disease."

23. *keeps issuing predictions about it:* van der Laan et al. 2011. See more on phantom limb syndrome at heam.info/phantom-1.

24. *likely to develop persistent pain:* Beggs et al. 2012. *heightened pain in later childhood:* Hermann et al. 2006; Walker et al. 2009. *routinely* not anesthetized: Wikipedia, s.v. "Pain in Babies," last modified February 23, 2016, http://en.wikipedia.org/wiki/Pain_in_babies. *linked to bad nociceptive predictions:* National Institute of Neurological Disorders and Stroke 2013; Maihöfner et al. 2005; Birklein 2005.

25. *scans will look somewhat different:* In chapter 1, we discussed the use of pattern classification to diagnose instances of different emotion categories (e.g., distinguishing instances of anger from fear). Each classifier is not a brain state for the emotion; the pattern that successfully diagnoses instances of an emotion is an abstract statistical representation that need not exist in any instance of the category. The same holds true for emotion and pain. My colleague Tor D. Wager has published a pattern classifier that successfully distinguishes between nociceptive pain and emotion (Wager et al. 2013; Chang, Gianaros, et al. 2015), and together we have published pattern

classifiers for anger, sadness, fear, disgust, and happiness (Wager et al. 2015). These classifiers are not neural essences for pain and emotion but are statistical summaries of highly variable instances of each category. *they look somewhat different too:* Wilson-Mendenhall et al. 2011.

26. *make sense of bodily sensations:* See heam.info/pain-8. *or threat to your tissue:* See heam.info/pain-5. *misleading data from your body:* Chronic pain spits in the face of the classical view of human nature; see heam.info/pain-6.

27. *"no longer be borne":* Styron 2010.

28. *a disease of the mind:* For a comparison of which diseases are "neurological" versus "psychiatric," Neuroskeptic (2011) tallied the number of scholarly papers, by topic, published in the journals *Neurology* versus the *American Journal of Psychiatry* from 1990 to 2011. See also heam.info/neurology-1. *your genes make you vulnerable:* Certain genes make you more or less sensitive to the environment (Ellis and Boyce 2008). For an informative lecture, see Akil 2015. See also heam.info/depression-1.

29. *not effective for everyone either:* Olfson and Marcus 2009; Kirsch 2010. See also heam.info/depression-5. *and then recur throughout life:* Curry et al. 2011. *war, or accidents:* Mathers et al. 2008.

30. *is not just one thing:* This is true because most human phenomena and characteristics are caused by degenerate gene combinations that are so variable that a detailed genetic explanation (involving the exact genes and mechanisms by which they influence one another) for any of them is unlikely, even when they have high heritability quotients, meaning that much of the observed variation in that characteristic is due to genetic variability (Turkheimer et al. 2014).

31. *sensory information from your body:* Your muscles contain energy sensors, for example, that send feedback about energy usage back to your brain (Craig 2015). *or other symptoms of depression:* Barrett and Simmons 2015. *heart disease, and cancer:* Your metabolism controls your immune system to some extent; fat cells emit proinflammatory cytokines (Mathis and Shoelson 2011), which means that obesity makes chronic inflammation worse. See, e.g., Spyridaki et al. 2014.

32. *scale that shuts you down:* Kaiser et al. 2015. When we look at the brains of people suffering from depression, we see activity and connectivity changes that are consistent with this hypothesis; see heam.info/depression-2.

33. *the parts of a machine:* In depression, dysregulation is widespread; see heam.info/depression-3. *built from toxic past experiences:* Ganzel et al. 2010; Dannlowski et al. 2012. Once a glucocorticoid gene becomes overexpressed at a young age (in rats), the brain pathways become set, creating a lifelong vulnerability to mood disorders and more lability, even if the gene turns off in adulthood (Wei et al. 2012). Toxic past experiences also lead to prolonged inflammation in childhood that increases the risk of depression and other illnesses later in life (Khandaker et al. 2014). *environment and every little problem:* Sometimes called "neuroticism" or "affective reactivity"; also see heam.info/depression-1. *post-traumatic stress disorder:* Risk is greatest with high levels of the ovarian hormone progesterone. This might help explain why the proportion of women suffering from mood disorders is so much higher than the proportion of men (Lokuge et al. 2011; Soni et al. 2013); e.g., Bryant et al. 2011. See also heam.info/women-1.

34. *your interoceptive network is restored:* Namely, the subgenual anterior cingulate cortex decreases in activity, and its connectivity with the rest of the interoceptive network increases, as does connectivity to the thalamus, which brings prediction error signals

(Riva-Posse et al. 2014; Seminowicz et al. 2004; Mayberg 2009; Goldapple et al. 2004; Nobler et al. 2001). For a meta-analytic review, see Fu et al. 2013. *for whom no treatments work:* McGrath et al. 2014.

35. *critical to anxiety as well:* On the connectivity of the interoceptive and control networks during anxiety, see McMenamin et al. 2014. On the similarity between anxiety and chronic pain, see Zhuo 2016, and Hunter and McEwen 2013. And for evidence consistent with the idea that anxiety enhances pain via prediction, see Ploghaus et al. 2001. *error across these two networks:* Paulus and Stein 2010. *stress, and depression:* E.g., Menon 2011; Crossley et al. 2014. Even fear and anxiety were once thought to be caused by separate circuits (Tovote et al. 2015). Also see heam.info/anxiety-1.

36. *is failing to regulate it:* Compare Suvak and Barrett 2011, and Etkin and Wager 2007. See also heam.info/anxiety-2.

37. *That's classic anxiety:* Anxiety followed by depression might be worse than depression followed by anxiety, because in the latter, a person might be starting to process prediction error again.

38. *sit in the control network:* van den Heuvel and Sporns 2013. *to learn effectively from experience:* Browning et al. 2015. *imprecisely or not at all:* A brain awash in prediction error is not always anxious; consider the infant's lantern of attention (chapter 6) or times when novelty and uncertainty are pleasant (e.g., meeting a new lover); see, e.g., Wilson et al. 2013. See heam.info/anxiety-3. *your brain ignores them:* Damasio and Carvalho 2013; Paulus and Stein 2010. *error that you can't resolve:* Specifically, from using prediction error as a "teaching signal" (McNally et al. 2011; Fields and Margolis 2015). *know their disease is permanent:* Six months after a serious operation (a colostomy), those who had a chance of having their colostomies reversed were less satisfied with life than those with permanent disability (Smith et al. 2009). Hope can be a cruel mistress.

39. *also with chronic fatigue syndrome:* To be clear, I am not saying that depression and chronic pain are the same phenomenon. I am saying that they have a set of common causes. There is a longstanding debate whether certain chronic pain syndromes are independent of depression, as opposed to being expressions of depression. In the past, this debate has been framed as a version of "it's all in your head," where spontaneously experienced pain in the absence of tissue damage is assumed to be a sign of mental illness. This line of argument assumes that depression is merely a mental illness, but this historical distinction is not meaningful in the light of modern neuroscience. Both depression and chronic pain can be considered neurodegenerative brain diseases that have metabolic and inflammatory roots. The fact that some prescription drugs are successful at reducing some instances of depression but not of chronic pain (or vice versa) does not mean the two are distinct biological categories, because depression has degenerate causes. Not everyone suffering from depression (i.e., the variable members of that category) is treated successfully with the same medication (i.e., variation is the norm). The same logic probably works for any category of chronic pain.

40. *are highly variable and malleable:* Barrett 2013.

41. *symptoms sounds just like autism:* The diagnostic symptoms of autism are consistent with my description; see heam.info/autism-1.

42. *have multiple, complex causes:* Jeste and Geschwind 2014. See also heam.info/autism-2.

43. *"An Inside View of Autism":* Grandin 1991. *"why she was not a cat":* Grandin 2009. *"and all is well":* Higashida 2013.

44. *is a failure of prediction:* Van de Cruys et al. 2014; Quattrocki and Friston 2014; Sinha et al. 2014. *the trajectory of brain development:* For a discussion, see heam.info/autism-3.

45. *sound of a laugh track:* There is now ample evidence that children and adolescents learn both physical and relational aggression from the media (Anderson et al. 2003). Situation comedies, both those designed for children and those for general audiences, contain some aggression in over 90% of programs sampled, compared with 71% of reality programs (Martins and Wilson 2011). In the fifty television shows that are most popular with children ages two to eleven years old, episodes contained, on average, fourteen different incidents of relational aggression per hour, or one every four or five minutes (Martins and Wilson 2012a). Young teenagers find relational and physical aggression funny (as opposed to upsetting) when it's performed by a likeable character in teenage ("tween") sitcoms; in addition, teens report that they are more likely to imitate the aggression themselves (Martins et al., in press). In younger school-aged children (K–5), girls are more likely to model relational aggression at school after having watched it on television (Martins and Wilson 2012b). Most concerning of all, these shows usually depict victims as experiencing no pain, particularly in the reality shows (Martins and Wilson 2011). Television shows influence not only how children and adolescents act but also their expectations of others. For example, after watching television clips with one character harming another in a physically or relationally aggressive way, children are more likely to predict that others have hostile intent (Martins 2013).

46. *midst of an opiate crisis:* Kolodny et al. 2015.

47. *when they are not hungry:* Mena et al. 2013. *act as a mild analgesic:* Mysels and Sullivan 2010. *might not be far off:* Avena et al. 2008.

48. *common underlying factors instead:* These observations led the U.S. National Institute of Mental Health (NIMH) to completely revamp its scientific approach in ways that are reminiscent of the theory of constructed emotion. Instead of considering each named illness as having a distinct essence, scientists now treat each one as a category full of variety and search for common, underlying causes (NIMH 2015).

11. Emotion and the Law

1. *is lowered into the ground:* Unless you are Dan Wegner, social psychologist and my good friend, who died in 2013 after bravely suffering amyotrophic lateral sclerosis (ALS). At Dan's memorial service, per his request, the speakers sauntered in wearing plastic Groucho Marx glasses with fake noses.

2. *responsible for your actions:* You are legally responsible for a criminal action but not necessarily for a civil action or negligent action like professional negligence, where the law requires a duty to another person, dereliction of that duty, proximate or legal cause, and compensable harm, for example. *an individual with free will:* One exception might be "fighting words," the idea that certain words spoken by another person are so offensive that you may be justified in harming the speaker.

3. *if you intended that harm:* The law distinguishes action, intent, and motivation; see heam.info/harm-1.

4. *the time of the crime:* People v. Patterson, 39 N.Y.2d 288 (1976).

5. *of destruction in its path:* Kahan and Nussbaum 1996; Percy et al. 2010. For wonderful metaphors, see Lakoff 1990. *person's responsibility for his actions:* Some legal scholars

acknowledge that emotions might not be a departure from rationality but rather a form of it; see heam.info/rational-1.

6. *sadness, and fear:* Kreibig 2010; Siegel et al., under review.

7. *wishing ill upon their oppressor:* Kuppens et al. 2007.

8. *deliberately into a frothing anger:* Kim et al. 2015. Knowing just when to get angry is a key aspect of emotional intelligence (Ford and Tamir 2012). See also heam.info/ anger-2. *"You have to go":* Zavadski 2015; Sanchez and Foster 2015.

9. *times are not necessarily emotional:* Barrett et al. 2004. See also heam.info/control-1.

10. *distinct systems in the brain:* Cisek and Kalaska 2010.

11. *due to the direct wiring:* Actually, it just seems as if there is one motor action. Many slightly different motor actions can be executed to perform the same behavior, as motor actions are degenerate. For a helpful summary, see Anderson 2014, Interlude 5. Also see Franklin and Wolpert 2011.

12. *decision-making:* Swanson 2012, following George Howard Parker (1919) and the neuroscientist and Nobel laureate Santiago Ramon y Cajal (1909–1911). See also heam.info/association-1.

13. *just that — an experience:* Your control network is always actively engaged whether you're aware or not; see heam.info/control-2.

14. *of your thoughts and actions:* The feeling of control is defined as awareness (you are able to report or reflect on your attempts at control), agency (you experience yourself as in control, as the agent), effort (you experience processing as effortful), and control (you are aware that automatic processes are occurring and are motivated to counteract them); see heam.info/control-3.

15. *the experience of having control:* I suspect the brain creates the experience of control like any other experience: you have a concept for "Agency," and you apply it as a prediction to a bunch of sensations. For a similar view, see Graziano 2013.

16. *are more stoic and analytical:* More on stereotypes of emotionality in men and women is at heam.info/stereo-1. *"that I considered important":* Albright 2003. See also heam .info/albright-1.

17. *there are no sex differences:* Barrett et al. 1998. *hardwired for stoicism or rationality:* Neuroscience evidence suggests that the "male brain" and "female brain" are myths; see heam.info/stereo-2.

18. *experiences of emotion while watching:* Kring and Gordon 1998; Dunsmore et al. 2009. Actually, women just move their facial muscles more in general, so they are not really more "expressive" (Kelly et al. 2006). Also, in studies that measure facial EMG, there are as many studies that find sex differences as those that don't (Barrett and Bliss-Moreau 2009b).

19. *the sex of the defendant:* Kahan and Nussbaum 1996.

20. *are supposed to be aggressors:* Tiedens 2001. *they're supposed to be afraid:* This belief exists even though all mammals attack during threat; see heam.info/attack-1. *and perhaps even their jobs:* Brescoll and Uhlmann 2008; Tiedens 2001. *be really competent and powerful:* Hillary Clinton is another example; see heam.info/clinton-1.

21. *who kill their intimate partners:* Percy et al. 2010; Miller 2010.

22. *passive, and helpless:* Morrison 2006; Moore 1994. See also "Developments in the Law" 1993, citing court opinions that portray battered women as "helpless, passive or psychologically disturbed" (1592).

23. *of second-degree murder:* Moore 1994. *manslaughter, a lesser charge:* African American women are in a catch-22; see heam.info/defense-1.

24. *the rapist a heavier sentence*: Schuster and Propen 2010, in Bandes, forthcoming. *just having a bad day*: Barrett and Bliss-Moreau 2009b.

25. *relief and happiness go unmentioned*: Abrams and Keren 2009. *people of the same sex*: Calhoun 1999.

26. *in and out of court*: For example, laws related to the "war on crime" put in place by Richard Nixon created a culture of fear against certain ethnic groups in the United States (Simon 2007). *the target of inconsistent rulings*: Abrams and Keren 2009, 2032.

27. *and her crime was possible*: Feresin 2011.

28. *findings in their defense strategy*: For a review, see Edersheim et al. 2012.

29. *neurons in the human brain*: Graziano 2016.

30. *to pain to math skills*: As shown by a meta-analysis of almost six thousand brain-imaging experiments; see heam.info/meta-1. *and impulsivity in some instances*: This is called the "reverse inference problem"; see heam.info/rev-1.

31. *aggression, let alone murder*: For more on brain region size and free will, see heam .info/size-1. *and cause severe personality changes*: Burns and Swerdlow 2003; Mobbs et al. 2007.

32. *automatically releases someone from responsibility*: The same argument could serve as a reason to keep Albertani locked up; see heam.info/albertani-1.

33. *"he has no regrets"*: McKelvey 2015. *"he is devoid of"*: Stevenson 2015.

34. *sex, or ethnicity*: Haney 2005, 189–209; Lynch and Haney 2011. See also heam.info/empathy-1. So much for the idea of being judged by a jury of your peers (which is enshrined in the Magna Carta and the U.S. Bill of Rights).

35. *the "Chechen wolf"*: Wikipedia, s.v. "Chechen Wolf," last modified March 18, 2015, http://en.wikipedia.org/wiki/Chechen_wolf.

36. *painful to shame your family*: Nisbett and Cohen 1996.

37. *leading to his death sentence*: Imagine if a defendant in a murder case smiled through the proceedings; see heam.info/trial-1.

38. *as evidence from the trial*: Keefe 2015. See also Gertner 2015.

39. *decision between imprisonment and death*: In fact, the jury's perception of whether or not a defendant is remorseful largely determines whether it recommends the death penalty (Lynch and Haney 2011).

40. *of the parole board resign*: Some reports say six members resigned; see heam.info/tsarnaev-1.

41. *to have a fair trial*: Riggins v. Nevada, 504 U.S. 127, 142 (1992) (Kennedy, J., concurring). Presumably, defendants are deprived of a fair trial by those things that interfere with a jury's perceiving remorse.

42. *cascade of predictions (chapter 6)*: It's so ubiquitous in Western culture that scholars keep rediscovering it and calling it by different names, such as "mind perception," "person-perception," and "mentalizing." For an entertaining and insightful treatment on this issue, see Wegner and Gray 2016.

43. *mental inferences, that is, guesses*: Gilbert 1998.

44. *conservative subjects inferred violent intentions*: Kahan et al. 2012.

45. *to recommend more severe punishments*: Nadler and Rose 2002; Salerno and Bottoms 2009, both in Bandes, forthcoming. See also Bandes and Blumenthal 2012. *a jury-swaying masterpiece*: Kelly v. California, 555 US 1020 (2008).

46. *stuck while he was traveling*: Goodnough 2009.

47. *justification for Florida's law*: Montgomery 2012.

48. *but they were not neuroscientists:* For the full statement of the Second Amendment, see heam.info/second. *gun will make them safer:* Kohut 2015, in Blow 2015.

49. *literal readout of the world:* Loftus and Palmer 1974; Kassin et al. 2001.

50. *place in Australia in 1975:* Massachusetts General Hospital Center for Law, Brain, and Behavior 2013.

51. *convicted based on eyewitness testimony:* Innocence Project 2015; Arkowitz and Lilienfeld 2010.

52. *go wrong in eyewitness testimony:* New Jersey Courts 2012; *State v. Lawson,* 291 P.3d 673, 352 Or. 724 (2012); *Commonwealth v. Gomes,* 470 Mass. 352, 22 N.E.3d 897 (2015). *that they did not commit:* Schacter and Loftus 2013; Deffenbacher et al. 2004.

53. *"most especially their emotions":* Scalia and Garner 2008.

54. *"or Easter bunnies":* United States v. Ballard, 322 U.S. 78, 93–94 (1944) (Jackson, J., dissenting). *instead of to hunger (chapter 4):* Danziger et al. 2011. *more likeable or sympathetic people:* Wistrich et al. 2015.

55. *is more likely to lose:* Black et al. 2011. *affective connotations in the judges' words:* Ironically, the late Justice Antonin Scalia was known for his emotional style of discourse; see heam.info/scalia-1.

56. *half of the United States:* Wikipedia, s.v. "David Souter," last modified March 30, 2016, http://en.wikipedia.org/wiki/David_Souter. *under the fiction of equanimity:* The sociologist Arlie Hochschild calls it "emotional labor" (Hochschild 1983).

57. *sentencing portion of criminal cases:* In 1972, the Supreme Court decreed that "any decision to impose the death sentence be, and appear to be, based on reason rather than caprice or emotion" (*Furman v. Georgia,* 408 U.S. 238, 311 [1972], [Stewart, J., concurring], as cited in Pillsbury 1989, 655n2). Since then, the Supreme Court has worked hard to remove emotional considerations from sentencing. Presumably, they assume that if a judge follows the rules, without the aid of emotion, then the outcome will be fair. Of course, the brain's wiring reveals that no judgment is ever free of body-budgeting considerations, and therefore a judge can implement the rules with affective realism (chapter 4) without ever knowing it. Ironically, judges know they need affect to do their job. Here is a quote from one judge: "Now, there's two things that can happen to you. Either you're going to remain a decent person and become terribly upset by it all because your emotions — because your feelings are being pricked by all of this constantly or you're going to become — you're going to grow a skin on you as thick as a rhino, in which case I believe you're going to become an inadequate judicial officer because once you lose the human — the feeling for humanity you can't really — I don't believe you can do the job" (Anleu and Mack 2005, 612). See heam.info/judges-1. *"more to be nurtured than feared":* Brennan 1988, as cited in Wistrich et al. 2015. Brennan foreshadowed Antonio Damasio. Science is on Justice Brennan's side here: no one is immune to affective realism (chapter 4).

58. *Aurora, Colorado, in 2012:* Wikipedia, s.v. "2012 Aurora Shooting," last modified April 21, 2016, http://en.wikipedia.org/wiki/2012_Aurora_shooting. *construct an experience of anger:* We might say that anger is appropriate, and even useful, because it is a form of social reality that shows the judge to be committed to preserving moral order in a society that promotes respect for others. See Berns 1979, in Pillsbury 1989, 689n112; also see Ortony et al. 1990. *victim of some sort himself:* Pillsbury 1989. There is a longstanding controversy over the role of empathy and emotions in judicial practice. Interested readers should see heam.info/empathy-2. *ignorance of the defendant's*

perspective: Anger as ignorance comes from contemplative philosophies such as Buddhism. *punishing the offender during sentencing:* Pillsbury 1989. It is difficult for a judge to see himself as similar to a defendant, which might be why judges are more likely to hand out maximum sentences (ibid., 705n155). *of emotion in the courtroom:* See heam.info/empathy-3. For an example of how enhanced emotional granularity improves moral decision-making, see Cameron et al. 2013.

59. *a host of other illnesses:* Copeland et al. 2013.
60. *early adversity have shorter telomeres:* Kiecolt-Glaser et al. 2011.
61. *disease of prediction gone wrong:* Borsook 2012.
62. *"cruel and unusual punishment":* Convention (III) relative to the Treatment of Prisoners of War. Geneva, August 12, 1949. Prisoners of war "are entitled in all circumstances to respect for their persons and their honour" (article 14) and "must at all times be protected . . . against insults and public curiosity" (article 13). U.S. Constitution, Eighth Amendment.
63. *telomeres and potentially their lifespan:* Guarneri-White 2014. *verbal aggression and physical threats:* Wikipedia, s.v. "Suicide of Phoebe Prince," last modified January 30, 2016, https://en.wikipedia.org/wiki/Suicide_of_Phoebe_Prince. *playground in a legal context:* Matters surrounding bullying are made more complicated by the fact that our culture models bullying as normative; see heam.info/bully-1.
64. *reported involvement with electronic bullying:* During a two-month period in 2005, using a nationally representative sample of over seven thousand children from grades six to ten (Wang et al. 2009).
65. *contaminating its warehouse with feces:* Monyak 2015. *"distress and mental anguish":* The lawyer arguing the case asked the jury to send corporate America a message; see heam.info/atlanta-1. *consequently so does compensation:* Note that the large majority of civil cases reach a settlement out of court; see heam.info/harm-2.
66. *which is far more variable:* How do you quantify suffering in dollars? See heam.info/harm-3.
67. *withdrawal from an addictive drug:* Fisher et al. 2010.
68. *a defendant than others will:* Zaki et al. 2008. *this synchrony and cultivate empathy:* Schumann et al. 2014.
69. *deep dividing lines in nature:* Even biological sex is not a natural kind; for informative discussions, see Dreger 1998, and Dreger et al. 2005. See also Dreger 2015.
70. *(self-reports are not necessarily valid):* One useful approach during voir dire can be adapted from the research of U.S. attorney Dan Kahan; see heam.info/kahan-1.
71. *guilt was true or false:* I am not implying that objective evidence is error-free, nor that it is completely free of human judgment. *consistency produces a just outcome:* Judges and lawyers must have realized that consistency does not always deliver justice, meaning that there will be some false positives (innocent people who are convicted). Thinking about the implication — that some sacrifices must be made for the good of the system — is worrisome, even alarming. Who said *The Hunger Games* was complete fiction?
72. *to hand out maximum sentences:* Pillsbury 1989, 705n155.
73. *influences you were pickled in:* This wonderful phrase comes from my friend and colleague Judith Edersheim, codirector of the Center for Law, Brain, and Behavior at Massachusetts General Hospital. *an unarmed African American civilian:* Fachner et al. 2015, 27–30. *the symbols of your culture:* As another example: a Confederate battle flag, which symbolizes racism to many people, flying atop a statehouse building and even appearing as part of a couple of state flags; see heam.info/flag-1.

12. Is a Growling Dog Angry?

1. *scientific discoveries in animal emotion:* A quick search of *Time, Pacific Standard, Newsweek, Atlantic Monthly, Boston Globe, Chicago Tribune, USA Today, Los Angeles Times,* and the *New York Times* turned up twenty-six articles between 2009–2014 reporting that animals have emotions. *dogs get jealous:* Harris and Prouvost 2014. *rats experience regret:* Steiner and Redish 2014. *crayfish feel anxiety:* Fossat et al. 2014. *flies fear the incoming flyswatter:* Gibson et al. 2015. *"they're largely the same":* Safina 2015, 34.

2. *but not for emotion:* LeDoux 2014.

3. *same basic nervous system plan:* Swanson 2012; Donoghue and Purnell 2005.

4. *about 25 million years ago:* Goodman 1999. All of these species have evolved since then to suit their habitats, so our modern forms hardly count for an evolutionary comparison. But scientists do their best to take that into consideration when interpreting the experimental results. *that the human network does:* Touroutoglou et al. 2016. More generally, macaque and human brains are very similar to one another (Barbas 2015), with a few notable changes, mostly at the front of the brain (Hill et al. 2010); see also heam.info/macaque-1.

5. *watching negative behaviors like cowering:* Bliss-Moreau et al. 2013. See also heam.info/macaque-2.

6. *are paired with electric shock:* Malik and Hodge 2014.

7. *can feel pleasure or pain:* Bentham believed in utilitarianism; see heam.info/bentham-1.

8. *more things matter to us:* Globalization is just a massive expansion of your affective niche; see heam.info/niche-1.

9. *"baby talk" tone of voice:* Amso and Scerif 2015. The infant and her caregiver are sharing attention; see heam.info/sharing-1.

10. *what is in her mind:* Okamoto-Barth and Tomonaga 2006; see also heam.info/gaze-1.

11. *large as a macaque brain:* Passingham 2009. *to learn purely mental concepts:* Most of the evolutionary changes have occurred in the cortical areas that have many neurons for processing prediction errors; see heam.info/evolution-2.

12. *animals learn concepts by smell:* Animals have concepts (Lea 2010). Primary olfactory cortex has a limbic structure that is closely connected to visceromotor limbic regions. For a review, see Chanes and Barrett 2016. *sight or sound as well:* While mammals are more dominated by olfactory concepts, birds are more visually dominated. Mammals and birds split from a common ancestor about 200 million years ago. *goats by vocal bleats:* Lea 2010.

13. *reward them with food or drink:* Mareschal et al. 2010. See also heam.info/animals-1. *regardless of font:* Vauclair and Fagot 1996. *animal images from food images:* Fabre-Thorpe 2010. *differ only by color:* Yoshikubo 1985; Marmi et al. 2004. For more examples, see Fabre-Thorpe 2010. *van Gogh, and Salvador Dalí:* Four macaques were trained to classify parts of paintings from these three painters and a fourth, Jean-Léon Gérôme. These parts contained no faces or full objects that could be memorized; monkeys were required to attend to the style of painting (Altschul et al. 2015).

14. *more of this critical wiring:* Goodman 1999. See also heam.info/evolution-2.

15. *making a mental inference:* Vallacher and Wegner 1987; Gilbert 1998. *thinking, desiring, or feeling:* Martin and Santos 2014.

16. *mental similarities amid perceptual differences:* For example, Tomasello 2014; Hare
     and Woods 2013. *just an action; it's a goal:* According to Michael Tomasello (2014,
     27–29), great apes create concepts that go beyond mere perceptual similarities, and
     they represent information about the situation (e.g., whether food is present or not).
     Most likely, they also create concepts in a generative way, meaning they can use bits
     and pieces of prior experience to create a novel prediction, up to a point (ibid., 28).
     A discussion of the concept "To Climb" can be found in ibid., 29. *have a shared men-
     tal goal:* The default mode networks in human and chimp brains are similar in the
     brain regions that are connected to one another but not in the microscopic wiring; see
     heam.info/chimp-1. *way that human infants do:* Scientists actively debate the brain
     mechanisms for human language; see heam.info/language-2.
17. *"wanting to have some":* Tomasello 2014, 105. See also heam.info/animals-2. *in order
     to request rewards:* Famous attempts to teach language to apes are described at heam
     .info/animals-3.
18. *use symbols on their own:* That is, just by exposing chimps to symbol-based language,
     without explicit rewards (e.g., Matsuzawa 2010; Hillix and Rumbaugh 2004). *the
     symbol to unfamiliar tools:* Tanaka 2011. Chimps seem to be able to recognize that
     different-looking objects can achieve the same function, as long as that function in-
     volves some sort of direct motor action. For example, chimps may understand that a
     stick can be used to obtain food in multiple ways: retrieving termites from the ground,
     opening a can of food, or shaking fruit from a tree. They might even understand
     that a ladder is a "Tool" to shake fruit from a tree. But would they understand that
     completely dissimilar objects, when employed with very dissimilar actions, are both
     "Tools," like a rock for cracking nuts and a ladder for reaching fruit in a tree? Would
     they understand that the same rock is also a "Tool" when used for non-food-related
     purposes, like weighting down light objects to keep them from blowing away in the
     wind? If a chimp uses a stick to threaten a subordinate, or if the chimp requests food
     from a human, would it understand that the stick and the human are "Tools" as well?
19. *waiting at the other end:* Herb Terrace, personal communication, June 6, 2015. *alone
     are not worth learning:* If an event or object does not perturb an animal's body budget,
     and is not relevant to energy regulation, then there is less need to invest the resources
     to build a concept for it. Research by the cognitive psychologist Patricia K. Kuhl sug-
     gests that language learning requires a brain's body-budgeting regions to be engaged,
     for example; see Kuhl 2014.
20. *cooperative than common chimps:* Chimps and bonobos last shared a common ances-
     tor about 1 million years ago (Becquet et al. 2007; Hey 2010). *the meaning of concrete
     words:* A comparison of chimps and bonobos is at heam.info/chimp-2.
21. *the results of the experiments:* Tetsuro Matsuzawa, personal communication, June 12,
     2015. See also heam.info/chimp-3.
22. *equally well under these conditions:* Murai et al. 2005.
23. *a flying leopard:* Tomasello 2014, 29. *from different points of view:* Ibid. This requires a
     type of simulation (Mesulam 2002) that a chimp brain does not seem wired to do. *the
     heads of other creatures:* Infant chimps stop following their mother's gaze during the
     first year of life (Matsuzawa 2010). Adult chimps can follow gaze under some circum-
     stances; see heam.info/chimp-4.
24. *exchangeable for goods in general:* Sousa and Matsuzawa 2006. Chimps are capable of
     constructing and using tools in complex ways. See also heam.info/chimp-5.

25. *had picked up the practice:* Trivedi 2004. For discussion, see Jablonka et al. 2014.

26. *unique in the animal kingdom:* Other scientists have similar views; see heam.info/reality-2.

27. *"letting the infant nurse":* Morell 2013, 222–223.

28. *motivation to interact with humans:* For more on Belyaev's story, see Hare and Woods 2013.

29. *can regulate ours in turn:* Learn more about the experiments showing human-dog body-budget regulation at heam.info/dogs-1.

30. *and vice versa:* Quaranta et al. 2007.

31. *heart rate and other factors:* Siniscalchi et al. 2013. For commentary, see heam.info/sides-1. *faces and voices of humans:* Turcsán et al. 2015.

32. *if trained to do so:* Range et al. 2008.

33. *the smells of other humans:* Settle et al. 1994.

34. *gestures and following human gaze:* Hare and Woods 2013, 50–51. *our mind in our eyes:* For a thoughtful discussion, see Bradshaw 2014, 200. *get information about the world:* Hare and Woods 2013, 50.

35. *more sophisticated than playing fetch:* Kaminski et al. 2009; Hare and Woods 2013, 129. *(affect) in the acoustic signal:* Owren and Rendall 2001. *food, and her crate:* Rossi and Ades 2008.

36. *clever study investigated this question:* Horowitz 2009.

37. *anus of the toy dog:* Harris and Prouvost 2014. *in only one condition:* An owner's subtle movements can have a large effect on an animal's behavior (due to statistical learning); see heam.info/animals-4.

38. *in distress, for example:* The act lifts a burden on their body budgets (e.g., Bartal et al. 2011). For more, see heam.info/burden-1. *infant who is in distress:* Dunfield and Kuhlmeier 2013; see heam.info/burden-2.

39. *with a bunch of strangers:* For an enlightening discussion of why wolves are not aggressive creatures by nature, read Bradshaw 2014. See also heam.info/wolves-1.

40. *experience some kind of grief:* Morell 2013, 148; Bekoff and Goodall 2008, 66. *operates similarly to drug withdrawal:* Vernon et al. 2016. *love is a drug:* Fisher et al. 2010.

41. *why isn't it "anger learning":* A similar point was made by Jerome Kagan (Kagan 2007).

42. *the "triune brain":* "Fear learning" studies, which assume a triune brain, have also been performed on humans, in support of the classical view (e.g., LaBar et al. 1998).

43. *this circuitry in elegant detail:* E.g., the neuroscientist Joseph LeDoux's groundbreaking research illustrates how synapses change within key sites of the amygdala, allowing neutral sensory inputs, like sounds, to automatically elicit an inborn defense response, like freezing (LeDoux 2015).

44. *automatically and effortlessly:* For an accessible introduction, see Wegner and Gray 2016. Mental inference is so ubiquitous in Western culture that scholars keep discovering it again and again and calling it by different names; see heam.info/inference-1.

45. *meaningful by making an inference:* This began with the first psychology experiment, which was conducted by Wilhelm Wundt in the late 1800s; see heam.info/wundt-2.

46. *into an industry of fear:* This confusion became institutionalized in psychology during behaviorism; see heam.info/behaviorism-1.

47. *rats run away:* E.g., Berlau and McGaugh 2003; see heam.info/rats-1. *in which case they attack:* Reynolds and Berridge 2008. See heam.info/rats-2. *goes down instead of up:* Iwata and LeDoux 1988. *not all of these varied behaviors require the amygdala:* Fear learning does not necessarily involve the amygdala. Aggression toward a preda-

tor (called "defensive treading" or "burying") does not depend on the amygdala (De Boer and Koolhaas 2003; Kopchia et al. 1992). The amygdala is involved when the threat is maximally ambiguous and learning is required (i.e., when prediction error must be processed [Li and McNally 2014]). Even if amygdala neurons are routinely involved in learning, they may not be necessary for learning to occur. For example, infant monkeys who have their amygdalae removed about two weeks after birth are able to learn about aversive things; a body-budgeting region (the anterior cingulate cortex) had expanded in these monkeys during brain development, and this region also supports aversive learning (Bliss-Moreau and Amaral, under review). *of the mental inference fallacy:* Gross and Canteras 2012; Silva et al. 2013. See also heam.info/ inference-2. *specific to freezing or fear:* Tovote et al. 2015; see heam.info/inference-3.

48. *be the circuitry for distress:* Blumberg et al. 2000. According to the neuroscientist Jaak Panksepp (Panksepp 1998), "distress/panic" calls are made by infant rats and occur following social isolation. For example, in a recent paper, he writes, "Distinct emotional powers that engender crying, allow young animals to signal their desperate need for care, especially when lost or isolated from caretakers by experimenters. These separation calls alert caretakers to seek out, retrieve, and attend to the needs of the offspring" (Panksepp 2011, 1799). *done by their absent mothers:* Blumberg and Sokoloff 2001. For a discussion, see Barrett, Lindquist, Bliss-Moreau, et al. 2007.

49. *evidence and revised his position:* His recent theoretical papers clearly distinguish an instance of the emotion "Fear" from freezing behavior (LeDoux 2015).

50. *the rodent is feeling empathy:* Burkett et al. 2016; Panksepp and Panksepp 2013. Don't get me wrong—rodents are social animals that regulate each other's body budgets, which means they can feel distress and perceive it in others of their species. Social insects regulate each other's body budgets with chemicals. Mammals also do it with touch and perhaps with sound. Humans use all these means, plus words. But the question remains, do all these animals feel empathy? Or do only humans have the goal-based concept necessary to impose additional functions to transform body-budget regulation into empathy?

51. *question is similar to yourself:* Mitchell et al. 1997. For other reasons, see Epley et al. 2007; Wegner and Gray 2016. *babies on her own flesh:* Kupfer et al. 2006. *of people chasing one another:* The similarities to humans can be simple; see heam.info/ inference-4.

52. *itself—it's completely normal:* I've avoided the term "anthropomorphism"; see heam .info/anthro-1.

53. *diminished versions of ourselves:* The classical view encourages this conceit, fueled by the "triune brain" myth of a simple brain evolving into something more complex; see heam.info/evolution-4. *not to build mental similarities:* Matsuzawa 2010.

54. *"crying for their mothers":* See more on Panksepp's circuits at heam.info/panksepp-1. *exist in any animal brain:* Barrett, Lindquist, Bliss-Moreau, et al. 2007. *are not dedicated to emotion:* Survival circuits are not one-to-one with emotion concepts; see heam.info/survival-1.

13. From Brain to Mind: The New Frontier

1. *distinctions between thinking and feeling:* Some cultures have a single word best translated as "thought-feeling" (e.g., Danziger 1997, chapter 1; William Reddy, personal communication, September 16, 2007; Wikan 1990); also see heam.info/balinese-1.

2. *are structured completely alike:* Van Essen and Dierker 2007; Finn et al. 2015; Hathaway 2015.

3. *neurons in certain brain regions:* Opendak and Gould 2015; Ernst and Frisén 2015. *also occurs with experience:* See heam.info/plasticity-1.

4. *neurotransmitters make this possible:* Bargmann 2012. Neurotransmitters change how efficiently your neurons communicate and more; see heam.info/neuro-1. *information flows along different paths:* Sporns 2011, 272. *greater than the sum of the parts:* For a review, see Park and Friston 2013; e.g., networks reconfigure as cognitive demands increase (Kitzbichler et al. 2011). For more, see heam.info/wiring-2.

5. *or even vision or hearing:* A single brain cell can be multipurpose, as we discussed in chapters 1 and 2, contributing to multiple psychological states; see heam.info/neurons-2.

6. *and other scholarly disciplines:* Bullmore and Sporns 2012. The brain is a complex, adaptive system, meaning that it constantly reconfigures the connectivity strength of its neurons to anticipate changes in the environment (which includes the body and outside world). Complex systems produce emergence, i.e., products of the system as a whole that cannot be reduced to the components of the system alone; they are "more than the sum of their parts" (Simon 1962). Complexity means that variation is the norm in patterns of brain activity; see heam.info/complexity-1.

7. *that it can support consciousness:* Tononi and Edelman 1998; Edelman and Tononi 2000. *its single function by itself:* A brain full of uniquely purposed neurons would also have low complexity, as would a fully synchronized brain, because in both cases, the majority of neurons do not share information (they all act differently in the former case and identically in the latter case).

8. *get to the same end:* Whitacre and Bender 2010, figure 10; see also heam.info/whitacre-1. *genes to the next generation:* Edelman and Gally 2001. Degeneracy accompanies natural selection. It makes the brain more resilient to injury, which is why natural selection favors a brain built with degeneracy. The variation that degeneracy provides is a prerequisite for natural selection in the first place; see heam.info/degeneracy-4.

9. *favors a complex brain:* The evolutionary success of a brain depends on its ability to model the ever-changing environment in a metabolically efficient way (Edelman and Gally 2001; Whitacre and Bender 2010). Evolution must select for individuals with a combination of genes that produce this kind of brain (and that genetic combination is, itself, degenerate and complex). The more important a system is to the survival of a species, the more degeneracy and complexity will exist in the genes that support that system. Therefore, degeneracy and complexity are prerequisites for and an inescapable product of natural selection. I am not claiming that natural selection favors ever-increasing complexity; natural selection does favor complex adaptive systems.

10. *and other properties of consciousness:* And perhaps a few other concepts as well; see heam.info/properties-1. *practices to address that dilemma:* See heam.info/world-1.

11. *perhaps a Jackson Pollock:* The brain doesn't construct a representation of an object like a bee or a car and then evaluate its significance for the self. The significance for your body budget is built into the construction in the first place, via interoceptive predictions. Note that this is at odds with a version of the classical view called causal appraisal theories of emotion, which assume that first you perceive an object and then you evaluate it for its self-relevance, novelty, etc.

12. *out for several thousand years:* Many other worldviews exist; see heam.info/world-1.

13. *has exactly the same function:* Pinker 2002, 40. *the next generation of humans:* Durham 1991; Jablonka et al. 2014; Richerson and Boyd 2008.
14. *mindful enough to cultivate doubt:* Firestein 2012.
15. *for parent-infant bonding:* See heam.info/synchrony-1.
16. *in the world just fine:* The activist Caroline Casey didn't know she was blind until age seventeen, when she proposed to learn to drive (Casey 2010).
17. *more aliases than Sherlock Holmes:* The default mode and salience networks go by many names (Barrett and Satpute 2013); see heam.info/dmn-5.
18. *prefrontal cortex (PFC):* Neurons in the upper layers of cortex are born last during the prenatal period and continue to mature and develop their connectivity after birth, during infancy and childhood (Kostović and Judaš 2015). Poverty is similarly toxic for other aspects of brain development (Noble et al. 2015). *(prediction error) and control:* Barrett and Simmons 2015; Finlay and Uchiyama 2015. *leads back to poverty:* See heam.info/children-1.
19. *accurate than we might think:* Jussim, Cain, et al. 2009; Jussim, Crawford, et al. 2009. *when compared to census figures:* Pinker 2002, 204. *muddled assumptions about human nature:* Jussim 2012; Pinker 2002.
20. *"next generation of tools":* Firestein 2012, 21.
21. *new government and social order:* Even the concept of a "Revolution" is social reality; see heam.info/revolution-1.

Appendix A

1. *2004 as a charity event:* "Fright Night" 2012.
2. *changing its rate of firing:* Marder 2012. The transmission is made more or less efficient by glial cells (Ji et al. 2013; Salter and Beggs 2014); see heam.info/glial-2.
3. *wired into circuits and networks:* The transition between the cortex and subcortical regions is called allocortex, and it ranges from having barely visible columns to three layers (Zilles et al. 2015).
4. *organized as clumps of neurons:* The word "cortical" means "in the cortex," hence "subcortical" is "below the cortex."
5. *important for coordinating physical movements:* The cerebellum's main role is to anticipate how the body's movements in time and space will influence the predictions and pattern completion going on in the cortex (Pisotta and Molinari 2014; Shadmehr et al. 2010).
6. *that replenish those resources:* There are three branches of the autonomic nervous system. The sympathetic nervous system, sometimes called the "fight or flight" system, tells the body to spend its energy resources. It sends information to the sweat glands in your skin, to the smooth muscles that surround your blood vessels, to your internal body organs, to the muscles that dilate your pupils, to the parts of the body that generate your immune cells, and so on. The parasympathetic nervous system, also known as the "rest and digest" system, tells the body to replenish its energy resources. It tells your pupillary muscles to contract, your body to secrete saliva and insulin, and other functions related to digesting food, in part by communicating with the third branch, called the enteric nervous system. See heam.info/nervous-1.

Appendix D

1. *(hearing, etc.) operate by prediction:* For a summary, see Chanes and Barrett 2016; details are at heam.info/prediction-12. *structured to function this way:* Barrett and Simmons 2015.

2. *cascade within the visual system:* Grill-Spector and Weiner 2014; Gilbert and Li 2013. *across the structure of cortex:* Barbas and Rempel-Clower 1997; Barbas 2015. *multisensory summaries in chapter 6:* Many neurons pass information to fewer, more densely connected neurons, meaning compression and dimension reduction must happen (Finlay and Uchiyama 2015).

3. *some of the same neurons:* A recent discovery is that conceptually similar visual instances are stored closer to one another in cortical space; for an example in visual cortex, see Grill-Spector and Weiner 2014.

4. *subjects were lying at rest:* Ironically, because scientists assumed that the brain was "off" when not stimulated by the external world, they missed evidence of this network several times. For more on how the default mode network was discovered, see Buckner 2012. *stimulated by an experimental procedure:* Obviously, intrinsic brain activity is not important only when the brain is not being probed explicitly in an experiment. Those who originally named the network probably did not appreciate the importance of this network (or intrinsic activity) to everyday thoughts, feelings, and perceptions when they named the network. *networks have since been discovered:* Yeo et al. 2011; Barrett and Satpute 2013. *name fit this network nicely:* The default mode network goes by many names; see heam.info/dmn-1.

5. *default mode network, as predicted:* Binder showed that conceptual processing occurs even when people are not explicitly asked about concepts (Binder et al. 1999). For more details on this experiment, see heam.info/binder-2. *similar brain-imaging experiments:* Binder et al. 2009.

6. *in the default mode network:* Spunt et al. 2010.

7. *way it is right now:* E.g., Barrett 2009; Bar 2007. For a review, see Buckner 2012.

8. *a key role in categorization:* Barrett 2012; Lindquist and Barrett 2012. For a similar but not identical point of view, see Edelman 1990, and Binder and Desai 2011.

9. *to construct instances of concepts:* The cognitive neuroscientist Eleanor A. Maguire comes close to this idea (Hassabis and Maguire 2009); see heam.info/maguire-1.

10. *"lantern" of attention:* Gao, Alcauter, et al. 2014.

11. *category to any brain region:* Lindquist et al. 2012. *the interoceptive and control networks:* Kober et al. 2008.

12. *clearly for happiness and sadness:* Wager et al. 2015. Further details are in chapter 1 and at heam.info/patterns-1. *exactly like its associated summary:* Clark-Polner, Johnson, et al., in press; Clark-Polner, Wager, et al., in press.

13. *changes in the interoceptive network:* Wilson-Mendenhall et al. 2013. Even more striking, when volunteers imagined physical danger, a relatively greater increase in neural activity was observed in a network that tracks and locates physical objects in space, but when they imagined social scenarios, the increase occurred in a network that helps infer the thoughts and feelings of others (Wilson-Mendenhall et al. 2011).

14. *is exactly what we observed:* Wilson-Mendenhall et al. 2015. See also Oosterwijk et al. 2015. For other brain-imaging studies that support the theory of constructed emotion, see heam.info/TCE-1.

15. *reported more intense emotional experiences:* Raz et al. 2016. More details are at heam. info/movies-1.

16. *similar case for emotion perception:* See research by the cognitive neuroscientist Robert Spunt and colleagues (e.g., Spunt and Lieberman 2012). See also Peelen et al. 2010, and Skerry and Saxe 2015, discussed in more detail in heam.info/dmn-3.

17. *represented by the entire brain:* Some scientists try to find a compromise between these two views of concepts (that they involve sensory and motor representations versus that they are "abstract," meaning they are stored without reference to sensory and motor details); see heam.info/dmn-4. *movements have increased their firing:* Chao and Martin 2000. See Barsalou 2008b for a review. *the name of the object ("hammer"):* Tucker and Ellis 2004. *gripping motion with your hand:* Klatzky et al. 1989; Tucker and Ellis 2001.

18. *represented throughout the entire brain:* For a review, see Barsalou 2009.

19. *of neurons for each goal:* Further details on this misconception are at heam.info/concepts-20. *see nothing of the kind:* For a discussion of evidence, see Lebois et al. 2015.

20. *can be different each time:* Within a concept, there can be several different goals, none of which is core; see heam.info/concepts-21.

21. *dark, empty bucket:* Years later, I finally forgave myself for this embarrassing error after reading Brian Greene's 2007 book *The Fabric of the Cosmos,* whose second chapter is titled "The Universe and the Bucket: Is Space a Human Abstraction or a Physical Entity?" (Greene 2007). *"eye of the beholder":* Ibid., 47.

22. *"memories" stored in your brain:* Schacter 1996.

Illustration Credits

Fig. 1-1: Illustration by Aaron Scott.
Fig. 1-2: Photos courtesy of Paul Ekman. Design layout by the author.
Fig. 1-3: Photo courtesy of Paul Ekman. Design layout by the author.
Fig. 1-4: Photos courtesy of Paul Ekman. Design layout by the author.
Fig. 1-5: Photo by Aaron Scott.
Fig. 1-6: Portrait of Martin Landau (center) by Howard Schatz from *In Character: Actors Acting* (Boston: Bulfinch Press, 2006). Other photos courtesy of Paul Ekman.
Fig. 1-7: Illustration by Aaron Scott.
Fig. 2-1: Photo courtesy of Richard Enfield. Modification courtesy of Daniel J. Barrett.
Fig. 3-1: Photo courtesy of Barton Silverman/New York Times/Redux.
Fig. 3-2: Photo courtesy of Paul Ekman. Design layout by the author.
Fig. 3-3: Photo courtesy of Paul Ekman. Design layout by the author.
Fig. 3-4: Photos courtesy of Paul Ekman. Design layout by the author.
Fig. 3-5: Photo courtesy of Debi Roberson.
Fig. 4-1: Illustration by Aaron Scott.
Fig. 4-2: Illustration by Aaron Scott.
Fig. 4-3: Illustration by Aaron Scott.
Fig. 4-4: Illustration by Aaron Scott.
Fig. 4-5: Illustration by Aaron Scott.
Fig. 4-6: Photo courtesy of Helen Mayberg.
Fig. 4-7: Illustration by Aaron Scott.
Fig. 5-1: Illustration by Aaron Scott.
Fig. 5-2: Illustration by Aaron Scott.
Fig. 5-3: Illustration by Aaron Scott.
Fig. 6-1: Illustration by Aaron Scott.
Fig. 6-2: Illustration by Aaron Scott.
Fig. 7-1: Photo courtesy of the author.
Fig. 7-2: Illustration by Aaron Scott.
Fig. 12-1: Photo courtesy of Ann Kring and Angie Hawk.
Fig. 12-2: Illustration by Aaron Scott.
Fig. AA-1: Illustration by Aaron Scott.
Fig. AA-2: Illustration by Aaron Scott.
Fig. AA-3: Illustration by Aaron Scott.
Fig. AA-4: Illustration by Aaron Scott.
Fig. AA-5: Illustration by Aaron Scott.
Fig. AA-6: Illustration by Aaron Scott.
Fig. AB-1: Photo (top) courtesy of Richard Enfield. Modification (bottom) courtesy of Daniel J. Barrett.
Fig. AC-1: Photo courtesy of Barton Silverman/New York Times/Redux.
Fig. AD-1: Illustration by Aaron Scott.
Fig. AD-2: Photo courtesy of Dr. Tor Wager and the author.
Fig. AD-3: Illustration by Aaron Scott.

Index